Population Genetics for Animal Conserva

It is widely accepted among conservation biolog. 　 ᵧenetics is, more than ever, an essential and efficient tool for wild and captive population management and reserve design. However, a true synergy between population genetics and conservation biology is lacking. Following the first International Workshop on Population Genetics for Animal Conservation in 2003 at the Centro di Ecologia Alpina, Trento, Italy (recently incorporated into the Edmund Mach Foundation), the scientific committee felt that, given the global urgency of animal conservation, it was imperative that discussions at the conference were made accessible to graduate students and wildlife managers. This book integrates 'the analytical methods approach' with the 'real problems approach' in conservation genetics. Each chapter is an exhaustive review of one area of expertise, and a special effort has been made to explain the statistical tools available for the analysis of molecular data as clearly as possible. The result is a comprehensive volume of the state of the art in conservation genetics, illustrating the power and utility of this synergy.

GIORGIO BERTORELLE currently teaches Biometry, Phylogeny Reconstruction and Conservation Genetics at the University of Ferrara, Italy. He is the President and Co-founder of the Italian Society for Evolutionary Biology.

MICHAEL W. BRUFORD, formerly Head of the Conservation Genetics Group at the Institute of Zoology, London, has been Professor and Research Group Leader at the Cardiff School of Biosciences since 1999, where he teaches undergraduate and graduate courses in conservation biology and molecular ecology.

HEIDI C. HAUFFE trained in Evolutionary Biology and established the first genetics laboratory at the Centro di Ecologia Alpina, Trento, Italy, in 1997. Now at the Edmund Mach Foundation, her research interests range from rodent-borne viruses to speciation to conservation genetics of alpine mammals. She is also currently affiliated to the University of York, UK, and the Institute of Vertebrate Biology, CZ.

ANNAPAOLA RIZZOLI is currently the Coordinator of the Environmental and Natural Resources Division and Head of the Wildlife Ecology and Epidemiology Group at the Edmund Mach Foundation. Her main research interests are host–parasite interactions and emerging zoonotic and vector-borne diseases.

CRISTIANO VERNESI is a researcher at the Edmund Mach Foundation, Trento, Italy. He is also one of the founders and Scientific Director of the Association 'Biosfera', a non-profit association devoted to research and teaching in conservation biology.

FONDAZIONE EDMUND MACH

ISTITUTO AGRARIO
DI SAN MICHELE ALL'ADIGE

Centro di Ecologia Alpina
Viote del Monte Bondone

UNIFE

Conservation Biology

This series aims to present internationally significant contributions from leading researchers in particularly active areas of conservation biology. It focuses on topics where basic theory is strong and where there are pressing problems for practical conservation. The series includes both authored and edited volumes and adopts a direct and accessible style targeted at interested undergraduates, postgraduates, researchers and university teachers.

1. *Conservation in a Changing World*, edited by Georgina Mace, Andrew Balmford and Joshua Ginsberg 0 521 63270 6 (hardcover), 0 521 63445 8 (paperback)
2. *Behaviour and Conservation*, edited by Morris Gosling and William Sutherland 0 521 66230 3 (hardcover), 0 521 66539 6 (paperback)
3. *Priorities for the Conservation of Mammalian Diversity*, edited by Abigail Entwistle and Nigel Dunstone 0 521 77279 6 (hardcover), 0 521 77536 1 (paperback)
4. *Genetics, Demography and Viability of Fragmented Populations*, edited by Andrew G. Young and Geoffrey M. Clarke 0 521 782074 (hardcover), 0 521 794218 (paperback)
5. *Carnivore Conservation*, edited by John L. Gittleman, Stephan M. Funk, David Macdonald and Robert K. Wayne 0 521 66232 X (hardcover), 0 521 66537 X (paperback)
6. *Conservation of Exploited Species*, edited by John D. Reynolds, Georgina M. Mace, Kent H. Redford, and John G. Robinson 0 521 78216 3 (hardcover), 0 521 78733 5 (paperback)
7. *Conserving Bird Biodiversity, and* edited by Ken Norris and Deborah J. Pain 0 521 78340 2 (hardcover), 0 521 78949 4 (paperback)
8. *Reproductive Science and Integrated Conservation*, edited by William V. Holt, Amanda R. Pickard, John C. Rodger and David E. Wildt 0 521 81215 1 (hardcover), 0 521 01110 8 (paperback)
9. *People and Wildlife*, edited by Rosie Woodroffe, Simon Thergood and Alan Rabinowitz 0 521 82505 9 (hardcover), 0 521 53203 5 (paperback)
10. *Phylogeny and Conservation*, edited by Andrew Purvis, John L. Gittleman and Thomas Brooks 0 521 82502 4 (hardcover), 0 521 53200 0 (paperback)
11. *Large Herbivore Ecology*, edited by Kjell Danell, Roger Bergstrom, Patrick Duncanand John Pastor 0 521 83005 2 (hardcover), 0 521 53687 1 (paperback)
12. *Top Predators in Marine Ecosysyems*, edited by Ian Boyd, Sarah Wanless and C.J. Camphuysen 0 521 84773 7 (hardcover), 0 521 61256 X (paperback)
13. *Coral Reef Conservation*, edited by Isbelle Côté and Jogn Reynolds 0521 85536 5 (hardcover), 0 521 67145 0 (paperback)
14. *Connectivity Conservation*, edited by Kevin R. Crooks and M. Sanjayan 0 521 85706 6 (hardcover), 0 521 67381 X (paperback)
15. *Zoos in the 21st Century*, edited by Alexandra Zimmermann, Matthew Hatchwell, Lesley Dicheie and Chris West 9780521853330 (hardcover) 9780521618588 (paperback)
16. *Setting Conservation Targets for Managed Forest Landscapes*, edited by Marc-André Villard and Bengt Gunnar Jonsson 9780521877091 (hardcover) 9780521700726 (paperback)

Population Genetics for Animal Conservation

Edited by

GIORGIO BERTORELLE
Department of Biology and Evolution, University of Ferrara, Italy

MICHAEL W. BRUFORD
School of Biosciences, Cardiff University, United Kingdom

HEIDI C. HAUFFE
Edmund Mach Foundation, Trento, Italy

ANNAPAOLA RIZZOLI
Edmund Mach Foundation, Trento, Italy

CRISTIANO VERNESI
Edmund Mach Foundation, Trento, Italy

CAMBRIDGE UNIVERSITY PRESS
Cambridge, New York, Melbourne, Madrid, Cape Town, Singapore, São Paulo, Delhi

Cambridge University Press
The Edinburgh Building, Cambridge CB2 8RU, UK

Published in the United States of America by Cambridge University Press, New York

www.cambridge.org
Information on this title: www.cambridge.org/9780521866309

First published 2009

Printed in the United Kingdom at the University Press, Cambridge

A catalogue record for this publication is available from the British Library

Library of Congress Cataloguing in Publication data
Population genetics for animal conservation / edited by Giorgio Bertorelle ... [et al.].
    p.   cm.
ISBN 978-0-521-86630-9
1. Conservation biology.   2. Animal genetics.   I. Bertorelle, Giorgio.   II. Title.
QH75.P64   2009
639.9–dc22
                                                                    2009005815

ISBN 978-0-521-86630-9 hardback
ISBN 978-0-521-68537-5 paperback

Conservation biology will succeed to the degree that its theoreticians, practitioners, and users acknowledge the larger context in which they exist, and to the degree that they respect one another's roles, contributions and problems.

Michael E. Soulé (1986)

Contents

List of contributors [ix]
Foreword [xiii]
Acknowledgements [xiv]

1 Introduction [1]
HEIDI C. HAUFFE AND VALERIO SBORDONI

Statistical approaches, data analysis and inference [23]

2 Statistical methods for identifying hybrids and groups [25]
ERIC C. ANDERSON

3 How to use MIGRATE or why are Markov chain Monte Carlo programs difficult to use? [42]
PETER BEERLI

4 Nested clade phylogeographic analysis for conservation genetics [80]
JENNIFER E. BUHAY, KEITH A. CRANDALL AND DAVID POSADA

5 A comparison of methods for constructing evolutionary networks from intraspecific DNA sequences [104]
PATRICK MARDULYN, INSA CASSENS AND MICHEL C. MILINKOVITCH

Molecular approaches and applications [121]

6 Challenges in assessing adaptive genetic diversity: overview of methods and empirical illustrations [123]
AURÉLIE BONIN AND LOUIS BERNATCHEZ

7 Monitoring and detecting translocations using genetic data [148]
GIORGIO BERTORELLE, CHIARA PAPETTI, HEIDI C. HAUFFE AND LUIGI BOITANI

8   Non-invasive genetic analysis in conservation   [167]
     BENOÎT GOOSSENS AND MICHAEL W. BRUFORD

9   The role of ancient DNA in conservation biology   [202]
     JON BEADELL, YVONNE CHAN AND ROBERT FLEISCHER

     From genetic data to practical management: issues and case studies   [225]

10   Future-proofing genetic units for conservation: time's up for
     subspecies as the debate gets out of neutral!   [227]
     MICHAEL W. BRUFORD

11   Genetic diversity and fitness-related traits in endangered
     salmonids   [241]
     KATRIINA TIIRA AND CRAIG R. PRIMMER

12   Genetics and conservation on islands: the Galápagos giant tortoise
     as a case study   [269]
     CLAUDIO CIOFI, ADALGISA CACCONE, LUCIANO B. BEHEREGARAY,
     MICHEL C. MILINKOVITCH, MICHAEL RUSSELLO AND JEFFREY R.
     POWELL

13   Evolution of population genetic structure in marine mammal
     species   [294]
     A. RUS HOELZEL

     Future directions in conservation genetics   [319]

14   Recent developments in molecular tools for conservation   [321]
     CRISTIANO VERNESI AND MICHAEL W. BRUFORD

15   Theoretical outlook   [345]
     MARK BEAUMONT

     Software index   [374]
     Species index (common name)   [376]
     Species index (Latin name)   [379]
     Subject index   [382]

The colour plates are situated between pages 210 and 211

Contributors

ERIC C. ANDERSON
Fisheries Ecology Division,
Southwest Fisheries Science Center,
110 Shaffer Road,
Santa Cruz,
CA 95060, USA

JON BEADELL
Center for Conservation and
Evolutionary Genetics,
National Zoological Park and National
Museum of Natural History,
Smithsonian Institution,
P.O. Box 37012 MRC 5503,
Washington,
DC 20013, USA
and
Department of Ecology and Evolutionary
Biology,
Yale University,
ESC 158b,
21 Sachem Street,
New Haven,
CT 06520, USA

MARK BEAUMONT
School of Biological Sciences,
Philip Lyle Research Building,
P.O. Box 68,
University of Reading,
Whiteknights,
Reading RG6 6BX,
United Kingdom

PETER BEERLI
Department of Scientific Computing
and Biological Sciences Department,
150-T Dirac Science Library,
Florida State University,
Tallahassee,
FL 32306, USA

LUCIANO B. BEHEREGARAY
Department of Biological Sciences,
Macquarie University,
Sydney,
NSW 2109, Australia

LOUIS BERNATCHEZ
Department of Biology,
Pavillon Charles-Eugène-Marchand
1030, Avenue de la Médecine,
Room 1145,
Université Laval,
Québec,
Québec G1V 0A6, Canada

GIORGIO BERTORELLE
Department of Biology and Evolution,
University of Ferrara,
44100 Ferrara, Italy

LUIGI BOITANI
Department of Animal and Human
Biology,
University of Rome 'La Sapienza',
viale Università 32,
00185 Roma, Italy

AURÉLIE BONIN
Laboratoire d'Ecologie Alpine,
CNRS-UMR 5553,
Université Joseph Fourier,
BP 53,
38041 Grenoble cedex 09,
France
and
Department of Biology,
Indiana University,
1001 E. Third Street,
Bloomington,
IN 47405, USA

MICHAEL W. BRUFORD
School of Biosciences,
Cardiff University,
Cathays Park,
Cardiff CF10 3AX,
United Kingdom

JENNIFER E. BUHAY
Integrative Biology Department,
Brigham Young University,
401 Widtsoe Building,
Provo,
UT 84602, USA

ADALGISA CACCONE
Department of Ecology and Evolutionary
Biology,
P.O. Box 208106,
Yale University,
New Haven,
CT 06520, USA

INSA CASSENS
Max-Planck-Institut für demografische
Forschung,
Konrad-Zuse-Str.1,
18057 Rostock, Germany

YVONNE CHAN
Department of Biological Sciences,
Stanford University,
Stanford,
CA 94305, USA
and
University of Hawaii–Manoa,
Hawaii Institute of Marine Biology,
P.O. Box 1346,
Kaneohe,
HI 96744, USA

CLAUDIO CIOFI
Department of Evolutionary Biology,
University of Florence,
via Romana 17,
50125 Firenze, Italy

KEITH A. CRANDALL
Integrative Biology Department,
Brigham Young University,
401 Widtsoe Building,
Provo,
UT 84602, USA

ROBERT FLEISCHER
Center for Conservation and
Evolutionary Genetics,
National Zoological Park and National
Museum of Natural History,
Smithsonian Institution,
P.O. Box 37012 MRC 5503,
Washington,
DC 20013, USA

BENOÎT GOOSSENS
School of Biosciences,
Cardiff University,
Cathays Park,
Cardiff CF10 3AX,
United Kingdom

HEIDI C. HAUFFE
Research and Innovation Centre,
Edmund Mach Foundation–IASMA,
Via Edmund Mach 1,
S. Michele all'Adige (TN), Italy

A. RUS HOELZEL
School of Biological and
Biomedical Sciences,
University of Durham,
South Road,
Durham DH1 3LE,
United Kingdom

PATRICK MARDULYN
Behavioral and Evolutionary Ecology,
CP 160/12,
Université Libre de Bruxelles,
Avenue FD Roosevelt 50,
1050 Brussels, Belgium

MICHEL C. MILINKOVITCH
Laboratory of Artificial and
Natural Evolution (LANE),
Department of Genetics and Evolution,
University of Geneva,
Sciences III, 30 Quai Ernest-Ansermet,
1211 Genève 4, Switzerland

CHIARA PAPETTI
Department of Biology,
University of Padova,
via U. Bassi 58/B,
35121 Padova, Italy

DAVID POSADA
Departamento de Bioquímica,
Genética e Inmunología,
Facultad de Biología,
Universidad de Vigo,
Vigo 36310, Spain

JEFFREY R. POWELL
Department of Ecology
and Evolutionary Biology,
P.O. Box 208106,
Yale University,
New Haven,
CT 06520, USA

CRAIG R. PRIMMER
Department of Biology,
University of Turku,
20014 Turku, Finland

MICHAEL RUSSELLO
Unit of Biology and Physical Geography,
University of British Columbia Okanagan,
3333 University Way, Kelowna,
British Columbia V1V 1V7, Canada

VALERIO SBORDONI
Department of Biology,
Tor Vergata University,
00133 Roma, Italy

KATRIINA TIIRA
Canine Genomics Research Group,
Department of Basic Veterinary
Sciences,
Department of Medical Genetics,
The Folkhälsen Institute
of Genetics,
Department of Molecular
Genetics,
P.O. Box 66,
University of Helsinki,
00014 Helsinki, Finland

CRISTIANO VERNESI
Research and Innovation Centre,
Edmund Mach Foundation–IASMA,
Via Edmund Mach 1,
S. Michele all'Adige (TN),
Italy

Foreword

The field of conservation genetics is evolving very quickly, due to both methodological and technical improvements. In this context, it is important that experts worldwide meet and share their knowledge about the latest theories and statistical developments. A group of scientists from the Centro di Ecologia Alpina (Trento) and the University of Ferrara decided to organize a meeting on Population Genetics for Animal Conservation. This meeting was held on 4–6 September 2003 at the Centro di Ecologia Alpina, and included many famous population geneticists and conservation biologists. This book is the outcome of this meeting. It does not correspond to a classical and comprehensive textbook, but focuses on the latest developments in conservation genetics, in a well-organized and integrated unity.

When going through the different chapters, I realized that this book does not simply reflect the situation of conservation genetics in 2003. All the most recent developments are included. I particularly appreciate that the authors of some chapters did not hesitate to share their feeling about the future of conservation genetics. Such views are usually excluded from scientific papers due to the peer-review process that promotes only well-accepted theories and methods. This is unusual in such a manual, but will be extremely useful to scientists having to design a research strategy for the next few years.

I am convinced that this volume complements extremely well the existing general textbooks on conservation genetics, and will stimulate the development of innovative studies in population genetics applied to the conservation of threatened populations or species.

Pierre Taberlet

Acknowledgements

Funding was received for the preparation of this book and the organization of the conference on Population Genetics for Animal Conservation from the Ministry of Research and Innovation of the Autonomous Province of Trento, the Edmund Mach Foundation, the Centro di Ecologia Alpina, the Department of Biology and Evolution of the University of Ferrara and the Association 'Biosfera', Florence, for which we are deeply grateful.

We are indebted to our many colleagues who agreed to referee chapters and spent valuable time making detailed suggestions, greatly improving the final text.

We would also like to heartily thank all the invited speakers, participants, organizers, students and staff of the ex-Centro di Ecologia Alpina (especially Elena Pecchioli, Barbara Crestanello, Annalisa Losa, Federico Lenzi, Alfonso Voltalini and Michela Manzi), whose tireless contributions made the first International Workshop on Population Genetics for Animal Conservation (PGAC) such a rewarding experience.

A special thanks goes to the former directors of the Centro di Ecologia Alpina, Gianni Nicolini and Claudio Chemini, for their endless faith in our abilities, and for backing the PGAC workshops from the outset.

And finally, the publication of this book would not have been possible without the enthusiasm, patience and guidance at various stages of production of Alison Evans, Alan Crowden, Betty Fulford, Anna Hodson, Dominic Lewis, Anna-Marie Lovett and Tracey Sanderson of Cambridge University Press.

$$\textcircled{1}$$

Introduction

HEIDI C. HAUFFE AND VALERIO SBORDONI

When it comes to advocating animal conservation, it is difficult to be convincing without becoming alarmist. The fact is, time is running out for many of the world's animal species. Habitat loss, introduced species, over-exploitation and pollution, all caused by human activities, combine with stochastic factors to place ever-increasing pressure on natural populations (Primack 2002). The estimates of the mid-1990s, predicting that thousands of species and millions of unique populations would go extinct in the following decades (Ehrlich and Wilson 1991; Smith *et al.* 1993; Lawton and May 1995), remain as relevant as when they were first made, and we are still living in an era of unprecedented biodiversity loss, with current extinction rates 100–1000 times the background rate (Primack 2002) and 5000–25 000 times that recorded in the fossil record (Frankham *et al.* 2002; but see Mace *et al.* 1996). Recently, however, there have been some positive signs in the media that biologists' warnings are being received (e.g. Gianni 2004; Devine *et al.* 2006; Black 2006; Gabriel 2007; Stern 2007), and a rapid and efficient approach in providing information pertinent to biodiversity preservation could be pivotal in policy decision and in optimizing resource allocation (Naidoo and Ricketts 2006; Marsh *et al.* 2007). Since the foundation of the field of conservation biology, it has been argued that a synergy between conservation biology and advanced population genetics could provide important information that policy-makers need. As should be obvious by its title, the purpose of this book is an attempt to go some way towards maturing such a synergy; hence, this introduction presents a brief history and the current state of this partnership.

THE EXTINCTION CRISIS

In order to be convinced of the urgency for animal conservation and the information necessary to practise it, an update on the current extinction

Population Genetics for Animal Conservation, eds. G. Bertorelle, M. W. Bruford, H. C. Hauffe, A. Rizzoli and C. Vernesi. Published by Cambridge University Press. © Cambridge University Press 2009.

crisis is pertinent. The IUCN Red List of Threatened Species is one of the most powerful tools available for assessing the extent of this crisis (Butchart *et al.* 2005; Rodrigues *et al.* 2006; Cardillo *et al.* 2006; but see Marsh *et al.* 2007). Of the 1.1–1.8 million known species of animals (Lecointre and Le Guyader 2001; Primack 2002; Halliday 2006; IUCN 2006), 7725 are listed by the IUCN (2006) as vulnerable, endangered or critically endangered. These include highly publicized and charismatic megafauna, such as the blue whale (*Balaenoptera musculus*) and the giant panda (*Ailuropoda melanoleuca*), a much longer list of smaller obscure creatures with curious names, like the booroolong frog (*Litoria booroolongensis*) and the pale lilliput naiad (*Toxolasma cylindrellus*), and, sadly, our closest nonhuman relatives the gorilla (*Gorilla gorilla*), the bonobo (*Pan paniscus*) and the orang-utan (*Pongo pygmaeus*).

Although the IUCN data have their flaws and gaps, in general they present an overwhelming picture of species decline. While amphibians are often quoted in the scientific and layman's literature as being the most threatened class of animals (one third of known species are at risk), with mammals, cartilaginous fishes and birds close behind (25, 13 and 12% of species in these groups, respectively, are listed as threatened; see Fig. 1.1, black bars), Fig. 1.1 (white bars) also shows that these four animal classes are the most thoroughly evaluated by the IUCN. In fact, the state of virtually all bird, mammal and amphibian species and more than 60% of cartilaginous fish species are listed as evaluated. Interestingly, a comparison of the number of species listed as *threatened* with the total number of *evaluated* species shows that 45% of evaluated ray-finned bony fish and mollusc species, 51% of reptiles and an astonishing 62% of arthropods are threatened (Fig. 1.1, black bars/grey bars). These numbers are almost certainly inflated since species from these taxonomic groups are probably more likely to be surveyed if they are noticeably threatened. However, if the evaluations of the less visible animal species are even vaguely representative, we can expect that many of these are also at serious risk of extinction and urgently need to be identified, as well as preserved and publicized. As a case in point, butterflies are among the best-studied arthropods and species extinction is reported in several cases (McLaughlin *et al.* 2002). A comparison of population and regional extinctions of birds, butterflies and vascular plants from Britain shows that butterflies experienced the greatest net losses in recent decades, disappearing on average from 13% of their previously occupied 10-kilometre squares (Thomas *et al.* 2004). If insects elsewhere in the world are similarly sensitive, the known global extinction rates of vertebrate species have an unrecorded parallel among the invertebrates.

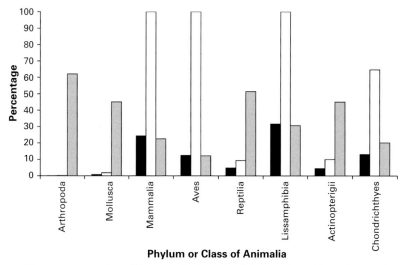

Figure 1.1. Percentage of *known* animal species listed as vulnerable, endangered or critically endangered on the IUCN Red List (2006) (black bars) compared with the total percentage *surveyed* by the IUCN (white bars). Grey bars show the *percentage* of IUCN-surveyed species listed as vulnerable, endangered or critically endangered in each taxonomic category.

THE SYNERGY BETWEEN CONSERVATION BIOLOGY AND GENETICS

Given the scale of the problem and the number of species involved, it is not surprising that conservation biology has emerged as a 'crisis discipline', a multidisciplinary approach to stemming the rapid rise in species' extinctions combining ecology, genetics and wildlife biology. However, the synergy of conservation biology and genetics has been a longer and more painful process than the initial optimism of its founding fathers would have led us to believe.

Conservation biology itself emerged as a field less than 30 years ago when, in 1978, Michael Soulé and Bruce Wilcox organized the 1st International Conference on Conservation Biology at the San Diego Wild Animal Park (Soulé and Wilcox 1980). Seven years later, in May 1985, Soulé, together with Paul Ehrlich and Jared Diamond, founded the Society for Conservation Biology. At that time, Soulé outlined the ethical principles of the field, which included the preservation of species and community diversity, and the maintenance of ecological and evolutionary processes (Soulé 1985). Upholding the convictions of philosophers such as Emerson (1836), Muir (1916), and Naess (Naess 1973; Naess and Sessions 1984), Soulé also maintained that biological diversity has its own intrinsic value. What makes this field unique, and ultimately inspiring, is that it is driven by a wide assortment of biologists from these various disciplines who are extraordinarily dedicated to preserving biological diversity. In fact, Soulé (1986) attributed the upsurge in conservation awareness, in part, to the need for academics to feel 'relevant' or altruistic in the environmental movement of the 1970s.

Conservation biologists realized they needed rapid, efficient, and relatively cheap methods for acquiring the relevant information for planning and implementing conservation strategies. One of the most powerful instruments for the conservation toolbox was immediately identified as population genetics. As far back as 1970, Frankel recognized the importance of genetics to conservation biology (Frankel 1970, 1974), and Frankel and Soulé produced the seminal text on the subject in 1981 (Frankel and Soulé 1981; Bruford 1998). Soulé went on to advocate genetics as the 'future' of conservation biology in later publications (Soulé 1986, 1987; Soulé and Simberloff 1986), as did others throughout the 1980s (e.g. Schonewald-Cox et al. 1983).

At that time, the confidence placed in genetics did not seem far-fetched since the theoretical foundations of this field had been set down some years before by Sewall Wright, John B. S. Haldane and Ronald A. Fisher (Fisher 1930; Wright 1931, 1943; Haldane 1956), and these authors also defined mathematically many of the standard variables we still use today, such as effective population size, mutation rates, and levels of inbreeding. In addition, Fisher established the maximum likelihood approach, and Wright (1940, 1943, 1965) proposed the use of F-statistics (in particular, F_{ST}) to express the partitioning of genetic differentiation (see also Weir and Cockerham 1984; Hartl and Clark 1989), as well as the mathematical basis for metapopulation analysis (the understanding of the spatial distribution of populations, and the relative importance of migration and of each population in maintaining the species); this latter theory was rearticulated by Andrewartha and Birch (1954), as well as Levins (1969, 1970).

But despite significant steps in theoretical population biology, population genetics only became much more useful to conservation biology with the development of a fantastic array of fast, accurate, relatively cheap and non-invasive genetic techniques that allowed the characterization of a sufficient number of loci for statistical analyses. Younger readers may not appreciate the remarkably rapid revolution that many of the authors of this book have witnessed. One of the first major advances, allozyme electrophoresis, only became possible for humans, model laboratory organisms such as *Drosophila* and some wild populations in the mid-1960s (Harris

1966; Hubby and Lewontin 1966; Selander and Yang 1969), and quickly became extremely widespread, especially for testing population genetic theory developed at this time (e.g. Nevo 1984). The advent of allozyme markers was paralleled by a plethora of activity in theoretical population genetics (e.g. Crow and Kimura 1970; Dobzhansky 1970; Lewontin 1974; Nei 1975; Hartl and Clark 1989 and references therein) and island biogeography (MacArthur and Wilson 1963, 1967), leading to hypotheses on the impact of changes in population size and structure, migration and dispersal, selection and drift. However, although electrophoretic methods are extremely cheap, rapid and reliable, and are still in use today for screening many organisms, only a few biologists attempted to apply these methods to endangered species, since large tissue samples were rarely available for these taxa (e.g. Bonnell and Selander 1974). In addition, from a conservation point of view, these protein markers gave somewhat unsatisfactory results because they required the killing of individuals, evolve relatively slowly and are hence non-polymorphic in many small populations (Carvalho 1998).

Successive advances in molecular biology continued to provide hope for conservationists. Painstaking work during the late 1960s and early 1970s to enable the manipulation of DNA molecules saw the advent of RFLP analysis (Linn and Arber 1968) and Southern blotting (Southern 1975), and led to the first comparison of several mitochondrial DNA (mtDNA) sequences (e.g. Brown *et al.* 1979). Further breakthroughs included cloning and manual DNA sequencing techniques (Maxam and Gilbert 1977; Sanger *et al.* 1977; Maniatus *et al.* 1978), but the first study of sequence variation in a natural population was only published by Kreitman in 1983, in *Drosophila*. The amplification and visualization of tandem repeats DNA, or DNA fingerprinting (Jeffreys *et al.* 1985a, b), was also hailed as a technological breakthrough and was applied to the study of some natural populations (e.g. Burke and Bruford 1987; Hoelzel and Amos 1988). However, even throughout the 1980s, laboratory workers were still taking days to extract a few copies of mtDNA using syringes and room-sized centrifuges, and struggling with manual reading of sequences from blurry, radioactive gels.

Then, finally, what conservation biology had been waiting for: PCR, or the polymerase chain reaction, for DNA amplification. Originally developed by Mullis *et al.* (1986) and Saiki *et al.* (1988), this versatile technique soon evolved from coaxing PCR products out of a series of manually controlled waterbaths to relatively cheap, incredibly rapid automated analysis that we see today (Olsen *et al.* 1996; O'Reilly and Wright 1995). These technological advancements in molecular techniques have made possible the characterization of a high number of DNA fragments and thus genetic markers in a large number of individuals in a relatively short period of time (see **Vernesi and Bruford**, this volume).

Conservation biologists were quick to realize the benefits of this simple, robust technique for which only minute tissue samples were required (e.g. Garza and Woodruff 1992; Hoelzel 1994; Morin *et al.* 1994; O'Brien 1994a, b; Mace *et al.* 1996; Mills *et al.* 2000). Robust statistical techniques were soon developed to match the extensive molecular data being produced, and those for a number of DNA sequences became available from the 1980s (Felsenstein 1981, 1995; Saitou and Nei 1987; Miyamoto and Cracraft 1991; Hillis *et al.* 1996). In fact, most conservation biology texts from the 1980s onwards included a chapter on the 'field' of conservation genetics (e.g. Western and Pearl 1989; Fiedler and Jain 1992; Meffe and Carroll 1994; Spellerberg 1996; Steinberg and Jordan 1998), and a number of overviews of the application of these new molecular methods to conservation were published (e.g. Avise 1989, 1994; Hedrick and Miller 1992; Loeschcke *et al.* 1994; Moritz 1994a; Schierwater *et al.* 1994; Avise and Hamrick 1996; McCullough 1996; Smith and Wayne 1996; Hoelzel 1998; Frankham 1999; King and Burke 1999; Landweber and Dobson 1999; Young and Clarke 2000; Hedrick 2002, 2003; Pearse and Crandall 2004).

However, even up to the mid-1990s, the application of genetics to conservation left many deluded, as it still wasn't entirely clear how much population genetics would concretely contribute to conservation. One of the principal reasons for this disappointment was that it had been fervently believed that the estimation of genetic diversity using neutral markers would *in itself* lead to an assessment of the loss of adaptive potential and an understanding of the genetic 'health' of populations and species. In fact, one of the basic premises of conservation genetics was that loss of hetero-zygosity, or genetic variability, could be correlated with a loss in reproduc-tive fitness. However, although Frankel and Soulé (1981) made a lengthy and compelling theoretical and empirical argument to support this corre-lation, only one recent meta-analysis supports it (Reed and Frankham 2003), and this basic premise of conservation genetics always was, and is still hotly debated (Caughley 1994; Gray 1996; Frankham 1999; Hedrick 2001; Reed and Frankham 2001; Moss *et al.* 2003). In addition, although several comparative studies have shown that population size varies directly with neutral genetic diversity (Frankham 1996, 1997, 1998), there has always been some scepticism that the genetic effects of small populations are deleterious, or that they lead to extinction (Berry 1983; Amos and

Balmford 2001; Jamieson 2007; Reed 2007). Lande (1988) also claimed that demographic factors were probably more important than genetics for promoting extinctions in small populations.

It is clear that not all conservationists or geneticists were convinced that measures of neutral genetic diversity would lead to the promised land. In addition to doubts about the interpretation of neutral genetic variation, Caughley (1994) and Meffe and Carroll (1994) expressed concern that overconfidence in genetic theory would lead some conservationists to ignore factors such as habitat destruction and disease. In general these criticisms were useful, leading to better definitions of genetic diversity (Frankham 1995; DeSalle and Amato 2004), inducing theoreticians to develop more accurate estimates and conduct more careful meta-analyses (Spielman *et al.* 2004), and encouraging conservation geneticists to interpret the results of genetical analysis within the wider historical scenario of a population or species (Avise 1996; Moritz 2002; DeSalle 2005). In a way, these arguments also brought conservationists and geneticists to consider a closer collaboration, since it was realized that a thorough knowledge of the natural history of a species was essential for interpreting genetic data (Steinberg and Jordan 1998). However, although measures of neutral genetic diversity currently remain a theoretical concern in themselves as a measure of inbreeding (mainly used for captive breeding programmes), and only secondarily as a (poor) surrogate measure of the loss of adaptive potential, many authors agree that a more accurate estimate of adaptive potential can only be made using genetic variability in quantitative trait loci (QTLs; Knott and Haley 1998; Lynch and Walsh 1998; Beebee 2005; Fitzpatrick *et al.* 2005), although these markers have not yet reached their full potential (Erickson 2005; see chapters in this volume by **Bonin and Bernachez, Tiira and Primmer, Vernesi and Bruford**).

More importantly, other geneticists, such as Avise (1996) have stepped past the genetic diversity debate and pointed out that the most important contribution of genetics to conservation is to significantly increase our knowledge of various aspects of particular species, including behaviour, ecology and evolution (see also Holsinger 1996; Mace *et al.* 1996; Reed 2007). In this regard, the advent of large numbers of nuclear markers and their associated analytical techniques have truly matured the synergy between population genetics and conservation by providing the necessary set of powerful tools for estimating basic ecological and demographic variables. While the amplification and automated sequencing of large, mainly neutral mtDNA molecules is still essential for taxonomic studies and macrogeographic pattern analysis, amplification and analysis of short tandem repeats (STRs or microsatellites) have become popular for conservation geneticists for their versatility, ease of amplification and high heterozygosity and mutation rates, useful for answering population-level questions (Pena *et al.* 1993 and articles therein; Bruford *et al.* 1996; Jarne and Lagoda 1996; Estoup and Angers 1998; Schlötterer 2004; see chapter by **Vernesi and Bruford**, this volume), although single nucleotide polymorphisms (SNPs) are now also used for many studies (Morin *et al.* 2004).

The statistical analysis of large numbers of nuclear markers is under constant development (Bertorelle *et al.* 2004; Pearse and Crandall 2004; see chapters in this volume by: **Anderson, Beaumont, Beerli, Bonin and Bernachez, Buhay *et al.*, Mardulyn *et al.**). For example, mismatch analysis (Slatkin and Hudson 1991; Rogers and Harpending 1992) was the first instrument of this kind to be extensively applied to identify the genetic signatures of past demographic events, while a parallel and more sophisticated approach relies on reconstruction of lineages-through-time plots, which are then compared to expectations from the coalescent theory to reveal past demographic trends (Kingman 1982; Griffiths and Tavare 1994; Harvey *et al.* 1994; Harvey and Steers 1999). More recently, the application of Bayesian frameworks to such coalescent-based approaches is greatly improving the power and accuracy of parameter estimation (Drummond *et al.* 2005 and references therein), and is fostering the current exponential growth in their application to empirical data sets from animal species. In addition, approaches to comprehensive multidisciplinary data analysis have been and are being developed, such as landscape or spatial genetics approaches (Templeton and Georgiadis 1996; Manel *et al.* 2003; Scribner *et al.* 2005; Epps *et al.* 2007), evolutionary conservation genetics (Ferrière *et al.* 2004), as well as the simultaneous analysis of molecular and quantitative genetic data (Moran 2002), simulation modelling (Steinberg and Jordan 1998), and the evaluation of neutral, detrimental and adaptive variation using surveys of genomic data (Kohn *et al.* 2006; see chapter by **Bonin and Bernachez**, this volume).

The development of this 'new synergy' means that characterization of numerous molecular markers combined with theoretical population genetics can now be used to detect and suggest practical solutions, not only to inbreeding and loss of genetic diversity, but also to a long list of real conservation concerns, such as the hybridization of native or captive species with allochthonous individuals by identifying the origin(s), structure, connectivity, taxonomic status and conservation importance of populations (see chapter by **Bruford**, this volume); identification of and/or the effect of wildlife corridors; the definition of sites of reintroduction or restocking and appropriate genotypes and source populations for such interventions; and the detection of illegal harvesting (e.g. Ryder 1986; Moritz 1994b, 1999, 2002; Ballou and Lacy 1995; Avise 1996; Mace *et al.* 1996; Smith and Wayne 1996; Bowen 1999; King and Burke 1999; Carvajal-Carmona *et al.* 2000; Pritchard *et al.* 2000; Dawson and Belkhir 2001; Frankham *et al.* 2002, Wilson and Rannala 2003; DeSalle and Amato 2004; Gaggiotti *et al.* 2004; Cassidy and Gonzales 2005; see chapters in this volume by **Bertorelle** *et al.*, **Bruford, Ciofi** *et al.*, **Hoelzel**). In addition, using mathematical models and molecular genetic data, it is now possible to greatly increase our knowledge of the biology of threatened species, since it is possible to estimate parameters such as effective population size, abundance, fragmentation, gene flow, genetic drift, genetic diversity, sex ratio, patterns of mate choice, pedigree (parentage or relatedness), effective and sex-specific dispersal rates, levels of inbreeding and introgressive exchange, viable population size, breeding system, and effects of bottlenecks and structure (e.g. Allendorf and Leary 1986; Templeton 1986, 1998; Latta and Mitton 1997; Luikart and Cornuet 1998; Ellegren 1999; Hedrick and Kalinowski 2000; Waits and Paetkau 2005).

Happily, most of these studies can now be completed using samples collected non-invasively (such as faeces, hair, skin and body fluids abandoned naturally in the field by individual animals; see chapter by **Goossens and Bruford**, this volume). The molecular analysis of extremely small quantities of DNA (incredibly, from a single cell), even allows the amplification of DNA from fossilized or semi-fossilized museum material (ancient DNA or aDNA), permitting changes in genetic diversity through time, the origin of current haplotypes and/or past dispersal patterns to be assessed (e.g. Roy *et al.* 1994; Rosenbaum *et al.* 2000; Barnes *et al.* 2002; Hedrick and Waits 2005; see chapter by **Beadell** *et al.*, this volume). At higher ecological levels, biodiversity can be estimated (Avise 1994; Mace *et al.* 1996); most recently, DNA barcoding has become a widely used method for species identification in such studies (Flintoft 2004).

POPULATION GENETICS FOR ANIMAL CONSERVATION (PGAC) WORKSHOP

The use of population genetics to provide demographic and ecological information to conservationists means that, gradually, the theories are being applied to many fields of wildlife ecology and management. It is widely accepted among conservation biologists that genetics is, more than ever, an essential and efficient tool for wild and captive population management and reserve design (Gray 1996). Vrijenhoek (1989) and Hedrick (2005) add that conservation genetics has the potential to 'set things straight', or restore what we can rather than just preserve what's left after our destructive activities (sometimes referred to as 'restoration genetics'). However, there are continuous calls for genetics to be applied more often and more rigorously to conservation problems (e.g. Milligan *et al.* 1994; Haig 1998; Hedrick 2004; Wayne and Morin 2004; DeYoung and Brennan 2005; DeYoung and Honeycutt 2005; Hogg *et al.* 2006), and especially to under-represented taxa, which, despite optimistic predictions by Burke (see King and Burke 1999), still include almost all invertebrates (e.g. Darvill *et al.* 2005), and microbial communities (Muyzer *et al.* 1993).

Therefore, to further encourage the 'new' synergy of population genetics and animal conservation, and to promote the exchange of ideas and expertise, the first PGAC conference was organized at the Centro di Ecologia Alpina (recently incorporated into the Fondazione Edmund Mach), nestled in the Dolomites near Trento, Italy in September 2003. It was designed as an intensive, international workshop, to discuss the latest theories, software, case studies and controversial issues concerning the genetics of animal conservation. It did not set out to be a 'conservation genetics' gathering as such, but a meeting of theoretical population geneticists interested in conservation genetics, and conservation biologists interested in population genetics methodology. The organizers reasoned that while the theoreticians publish detailed theoretical or statistical methods, they generally contemplate the practical implications of their results superficially; on the other hand, the researchers involved in the practical problems of conservation do not have always the tools or the time to follow and understand recent developments in theoretical population genetics, thereby making inefficient use of their valuable and often hard-won data. The PGAC workshop was designed to bridge this gulf.

Given the global urgency of animal conservation we felt it was imperative that results of the discussions at the PGAC conference were made accessible to graduate students and wildlife managers. Hence, the present volume is an advanced textbook that integrates 'the analytical methods approach' with the 'real problems approach' in conservation genetics. Most chapters are based on presentations made by speakers at the PGAC workshop, but several chapters have also been added to fill obvious gaps. Each author was encouraged to collaborate with other contributors in order to produce a comprehensive review of their area of expertise. As an advanced textbook, this volume does not intend to provide an inclusive introduction to either population genetics or conservation genetics. For this, readers are directed to several excellent manuals: a comprehensive undergraduate text on conservation genetics was published in 2002 (Frankham *et al.* 2002), while essential guides to using genetic information to develop conservation plans can be found in Allendorf and Luikart (2007) and Taberlet and Luikart (in press). Amato *et al.* (2006) also supply the results of their symposium *Conservation Genetics in the Age of Genomics* (American Museum of Natural History, 2001) in a book of the same name. Instead, this book aims to review advanced analytical approaches to animal conservation and how they have been successfully applied.

As you will have noted from the title, and from the chapters indicated in **bold** above, this book primarily provides reviews of recent developments in the field of population genetics as applied to conservation. This book is divided into four sections: Statistical Approaches, Data Analysis and Inference; Molecular Approaches and Applications; From Genetic Data to Practical Management: Issues and Case Studies and Future Directions in Conservation Genetics. The statistical and methodological contributions of the first two sections assess, in an approachable language, recent theoretical and methodological techniques for analysing genetic data for conservation and management purposes, providing examples describing their potential applications and, when possible, the available computer software. The issues and case studies section covers problems and controversies in the conservation of different taxonomic groups, and describes how novel analyses of genetic data have contributed or could contribute to their resolution. The last section provides a prospective on future methodological and theoretical developments.

We trust that this book goes some small but significant way toward creating a mutually beneficial synergy between population genetics and conservation biology, creating an overview of currently available technologies and analytical approaches to academics and policy-makers alike. We will have achieved our purpose if you the reader take these concepts and apply them to practical management problems. However, we would add that these tools are no substitute for awareness and conscientious reduction of the impact of human activities on natural environments. As the conservation geneticist John C. Avise (1996) asserted:

genetic perspectives truly have enriched our understanding of nature. Let us relish these important contributions, but at the same time retain sight of the underlying root of the conservation predicament.

We too encourage such an open, broad-minded, and multidisciplinary approach.

ACKNOWLEDGEMENTS

The authors would like to thank Michael W. Bruford, Gabriele Gentile and Kathryn M. Rodrigúez-Clark whose comments and suggestions greatly improved the final text. HCH was supported by the Edmund Mach Foundation and the Centro di Ecologia Alpina during the preparation of this manuscript.

REFERENCES

Allendorf, F. W. and Leary, R. F. (1986). Heterozygosity and fitness in natural populations of animals. In *Conservation Biology: The Science of Scarcity and Diversity*, ed. M. E. Soulé. Sunderland: Sinauer Associates, pp. 57–76.

Allendorf, F. W. and Luikart, G. (2007). *Conservation and the Genetics of Populations*. Oxford: Blackwell Publishing.

Amato, G., Ryder, O. A., Rosenbaum, H. C. and DeSalle, R., eds. (2006). *Conservation Genetics in the Age of Genomics*. New York: Columbia University Press.

Amos, W. and Balmford, A. (2001). When does conservation genetics matter? *Heredity*, **87**, 257–265.

Andrewartha H. G. and Birch, L. C. (1954). *The Distribution and Abundance of Animals*. Chicago: University of Chicago Press.

Avise, J. C. (1989). A role for molecular genetics in the recognition and conservation of endangered species. *Trends in Ecology and Evolution*, **4**, 279–281.

Avise, J. C. (1994). *Molecular Markers, Natural History and Evolution*. New York: Chapman and Hall.

Avise, J. C. (1996). Introduction: the scope of conservation genetics. In *Conservation Genetics: Case Histories from Nature*, ed. J. C. Avise and J. L. Hamrick. New York: Chapman and Hall, pp. 1–9.

Avise, J. C. and Hamrick, J. L., eds. (1996). *Conservation Genetics: Case Histories from Nature*. New York: Chapman and Hall.

Ballou, J. D. and Lacy, R. C. (1995). Identifying genetically important individuals for management of genetic variation in pedigreed populations. In *Management for Survival and Recovery*, ed. J. D. Ballou, M. Gilpin and T. J. Foose. New York: Columbia University Press, pp. 76–111.

Barnes, I., Matheus, P., Shapiro, B., Jensen, D. and Cooper, A. (2002). Dynamics of Pleistocene population extinctions in Beringian brown bears. *Science*, **295**, 2267–2270.

Beebee, T. J. C. (2005). Conservation genetics of amphibians. *Heredity*, **95**, 423–427.

Berry, R. J. (1983). Genetics and conservation. In *Conservation in Perspective*, ed. A. Warren and F. B. Goldsmith. Chichester: John Wiley, pp. 141–156.

Bertorelle, G., Bruford M., Chemini, C., Vernesi C. and Hauffe, H. C. (2004). New, flexible Bayesian approaches to revolutionize conservation genetics. *Conservation Biology*, **18**, 1–2.

Black, R. (2006). 'Only 50 years left' for sea fish. BBC News Online http://news.bbc.co.uk/go/pr/fr/-/2/hi/science/nature/6108414.stm.

Bonnell, M. L. and Selander, R. K. (1974). Elephant seals: genetic variation and near extinction. *Science*, **184**, 980–989.

Bowen, B. W. (1999). Preserving genes, species, or ecosystems? Healing the fractured foundations of conservation policy. *Molecular Ecology*, **8**, S5–S10.

Brown, W. M., George, M. and Wilson, A. C. (1979). Rapid evolution of animal mitochondrial DNA. *Proceedings of the National Academy of Sciences USA*, **76**, 1967–1971.

Bruford, M. W. (1998). Conservation genetics finally brought to book – but is it really worth the money? *Biodiversity and Conservation*, **7**, 405–413.

Bruford, M. W., Cheesman, D. J., Coote, T. *et al.* (1996). Microsatellites and their application to conservation genetics. In *Molecular Genetic Approaches in Conservation*, ed. T. B. Smith and R. K. Wayne. Oxford: Oxford University Press, pp. 278–297.

Burke, T. and Bruford, M. W. (1987). DNA fingerprinting in birds. *Nature*, **327**, 149–52.

Butchart, S. H. M., Statterfield, A. J., Baillie, J. *et al.* (2005). Using Red List indices to measure progress towards the 2010 target and beyond. *Philosphical Transactions of the Royal Society Series B*, **360**, 255–268.

Cardillo, M., Mace, G. M., Gittleman, J. L. and Purvis, A. (2006). Latent extinction risk and the future battlegrounds of mammal conservation. *Proceedings of the National Academy of Sciences USA*, **103**, 4157–4161.

Carvajal-Carmona, L. G., Soto, I. D., Pineda, N. *et al.* (2000). Strong Amerind/white sex bias and a possible Sephardic contribution among the founders of a population in northwest Colombia. *American Journal of Human Genetics*, **67**, 1287–1295.

Carvalho, G. R., ed. (1998). *Advances in Molecular Ecology*, NATO Science Series A: Life Sciences vol. 306. Amsterdam: IOS Press.

Cassidy, B. G. and Gonzales, R. A. (2005). DNA testing in animal forensics. *Journal of Wildlife Management*, **69**, 1454–1462.

Caughley, G. (1994). Directions in conservation biology. *Journal of Animal Ecology*, **63**, 215–244.

Crow, J. F. and Kimura, M. (1970). *An Introduction to Population Genetics Theory*. New York: Harper and Row.

Darvill, B., Ellis, J. S., Lye, G. C. and Goulson, D. (2005). Population structure and inbreeding in a rare and declining bumblebee, *Bombus muscorum* (Hymenoptera: Apidae). *Molecular Ecology*, **15**, 601–611.

Dawson, K. J. and Belkhir, K. (2001). A Bayesian approach to the identification of panmictic populations and assignment of individuals. *Genetical Research*, **78**, 59–77.

DeSalle R. 2005. Genetics at the brink of extinction. *Heredity*, **94**, 386–387.

DeSalle, R. and Amato, G. (2004). The expansion of conservation genetics. *Nature Review Genetics*, **5**, 702–712.

Devine, J. A., Baker, K. D. and Haedrich, R. L. (2006). Fisheries: deep-sea fishes qualify as endangered. *Nature*, **439**, 29.

DeYoung, R. W. and Brennan, L. A. (2005). Molecular genetics in wildlife science, conservation and management. *Journal of Wildlife Management*, **69**, 1360–1361.

DeYoung, R. W. and Honeycutt, R. L. (2005). The molecular toolbox: genetic techniques in wildlife ecology and management. *Journal of Wildlife Management*, **69**, 1362–1384.

Dobzhansky, Th. (1970). *Genetics of the Evolutionary Process*. New York: Columbia University Press.

Drummond, A. J., Rambaut, A., Shapiro, B. and Pybus, O. G. (2005). Bayesian coalescent inference of past population dynamics from molecular sequences. *Molecular Biology and Evolution*, **22**, 1185–1192.

Ehrlich, P. R. and Wilson, E. O. (1991). Biodiversity studies: science and policy. *Science*, **253**, 758–762.

Ellegren, H. (1999). Inbreeding and relatedness in Scandinavian grey wolves *Canis lupus. Hereditas*, **103**, 239–244.

Emerson, R. W. (1836). *Nature*. Boston: James Monroe and Co.

Erickson, D. (2005). Mapping the future of QTLs. *Heredity*, **95**, 417–418.

Epps, C. W., Wehausen, J. D., Bleich, V. C., Torres, S. G. and Brashres, S. (2007). Optimizing dispersal and corridor models using landscape genetics. *Journal of Applied Ecology*, **44**, 714–724.

Estoup, A. and Angers, B. (1998). Microsatellites and minisatellites for molecular ecology: theoretical and empirical considerations. In *Advances in Molecular Ecology*. NATO Science Series A: Life Sciences vol. 306, ed. G. R. Carvalho. Amsterdam: IOS Press, pp. 55–86.

Felsenstein, J. (1981). Evolutionary trees from DNA sequences: a maximum likelihood approach. *Journal of Molecular Evolution*, **17**, 368–376.

Felsenstein, J. (1995). *Theoretical Evolutionary Genetics*. Seattle: ASUW Publishing.

Ferrière, R., Dieckmann, U. and Couvet, D., eds. (2004). *Evolutionary Conservation Biology*. Cambridge: Cambridge University Press.

Fiedler, P. L. and Jain, S. K., eds. (1992). *Conservation Biology*. New York: Chapman and Hall.

Fisher, R. A. (1930). *The Genetical Theory of Natural Selection*. Oxford: Clarendon Press.

Fitzpatrick M. J., Ben-Shahar, Y., Smid, H. M. *et al.* (2005). Candidate genes for behavioural ecology. *Trends in Ecology and Evolution*, **20**, 96–104.

Flintoft, L. (2004). A barcode for life? *Nature Review Genetics*, **5**, 805.

Frankel O. H. (1970). Variation, the essence of life. *Proceedings of the Linnean Society of New South Wales*, **95**, 158–169.

Frankel, O. H. (1974). Genetic conservation: our evolutionary responsibility. *Genetics*, **78**, 53–65.

Frankel, O. H. and Soulé, M. E. (1981). *Conservation and Evolution*. Cambridge: Cambridge University Press.

Frankham, R. (1995). Conservation genetics. *Annual Review of Genetics*, **29**, 305–327.

Frankham, R. (1996). Relationships of genetic variation to population size in wildlife. *Conservation Biology*, **10**, 1500–1508.

Frankham, R. (1997). Do island populations have less genetic variation than mainland populations? *Heredity*, **78**, 311–327.

Frankham, R. (1998). Inbreeding and extinction: Island populations. *Conservation Biology*, **12**, 665–675.

Frankham, R. (1999). Quantitative genetics in conservation biology. *Genetical Research Cambridge*, **74**, 237–244.

Frankham, R., Ballou, J. D. and Briscoe, D. A. (2002). *Introduction to Conservation Genetics*. Cambridge: Cambridge University Press.

Gabriel, S. (2007). Biodiversity 'fundamental' to economics. BBC News Online. http://news.bbc.co.uk/2/hi/science/nature/6432217.stm.

Gaggiotti, O. E., Brooks, S. P., Amos, W. and Harwood, J. (2004). Combining demographic, environmental and genetic data to test hypotheses about colonization events in metapopulations. *Molecular Ecology*, 1, 811–825.

Garza, J. C. and Woodruff, D. S. (1992). A phylogenetic study of the gibbons (*Hylobates*) using DNA obtained nondestructively from hair. *Molecular Phylogenetics and Evolution*, 1, 202–210.

Gianni, M. (2004). *High Seas Bottom Trawl Fisheries and their Impacts on the Biodiversity of Vulnerable Deep-Sea Ecosystems: Options for International Action.* Gland: IUCN – The World Conservation Union.

Gray, A. J. (1996). The genetic basis of conservation biology. In *Conservation Biology*, ed. I. F. Spellerberg. Edinburgh: Longman, pp. 115–121.

Griffiths, R. C. and Tavare, S. (1994). Sampling theory for neutral alleles in a varying environment. *Philosophical Transactions of the Royal Society Series B*, **344**, 403–410.

Haig, S. M. (1998). Molecular contributions to conservation. *Ecology*, **79**, 413–425.

Haldane, J. B. S. (1956). The estimation of viabilities. *Journal of Genetics*, **54**, 294–296.

Halliday, T. (2006). All is silent down at the pond. BBC News Viewpoint 27 January 2006. http://news.bbc.co.uk/go/pr/fr/-/2/hi/science/nature/4582024.stm.

Harris, H. (1966). Enzyme polymorphisms in man. *Proceedings of the Royal Society Series B*, **164**, 298–310.

Hartl, D. L. and Clark, A. G. (1989). *Principles of Population Genetics.* Sunderland: Sinauer Associates.

Harvey, P. H. and Steers, H. (1999). On use of phylogenies for conservation biologists: inferring population history from gene sequences. In *Genetics and the Extinction of Species*, ed. L. F. Landweber and A. P. Dobson. Princeton: Princeton University Press, pp. 101–120.

Harvey, P. H., May, R. R. and Nee, S. (1994). Phylogenies without fossils. *Evolution*, **48**, 523–529.

Hedrick, P. W. (2001). Conservation genetics: where are we now? *Trends in Ecology and Evolution*, **16**, 629–636.

Hedrick, P. W. (2002). Application of molecular genetics to managing endangered species. In *Population Viability Analysis: Assessing Models for Recovering Endangered Species*, ed. S. R. Beissinger and D. R. McCullough. Chicago: University of Chicago Press, pp. 367–387.

Hedrick, P. W. (2003). Conservation biology: The impact of population biology and a current perspective. In *The Evolution of Population Biology Modern Synthesis*, ed. R. Singh and M. Uyenoyama. Cambridge: Cambridge University Press, pp. 347–365.

Hedrick, P. W. (2004). Recent developments in conservation genetics. *Forest Ecology and Management*, **197**, 3–19.

Hedrick, P. W. (2005). 'Genetic restoration': a more comprehensive perspective than 'genetic rescue'. *Trends in Ecology and Evolution*, **20**, 109.

Hedrick, P. W. and Kalinowski, S. T. (2000). Inbreeding depression in conservation biology. *Annual Review of Ecology and Systematics*, **31**, 139–216.

Hedrick, P. W. and Miller, P. S. (1992). Conservation genetics: techniques and fundamentals. *Ecological Applications*, **2**, 30–46.

Hedrick, P. and Waits, L. (2005). What ancient DNA tells us. *Heredity*, **94**, 463–464.

Hillis, D. M., Moritz, C. and Mable, B. K., eds. (1996). *Molecular Systematics.* Sunderland: Sinauer Associates.

Hoelzel, A. R. (1994). Genetics and ecology of whales and dolphins. *Annual Review of Ecology and Systematics*, **25**, 377–399.

Hoelzel, A. R., ed. (1998). *Molecular Genetic Analysis of Populations: A Practical Approach*, 2nd edn. Oxford: Oxford University Press.

Hoelzel, A. R. and Amos, W. (1988). DNA fingerprinting and 'scientific' whaling. *Nature*, **333**, 305.

Hogg, J. T., Forbes, S. H., Steele, B. M. and Luikart, G. (2006). Genetic rescue of an insular population of large mammals. *Proceedings of the Royal Society Series B*, **273**, 1491–1499.

Holsinger, K. E. (1996). The scope and the limits of conservation genetics. *Evolution*, **50**, 2558–2561.

Hubby, J. L. and Lewontin, R. C. (1966). A molecular approach to the study of genic heterozygosity in natural populations. I. The number of alleles at different loci in *Drosophila pseudoobscura*. *Genetics*, **54**, 577–594.

IUCN (2006). *2006 IUCN Red List of Threatened Species*. www.iucnredlist.org.

Jamieson, I. G. (2007). Role of genetic factors in extinction of island endemics: complementary or competing explanations? *Animal Conservation*, **11**, 151–153.

Jarne, P. and Lagoda, P. J. L. (1996). Microsatellites, from molecules to populations and back. *Trends in Ecology and Evolution*, **11**, 424–429.

Jeffreys, A. J., Wilson, V. and Thein, S. L. (1985a). Hypervariable ministatellite regions in human DNA. *Nature*, **314**, 76–79.

Jeffreys, A. J., Wilson, V. and Thein, S. L. (1985b). Individual-specific fingerprints of human DNA. *Nature*, **316**, 76–79.

King, T. L. and Burke, T., eds. (1999). Special issue on gene conservation: identification and management of genetic diversity. *Molecular Ecology*, **8**, S1–S3.

Kingman, J. F. C. (1982). On the genealogy of large populations. *Journal of Applied Probability*, **19A**, 27–43.

Kohn, M. H., Murphy, W. J., Ostrander, E. A. and Wayne, R. K. (2006). Genomics and conservation genetics. *Trends in Ecology and Evolution*, **21**, 629–637.

Knott, S. A. and Haley, C. S. (1998). Simple multiple-marker sib pair analysis for mapping quantitative traits. *Heredity*, **81**, 48–54.

Kreitman, M. (1983). Nucleotide polymorphism at the alcohol dehydrogenase locus of *Drosophila melanogaster*. *Nature*, **304**, 412–417.

Lande, R. (1988). Genetics and demography in biological conservation. *Science*, **241**, 1455–1460.

Landweber, L. F. and Dobson, A. P., eds. (1999). *Genetics and the Extinction of Species: DNA and the Conservation of Biodiversity*. Princeton: Princeton University Press.

Latta, R. G. and Mitton, J. B. (1997). A comparison of population differentiation across four classes of gene markers in limber pine (*Pinus flexilis* James). *Genetics*, **146**, 1153–1163.

Lawton, J. H. and May, R. M., eds. (1995). *Extinction Rates*. Oxford: Oxford University Press.

Lecointre, G. and Le Guyader, H. (2001). *Classification phylogénétique du vivant*. Paris: Belin.

Levins, R. (1969). Some demographic and genetic consequences of environmental heterogeneity for biological control. *Bulletin of the Entomological Society of America*, **15**, 237–240.

Levins, R. (1970). Extinction. In *Some Mathematical Problems in Biology*, ed. M. Gesternhaber. Providence: American Mathematical Society, pp. 77–107.

Lewontin, R. (1974). *The Genetic Basis of Evolutionary Change*. New York: Columbia University Press.

Linn, S. and Arber, W. (1968). Host specificity of DNA produced by *Escherichia coli*. X. *In vitro* restriction of phage fd replicative form. *Proceedings of the National Academy of Sciences USA*, **59**, 1300–1306.

Loeschcke, V., Tomiuk, J. and Jain, S. K., eds. (1994). *Conservation Genetics*. Basel: Birkhäuser Verlag.

Luikart, G. and Cornuet, J.-M. (1998). Empirical evaluation of a test for identifying recently bottlenecked populations from allele frequency data. *Conservation Biology*, **12**, 228–237.

Lynch, M. and Walsh, B. (1998). *Genetics and Analysis of Quantitative Traits*. Sunderland: Sinauer Associates.

MacArthur, R. H. and Wilson, E. O. (1963). An equilibrium theory of insular zoogeography. *Evolution*, **17**, 373–387.

MacArthur, R. H. and Wilson, E. O. (1967). *The Theory of Island Biogeography*. Princeton: Princeton University Press.

Mace, G. M., Smith, T. B., Bruford, M. W. and Wayne, R. K. (1996). An overview of the issues. In *Molecular Genetic Approaches in Conservation*, ed. T. B. Smith and R. K. Wayne. Oxford: Oxford University Press, pp. 1–21.

Manel, S., Schwartz, M. K., Luikart, G. and Taberlet, P. (2003). Landscape genetics: combining landscape ecology and population genetics. *Trends in Ecology and Evolution*, **18**, 189–197.

Maniatis, T., Hardison, R. C., Lacy, E. *et al.* (1978). The isolation of structural genes from libraries of eukaryotic DNA. *Cell*, **15**, 687–701.

Marsh, H., Dennis, A., Hines, H. *et al.* (2007). Optimizing allocation of management resources for wildlife. *Conservation Biology*, **21**, 387–399.

Maxam, A. M. and Gilbert, W. (1977). A new method for sequencing DNA. *Proceedings of the National Academy of Sciences USA*, **74**, 560–564.

McCullough, D. R., ed. (1996). *Metapopulations and Wildlife Conservation*. Washington, DC: Island Press.

McLaughlin, J. F., Hellmann, J., Boggs, C. L. and Ehrlich, P. R. (2002). The route to extinction: population dynamics of a threatened butterfly. *Oecologia*, **132**, 538–548.

Meffe, G. K. and Carroll, C. R., eds. (1994). *Principles of Conservation Biology*. Sunderland: Sinauer Associates.

Milligan, B. G., Leebensmack, J. and Strand, A. E. (1994). Conservation genetics – beyond the maintenance of marker diversity. *Molecular Ecology*, **3**, 423–435.

Mills L. S., Citta, J. J., Lair, K. P., Schwartz, M. K. and Tallmon, D. A. (2000). Estimating animal abundance using noninvasive DNA sampling: promise and pitfalls. *Ecological Applications*, **10**, 283–294.

Miyamoto, M. M., and Cracraft, J., eds. (1991). *Phylogenetic Analysis of DNA Sequences*. New York: Oxford University Press.

Moran, P. (2002). Current conservation genetics: building an ecological approach to the synthesis of molecular and quantitative genetic methods. *Ecology of Freshwater Fish*, **11**, 30–55.

Morin, P. A., Luikart, G., Wayne, R. K. and The Single Nucleotide Polymorphism Workshop Group (2004). SNPs in ecology, evolution, and conservation. *Trends in Ecology and Evolution*, **19**, 208–216.

Morin, P. A., Moore, J. J., Chakraborty, R. *et al.* (1994). Kin selection, social structure, gene flow and the evolution of chimpanzees. *Science*, **265**, 1193–1201.

Moritz, C. (1994a). Application of mitochondrial DNA analysis on conservation: a critical review. *Molecular Ecology*, **3**, 401–411.

Moritz, C. (1994b). Defining 'Evolutionarily Significant Units' for conservation. *Trends in Ecology and Evolution*, **9**, 373–375.

Moritz, C. (1999). Conservation units and translocation: strategies for conserving evolutionary processes. *Hereditas*, **130**, 217–228.

Moritz, C. (2002). Strategies to protect biological diversity and the evolutionary processes that sustain it. *Systematic Biology*, **51**, 238–254.

Moss, R., Piertney, S. B. and Palmer, S. C. F. (2003). The use and abuse of microsatellite DNA markers in conservation biology. *Wildlife Biology*, **9**, 243–250.

Muir, J. (1916). *A Thousand-Mile Walk to the Gulf.* Boston: Houghton Mifflin.

Mullis, K., Faloona, F., Svharf, R. *et al.* (1986). Specific enzymatic amplification of DNA *in vitro*: the polymerase chain reaction. *Cold Spring Harbor Symposium on Quantitative Biology*, **51**, 263–273.

Muyzer, G., De Wall, E. C. and Uiterlinden, A. G. (1993). Profiling of complex microbial populations by denaturing gradient gel electrophoresis analysis of polymerase chain reaction amplified genes coding for 16S rRNA. *Applied and Environmental Microbiology*, **59**, 695–700.

Naess, A. (1973). The shallow and the deep, long-range ecology movements: a summary. *Inquiry*, **16**, 95–100.

Naess, A. and Sessions, G. (1984). Basic principles of deep ecology. *Ecophilosophy*, **6**, 3–7.

Naidoo, R. and Ricketts, T. H. (2006). Mapping the economic costs and benefits of conservation. *PLoS Biology*, **4**, 2153–2164.

Nei, M. (1975). *Molecular Population Genetics and Evolution.* Amsterdam: North-Holland.

Nevo, E. (1984). The evolutionary significance of genic diversity: ecological, demographic and life history correlates. *Lecture Notes in Biomathematics*, **53**, 13–213.

O'Brien, S. J. (1994a). Genetic and phylogenetic analyses of endangered species. *Annual Review of Genetics*, **28**, 467–489.

O'Brien, S. J. (1994b). A role for molecular genetics in biological conservation. *Proceedings of the National Academy of Sciences USA*, **91**, 5748–5755.

Olsen, J. B., Wenburg, J. K. and Benzen P. (1996). Semiautomated multilocus genotyping of pacific salmon (*Oncorhynchus* spp.) using microsatellites. *Molecular Marine Biology and Biotechnology*, **5**, 259–272.

O'Reilly, P. and Wright, J. M. (1995). The evolving technology of DNA fingerprinting and its application to fisheries and aquaculture. *Journal of Fish Biology*, **47** (Suppl. A), 29–55.

Pearse, D. E. and Crandall, K. A. (2004). Beyond F-ST: Analysis of population genetic data for conservation. *Conservation Genetics*, **5**, 585–602.

Pena, S. D. J., Chakraborty, R., Epplen, J. T. and Jeffreys A. J., eds. (1993). *DNA Fingerprinting: The State of the Science*. Basel: Birkhäuser Verlag.

Primack, R. B. (2002). *Essentials of Conservation Biology*, 3rd edn. Sunderland: Sinauer Associates.

Pritchard, J. K., Stephens, M. and Donnelly, P. (2000). Inference of population structure using multilocus genotype data. *Genetics*, **155**, 945–959.

Reed, D. H. (2007). Extinction of island endemics: it is not inbreeding depression. *Animal Conservation*, **10**, 145–148.

Reed, D. H. and Frankham, R. (2001). How closely correlated are molecular and quantitative measures of genetic variation? A meta-analysis. *Evolution*, **55**, 1095–1102.

Reed, D. H. and Frankham, R. (2003). Correlation between fitness and genetic diversity. *Conservation Biology*, **17**, 230–237.

Rodrigues, A. S. L., Pilgrim, J. D., Lamoreux, J. F., Hoffman, M. and Brooks, T. M. (2006). The value of the IUCN Red List for conservation. *Trends in Ecology and Evolution*, **21**, 71–76.

Rogers, A. R. and Harpending, H. (1992). Population growth makes waves in the distribution of pairwise genetic differences. *Molecular Biology and Evolution*, **9**, 552–569.

Rosenbaum, H. C., Egan, M. G., Clapham, P. J. *et al.* (2000). Utility of north Atlantic right whale museum specimens for assessing changes in genetic diversity. *Conservation Biology*, **14**, 1837–1842.

Roy, M. S., Girman, D. J., Taylor, A. C. and Wayne, R. K. (1994). The use of museum specimens to reconstruct the genetic-variability and relationships of extinct populations. *Experientia*, **50**, 551–557.

Ryder, O. A. (1986). Species conservation and systematics: the dilemma of subspecies. *Trends in Ecology and Evolution*, **1**, 9–10.

Saiki, R. K., Gelfand, D. H., Stoffel, S. *et al.* (1988). Primer-directed amplification of DNA with a thermostable DNA polymerase. *Science*, **239**, 487–491.

Saitou, N. and Nei, M. (1987). The neighbor-joining method: a new method for reconstructing phylogenetic trees. *Molecular Biology and Evolution*, **4**, 406–425.

Sanger, F., Nicklen, S. and Coulsen, A. R. (1977). DNA sequencing with chain-terminating inhibitors. *Proceedings of the National Academy of Sciences USA*, **74**, 5463–5467.

Schierwater, B., Streit, B., Wagner, G. P. and DeSalle, R., eds. (1994). *Molecular Ecology and Evolution: Approaches and Applications*. Basel: Birkhäuser Verlag.

Schlötterer, C. (2004). The evolution of molecular markers: just a matter of fashion? *Nature Review Genetics*, **5**, 63–69.

Schonewald-Cox, C. M., Chambers, S. M., MacBryde, B. and Thomas, L., eds. (1983). *Genetics and Conservation: A Reference for Managing Wild Animal and Plant Populations*. Menlo Park: Benjamin/Cummings.

Scribner, K. T., Blanchong, J. A., Bruggeman, D. J. *et al.* (2005). Geographical genetics: conceptual foundations and empirical applications of spatial genetic data in wildlife management. *Journal of Wildlife Management*, **69**, 1434–1453.

Selander, R. K. and Yang, S. Y. (1969). Protein polymorphism and genic heterozygosity in a wild population of the house mouse (*Mus musculus*). *Genetics*, **63**, 653–667.

Slatkin, M. and Hudson, R. R. (1991). Pairwise comparisons of mitochondrial DNA sequences in stable and exponentially growing populations. *Genetics*, **129**, 555–562.

Smith, F. D. M., May, R. M., Pellew, R., Johnson, T. H. and Walter, K. R. (1993). How much do we know about the current extinction rate? *Trends in Ecology and Evolution*, **8**, 375–378.

Smith, T. B. and Wayne, R. K., eds. (1996). *Molecular Genetic Approaches in Conservation*. New York: Oxford University Press.

Soulé, M. E. (1985). What is conservation biology? *BioScience*, **35**, 727–734.

Soulé, M. E., ed. (1986). *Conservation Biology: The Science of Scarcity and Diversity*. Sunderland: Sinauer Associates.

Soulé, M. E., ed. (1987). *Viable Populations for Conservation*. Cambridge: Cambridge University Press.

Soulé, M. E. and Simberloff, D. (1986). What do genetics and ecology tell us about the design of nature reserves? *Biological Conservation*, **35**, 19–40.

Soulé, M. E., and Wilcox, B. A., eds. (1980). *Conservation Biology: An Evolutionary–Ecological Perspective*. Sunderland: Sinauer Associates.

Southern, E. M. (1975). Detection of specific sequences among DNA fragments separated by gel electrophoresis. *Journal of Molecular Biology*, **98**, 503–517.

Spellerberg, I. F., ed. (1996). *Conservation Biology*. Edinburgh: Longman.

Spielman, D., Brook, B. W. and Frankham, R. (2004). Most species are not driven to extinction before genetic factors impact them. *Proceedings of the National Academy of Sciences USA*, **101**, 15261–15264.

Steinberg, E. K. and Jordan, C. E. (1998). Using molecular genetics to learn about the ecology of threatened species: the allure and the illusion of measuring genetic structure in natural populations. In *Conservation Biology for the Coming Decade*, (2nd edn) ed. P. L. Fiedler and P. M. Kareiva. New York: Chapman and Hall, pp. 440–460.

Stern, N. (2007). *The Economics of Climate Change: The Stern Review*. Cambridge: Cambridge University Press.

Taberlet, P. and Luikart, G. (2008). *Ecological and Conservation Genetics: A Handbook of Techniques*. Oxford: Oxford University Press.

Templeton, A. R. (1986). Coadaptation and outbreeding depression. In *Conservation Biology: The Science of Scarcity and Diversity*, ed. M. E. Soulé. Sunderland: Sinauer Associates, pp. 105–116.

Templeton, A. R. (1998). Nested clade analysis of phylogeographic data: testing hypotheses about gene flow and population history. *Molecular Ecology*, **7**, 381–397.

Templeton, A. R. and Georgiadis, N. J. (1996). A landscape genetic approach to conservation genetics: conserving evolutionary potential in the African Bovidae. In *Conservation Genetics: Case Histories from Nature*, ed. J. C. Avise and J. L. Hamrick. New York: Chapman and Hall, pp. 398–430.

Thomas, J. A., Telfer, M. G., Roy, D. B. *et al.* (2004). Comparative losses of British butterflies, birds, and plants and the global extinction crisis. *Science*, **303**, 1879–1881.

Vrijenhoek, R. C. (1989). Population genetics and conservation. In *Conservation for the Twenty-First Century*, ed. D. Western and M. C. Pearl. New York: Oxford University Press, pp. 89–98.

Waits, L. P. and Paetkau, D. (2005). New noninvasive genetic sampling tools for wildlife biologists: a review of applications and recommendations for accurate data collection. *Journal of Wildlife Management*, **69**, 1419–1433.

Wayne, R. K. and Morin, P. A. (2004). Conservation genetics in the new molecular age. *Frontiers in Ecology and the Environment*, **2**, 89–97.

Weir, B.S. and Cockerham, C.C. (1984). Estimating *F*-statistics for the analysis of population structure. *Evolution* **38**, 1358–1370.

Western, D. and Pearl, M.C., eds. (1989). *Conservation for the Twenty-First Century.* New York: Oxford University Press.

Wilson, G. A. and Rannala, B. (2003). Bayesian inference of recent migration rates using multilocus genotypes. *Genetics*, **163**, 1177–1191.

Wright, S. (1931). Evolution in Mendelian populations. *Genetics* **16**, 97–159.

Wright, S. (1940). Breeding structure of populations in relation to speciation. *American Naturalist*, **74**, 232–248.

Wright, S. (1943). Isolation by distance. *Genetics*, **28**, 114–138.

Wright, S. (1965). The interpretation of population structure by *F*-statistics with special regard to systems of mating. *Evolution*, **19**, 395–420.

Young, A. G. and Clarke, G. M. (2000). *Genetics, Demography and Viability of Fragmented Populations.* Cambridge: Cambridge University Press.

Statistical approaches, data analysis and inference

Statistical methods for identifying hybrids and groups

ERIC C. ANDERSON

INTRODUCTION

Recently, statistical geneticists have developed a number of model-based methods that use genetic data to infer the population of origin of the gene copies within an individual. In this chapter we focus on three of these methods which are known by the software that implements them: STRUCTURE (Pritchard *et al.* 2000), NEWHYBRIDS (Anderson and Thompson 2002) and BAYESASS+ (Wilson and Rannala 2003). These programs are increasingly used in animal conservation for population assignment, detection of hybridization and estimation of recent migration rates. Unlike more generic statistical approaches (Bowcock *et al.* 1994; Roques *et al.* 2001), the three methods we review here are all based on an underlying probability model that is intended to mimic the inheritance of genes and the sampling of individuals. Such model-based inference has a number of advantages. First, it typically uses more of the information in the data than approaches that are not based explicitly on genetic models, and second, the variables appearing in genetically based statistical models relate directly to genetic phenomena, so they are easily interpreted.

The statistical genetic models underlying STRUCTURE, NEWHYBRIDS and BAYESASS+ are simple and quite similar. The primary goal of this chapter is to describe these models with as few equations as possible. In lieu of mathematical equations we will explore the structure of these models in terms of simple, intuitive diagrams called directed acyclic graphs (DAGs), which show the relationship between variables in a model. This should allow users to better understand what the methods do, how they are similar, and the important ways in which they differ. Though the softwares implementing these techniques are user-friendly, they are certainly not 'plug-and-play' methods. I hope that this chapter will allow users to understand the

Population Genetics for Animal Conservation, eds. G. Bertorelle, M. W. Bruford, H. C. Hauffe, A. Rizzoli and C. Vernesi. Published by Cambridge University Press. © Cambridge University Press 2009.

methods enough to ensure they get reasonable results and they can interpret them appropriately.

After discussing the three different models, we focus on practical issues.
Because previous reviews (Pearse and Crandall 2004; Manel *et al.* 2005)
have summarized when these various methods (and many related ones, e.g.
Rannala and Mountain 1997; Dawson and Belkhir 2001; Corander *et al.*
2004; Piry *et al.* 2004) are useful, and have offered many general guidelines
for their use, this final section is devoted to the simple proposition that it is
important to assess the results of these programs by comparison to simulated data that look like your own.

CONCEPTUAL MODELS AND GRAPHICAL MODELS

All statistical inference depends in some way on a probability model. This
model may be completely specified in terms of the equations describing the
statistical distributions involved; though if you simply want to understand
the assumptions of the model, it is usually sufficient to understand the
verbal description of the model. As a model gets more complex, however, it
is helpful to have a visual roadmap as well as a verbal description. A DAG
is such a roadmap, providing a diagram of the relationship between components in a model and a comparison of the structure of different models.
We will illustrate our first DAG by considering the estimation of allele
frequencies in a closed population.

Let us imagine that we are interested in estimating the frequency of
alleles at a single locus in a lake population of fish. An obvious course of
action would be to draw a sample of M fish from the lake, genotype them at
the locus, and estimate the allele frequencies from the observed proportion
of alleles in the sample. The conceptual model underlying this procedure is
one in which each fish carries two gene copies drawn at random from a
large pool of alleles whose proportions are the unknown allele frequencies
in the lake.

The relationship between the allelic types in the fish we sample and the
population allele frequencies is captured in the DAG of Fig. 2.1a. The circles
(called nodes) in the graph represent the different variables in the model.
The frequencies of the alleles are denoted by θ. The node associated with θ
is unshaded, representing the fact that the values of the population allele
frequencies are unknown. The type of the allele is denoted by $Y_{i,1}$ for the
first gene copy of the ith sampled fish, and $Y_{i,2}$ for the second gene copy.
The nodes associated with these variables are shaded black to denote that
they are observed – i.e. the fish are genotyped. The allelic type of each gene

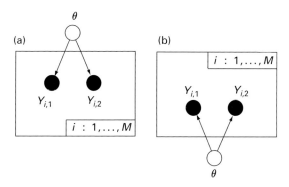

Figure 2.1. DAGs describing the sampling to estimate allele frequency θ: (a) and (b) describe identical models.

copy is independent (under the assumption of Hardy–Weinberg equilibrium) and depends only on the frequency of alleles in the population. Hence, there are distinct arrows drawn from the node at θ to the nodes for $Y_{i,1}$ and $Y_{i,2}$. The meaning of the arrow can be read as, for example, 'the allelic type, $Y_{i,1}$, depends on θ'. Finally, the two Y nodes are placed inside a box which is known as a *plate* (or, in this case, an M-plate). The legend at the lower right of the plate indicates that the variables within the plate are duplicated M times, over the subscript i. This shorthand expresses that M fish are sampled independently from the lake. The node for θ is not included on the plate because each of the M fish is assumed to be sampled from the same population with the same allele frequencies. Figure 2.1b represents exactly the same model. This figure is included to emphasize that the spatial position of variables in the graph is unimportant; only the orientation of the arrows, and the connections they make, are relevant.

Recall that the original problem was to estimate the allele frequencies, θ, given the observed genotypes of a sample of fish. This problem is also apparent in the DAG because θ is something we wish to know about, and yet it is unknown (as signified by its unshaded node). Generally, the problem of estimation can be interpreted in a DAG as the process of learning about variables or parameters with unshaded nodes given what is observed in the data (the shaded nodes).

As a final word on Fig. 2.1, we should keep in mind that we would have obtained the same DAG if we were sampling any objects, two at a time, from a large population of objects. In fact, it is often easier to think of the sampling process as that of randomly drawing coloured balls, two at a time, out of a large barrel. In this case, each ball is a gene copy, its colour is its allelic type, and the barrel full of balls is the population of gene copies carried by fish in the lake. We have explored this example in detail because the 'balls-in-barrels' conceptual model, and the DAG that goes with it, are basic building blocks for understanding more complex models. In the next section we use these building blocks to describe a class of models called mixture models.

MIXTURE MODELS

The problem recently called 'population assignment' in the molecular ecology literature is a special case of inference in a finite mixture model. In statistics, a finite mixture model is one in which the collection from which the sample is taken is a mixture of individuals from different populations. Such models were applied to the problem of population assignment and 'genetic stock identification' as early as 1981 in the fisheries management literature (Milner *et al.* 1981). The programs STRUCTURE, NEWHYBRIDS and BAYESASS+ are all elaborations of the basic mixture model. In fact, the version of STRUCTURE 'without admixture' employs the same mixture model as an earlier method used to estimate proportions of Columbia River tributary salmon caught in a mixed stock fishery (Smouse *et al.* 1990).

This salmon-fishery mixture model arises from a scenario such as the following: K separate spawning populations of salmon, each with its own unknown allele frequencies, reproduce in different tributaries of a river. Fish from each population migrate through the same place (for example, the mouth of the river), where they are subject to a fishery. By sampling M fish in the fishery and genotyping them at L loci we hope to estimate the proportion of fish from each of the K tributaries that were at the fishery site when the sample was taken. We might also want to infer the population of origin of each of the sampled fish.

Figure 2.2(a) shows the DAG for the mixture model described above. This DAG is composed of a number of elements that look suspiciously like the DAG of Fig. 2.1. Working our way through the graph, from top to bottom, we first have π, which denotes the unknown proportions of fish from the K different populations at the fishery site. W_i is a variable that denotes the population of origin of the ith fish. It can be thought of as a ball that is tied to the fin of the fish, with the colour of the ball telling us where the fish comes from. Under this interpretation, each fish is sampled from the fishery as if it were a coloured ball drawn from a barrel in which the different colours of balls are in the unknown proportions π. Of course, the node associated with W_i is unshaded because we don't know where the fish

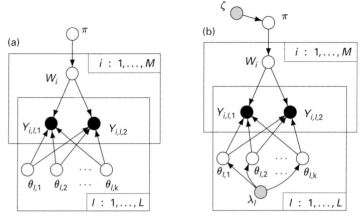

Figure 2.2. DAGs for the mixture model: (a) represents the likelihood model, (b) includes nodes associated with the prior distributions for a Bayesian specification of the problem.

come from – that is what we would like to learn. The remainder of the graph looks complicated, but we can break it down as follows: the L-plate (the lower of the two plates, with the legend 'ℓ : 1, ..., L') signifies that for each fish, there are L loci genotyped, and that their allelic types are independent, given the fish's population of origin, W_i. $\theta_{\ell,k}$ is the frequency of alleles at locus ℓ in population k (where k denotes any one of the K populations). $Y_{i,\ell,1}$ and $Y_{i,\ell,2}$ are the allelic types of the gene copies carried by fish i at locus ℓ. As the arrows in the graph show, the allelic types of these gene copies depend *both* on W_i and on the allele frequencies in the different populations. The nature of this dependence is straightforward: the two allelic types $Y_{i,\ell,1}$ and $Y_{i,\ell,2}$ are drawn from the population that the fish is from – which is denoted by W_i.

The whole sampling model can be summarized by thinking about generating a sample from it. The steps in doing so would be: (1) Draw a ball from a barrel with frequencies $\boldsymbol{\pi}$. (2) The colour of the ball tells you which population to sample a fish from. (3) To generate the genotype for that fish you draw two balls, representing the genes in that fish, from each of L barrels. Each barrel represents the alleles in the population at one of the L loci.

As before, the inference problems that can be tackled with this model can be seen in the DAG. The exercise of population assignment (Paetkau *et al.* 1995; Rannala and Mountain 1997) is merely that of inferring the values of the W_i variables. On the other hand, estimating the proportion of fish from different populations is just the process of inferring the value of $\boldsymbol{\pi}$. Finally, if desired, one could also pursue inference of the allele frequencies in the populations. These are all inference problems that are just different uses of the same underlying model. In actual practice, these different inference problems are typically tackled at the same time, but it is still useful to view them as separate inference problems.

Many times, individuals known to be from particular populations may be sampled. Such individuals constitute what are called learning samples or training samples. These would be represented in the DAG simply as individuals for whom the node associated with W_i was shaded. Inference then proceeds much as before – unknown quantities of interest are estimated given the observed data, which in this case includes the W_is of the individuals in the learning samples. Though with multiple loci mixture inference may be possible without learning samples, if there are many populations contributing to the mixture then accurate inference may be impossible without learning samples.

BAYESIAN INFERENCE

Structure, NewHybrids and BayesAss+ all use the Bayesian paradigm for inferring quantities of interest. This means that estimation is conducted by summarizing the posterior distribution of quantities of interest. The posterior distribution of an unknown variable is just its probability distribution conditional on the observed data. Computing the posterior distribution can be difficult, and, indeed, in Structure, NewHybrids and BayesAss+ it is approximated using Markov chain Monte Carlo. However, the fact that the inference is done in a Bayesian manner does not substantially alter the structure of the underlying models. This is illustrated in Fig. 2.2b, which shows the DAG for a Bayesian specification of the mixture model of Fig. 2.2a. It is apparent that the 'heart' of the model is unchanged. In fact, the only modification is the addition of prior distributions parametrized by ζ for π and λ_ℓ for the θ_ℓs. The nodes for ζ and λ_ℓ are shaded grey to denote that values of those parameters are assumed rather than observed. Prior distributions are necessary for Bayesian inference. Usually the parameters of the prior distribution are chosen to reflect prior knowledge – or in many cases, ignorance – about the associated variables.

A SURVEY OF METHODS

Having established the language of graphical models, we are now in position to quickly survey the models used in Structure, NewHybrids and BayesAss+.

The STRUCTURE model without admixture

As indicated above, the STRUCTURE model without admixture is identical to the model shown in Fig. 2.2(b), and the details of that model have already been described. It assumes that all individuals descend exclusively from one of K populations, where K can be set by the user. In other words, there is no facility in this model for explicitly dealing with hybrids or admixed individuals. Therefore, the method should be used with collections of organisms that are believed to be non-interbreeding. The data required are the multilocus genotypes of the individuals in a sample. The individuals may belong to 'cryptic' subpopulations. That is, it is not necessary to have prior knowledge of separate groups – the program will automatically infer K subpopulations; however, the inclusion of learning samples can be helpful in resolving groups, especially if K is large, or genetic differentiation between populations is limited. The program computes the posterior probability that each individual belongs to each of the K subpopulations, and, in the process it also estimates the allele frequencies in the K separate subpopulations.

It is worth noting that when STRUCTURE uses the model with no admixture, it assumes that the proportion of individuals from each subpopulation is equal (each subpopulation contributes a proportion $1/K$ to the mixture). This feature will cause STRUCTURE to overestimate the true posterior probability of group membership for individuals from subpopulations that are rare in the mixture. If this is a concern, then it may be preferable to use the program *BAYES* (Pella and Masuda 2001) which was developed for analysing large mixtures of salmon.

The STRUCTURE model with admixture

This model provides a flexible way of accommodating individuals of mixed ancestry. No longer must each individual be purely descended from one of the K subpopulations. Rather, each gene copy within an individual may come from a different one of the K subpopulations. The subpopulations of origin of the two gene copies at locus ℓ in the ith individual are indicated by the unobserved variables $W_{i,\ell,1}$ and $W_{i,\ell,2}$, and the expected proportion of ancestry of the ith individual from each of the K subpopulations is a variable to be inferred, denoted by Q_i. The DAG for this model appears in Fig. 2.3a. Here, α is a parameter that determines whether individuals tend to be mostly admixed (high values of α) or mostly purebred (low values of α). It is a value that can be assumed, or inferred. If it is inferred, its prior distribution is assumed to be uniform on the interval $(0, A)$.

We can use the DAG to follow how we would generate data under the model, given α and the allele frequencies: (1) Conditional on α we would

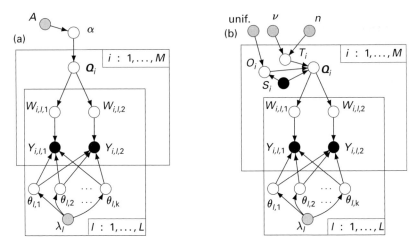

Figure 2.3. (a) The *structure* model with admixture. (b) The *structure* model with admixture *and* prior population information.

simulate a different, random Q_i for each individual i in the sample. Q_i can be thought of as the proportion of balls of K different colours filling a 'Q-barrel' associated with individual i. (2) For each locus, we would draw two balls from individual i's Q-barrel. The colours of the balls drawn tell us which of the K different subpopulations the two gene copies at each locus came from. (3) The allelic type of each gene copy would then be drawn from the allele frequencies in the gene copy's subpopulation of origin.

This is a flexible and general model. It applies generically to many different scenarios: estimating the hybrid index (i.e. Q_i) of individuals in hybrid zones, detecting recent gene flow between populations, and elucidating population structure (cryptic or otherwise). It also provides a facility for estimating the number of subpopulations in a structured population, without prior knowledge about population boundaries.

The data required are the multilocus genotypes of sampled individuals. Learning samples are not required, so it is possible to identify cryptic genetic population structure in a sample of individuals from a single location. However, the capacity to detect cryptic structure declines as the degree of admixture of the individuals in the sample increases (Falush *et al.* 2003). In other words, if most individuals in the sample are highly admixed members of a hybrid swarm, it will be more difficult to correctly infer the nature of the population structure than if some of the individuals in the sample retain the genotypes of pure subspecies, and others are admixed.

The STRUCTURE model with admixture and prior population information

A variant available with STRUCTURE is the model with 'prior population information' in which genotyped individuals have been sampled from K known, separate subpopulations. This model is used to identify individuals in each sample that are migrants from other subpopulations or that have recent immigrant ancestry. In this case, it is necessary to have prior knowledge that there are distinct subpopulations, and that K of them have been sampled. A subpopulation is typically comprised of individuals living in a particular locality; however, the definition of 'subpopulation' is flexible. For example, one might be able to define K subpopulations on the basis of distinct morphological traits possessed by different species or subspecies.

Figure 2.4a is a schematic of the population model in the case of $K = 3$ subpopulations of cats. The subpopulations are distinct, but there is migration between them. Immigration is assumed to be symmetrical and equal between all subpopulations. The model specifies that each individual has a probability $1 - v$ of being descended purely from ancestors belonging to the subpopulation from which it was sampled. With probability v, however, an individual has immigrant ancestry. If v is unknown (as it usually is) then it must be assumed.

If individual i has immigrant ancestry, then it is assumed that only one ancestor in the last n generations was a migrant, and that this migrant ancestor arrived from subpopulation O_i in the T_i^{th} generation before sampling. If $T_i = 0$ then the sampled individual i is itself the migrant; if $T_i = 1$ then one of individual i's two parents was a migrant; if $T_i = 2$ then one of i's four grandparents was a migrant, and so forth (Fig. 2.4c). O_i and T_i are unknown. We will let S_i denote the subpopulation from which the ith individual was sampled; S_i is an observed variable.

The DAG in Fig. 2.3b shows that this model with prior population information is identical to the original STRUCTURE model, except for the parts 'upstream' from the Q_j node. In effect, the model with admixture and prior population information just establishes a new, and more easily interpreted, prior probability distribution for Q_j that ultimately depends on v and n. The arrows in the DAG appear as they do because 1) the parameters v and n determine the probability that an individual has a migrant ancestor at time T_i; 2) if individual i has a migrant ancestor, then the origin of that migrant depends on S_i because the migrant must have come from somewhere *other* than S_i; and finally 3) given T_i, S_i and O_i, the value Q_j is determined (Fig. 2.4c).

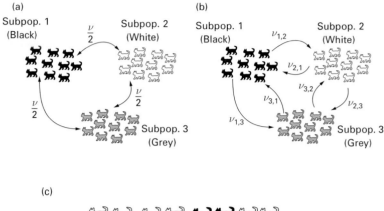

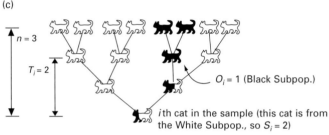

Figure 2.4. (a) A schematic of the *structure* model with prior population information assuming three subpopulations of cats. v is the fraction of individuals in any subpopulation having a single immigrant ancestor in the last n generations from any of the other subpopulations. The other subpopulations are assumed to contribute migrants at the same rate so the probability that an individual has an ancestor from a specific subpopulation is $v/(K-1)$, which in this case is $v/2$ because there are $K=3$ subpopulations. (b) The migration model in *BayesAss+*. $v_{j,k}$ is the fraction of individuals in subpopulation k having immigrant ancestors from subpopulation j in the last n generations. (c) Notation relating to migrants and their descendants. S_i is the location where cat i was sampled. T_i is the number of generations back in time that i had a single migrant ancestor. O_i is the origin of that migrant. n is the total number of generations in the past during which it is assumed an individual might have a migrant ancestor. The cat shown at the bottom of the pedigree was sampled from the White Subpopulation ($S_i=2$) and it has a single migrant ancestor from the Black Subpopulation ($O_i=1$) two generations ago ($T_i=2$). Correspondingly, it is expected to have $\frac{1}{4}$ of its ancestry from the Black Subpopulation, $\frac{3}{4}$ from the White Subpopulation, and no ancestry from the Grey Subpopulation, i.e. $Q_i = \left(\frac{1}{4}, \frac{3}{4}, 0\right)$.

This specialized model is tailored to provide more power than the generic STRUCTURE model for detecting individuals with recent immigrant ancestry. The data required are the multilocus genotypes of the sampled individuals *and* knowledge of the subpopulation each individual was sampled from. Being more specialized, this model also makes more assumptions. Specifically, it is assumed that migration occurs infrequently at a known rate, and that migration occurs at the same rate from and into all subpopulations. This is a model for detecting migrants; not for detecting non-migrants. It is important to note that given the way the model is set up, if there were no genetic data, the posterior probability that a individual is *not* a migrant is $1 - v$. Therefore, if you choose v to be 0.01 and run STRUCTURE to discover that the posterior probability that each individual in your sample is a non-migrant is 0.99, you *must not* infer that this is telling you anything about the power of your genetic data to distinguish the subpopulations – you would have obtained the same result even if you had no genetic data.

Looking at the DAG of Fig. 2.3b, one might not immediately see how the genetic data, $Y_{i,\ell,1}$ and $Y_{i,\ell,2}$, will influence the posterior distribution of Q_j – after all, there are no arrows from $Y_{i,\ell,1}$ or $Y_{i,\ell,2}$ to Q_j, so how can Q_j depend on $Y_{i,\ell,1}$ or $Y_{i,\ell,2}$? The answer is that, even though it is natural in the formulation of a probability model to speak of one variable depending on another – for example, the colour of a ball drawn from a barrel depends on the frequency of different-coloured balls in the barrel – the influence between variables runs in both directions along the arrow. This is, in fact, why it is possible to do inference: if most of the balls you draw from a barrel are orange, then you may infer that there is a high frequency of orange balls in the barrel. In other words, the observed data influence your belief about unobserved variables. In the case of the STRUCTURE model of Fig. 2.3b, knowing the allelic type $Y_{i,\ell,1}$ gives you some information about where that gene copy came from ($W_{i,\ell,1}$) if you have some idea about the allele frequencies. Information about $W_{i,\ell,1}$, in turn, influences your belief about Q_j which, in turn, influences your belief about T_i and O_i which are variables that describe whether an individual is a migrant or not. In other words, during the inference process information obtained from observed variables flows throughout the graph to influence one's belief about *all* the unobserved variables and parameters. A corollary is that with no data, the posterior distribution of variables or parameters will merely be their prior distribution, i.e. with no genetic data, the posterior probability that an individual is a migrant is merely its prior probability, v.

There are two important limitations of the STRUCTURE model with admixture and prior population information. The first is that it does not account for the fact that descendants of migrants will inherit genes in predictable *patterns* (not just in predictable *proportions*) from the different subpopulations (more details appear in the following section). The second limitation is the requirement that the migration rate v must be known, or assumed. It would be preferable to allow the estimation of v from the data.

The NEWHYBRIDS model

NEWHYBRIDS is designed to identify individuals that are recent hybrids between two species or populations. It can distinguish between genealogical classes like F_1, F_2, and backcrosses in a way that STRUCTURE cannot because the NEWHYBRIDS model takes account of the predictable *patterns* of gene inheritance in hybrids, while the STRUCTURE model does not. The simplest example occurs in comparing F_1 hybrids (the offspring of parents from different populations or species) with F_2 hybrids (the offspring of two parents who are themselves F_1 hybrids). F_1 hybrids will have *exactly* one gene copy from one population and one gene copy from the other population *at every locus*. An F_2 individual will also have, on average, half of its gene copies from one population and half from the other; however, only in half of its loci, on average, will there be exactly one gene copy from each population. The model in STRUCTURE is not able to detect differences between F_1s and F_2s because it models admixture strictly in terms of Q_j, which is the proportion of gene copies an individual will have, *on average*, from different subpopulations.

The DAG for the NEWHYBRIDS model (Fig. 2.5a), shows that it is a mixture model. In this case, however, the different components of the mixture are different genealogical classes, rather than simply different

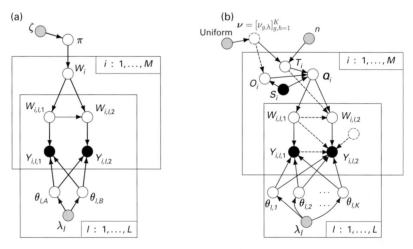

Figure 2.5. (a) The NEWHYBRIDS model. (b) The model used in BAYESASS+. The additions to the model that make it different from STRUCTURE with admixture and prior population information are depicted with dashed lines.

populations. π denotes the proportions of individuals of different genealogical classes present where the sample is drawn, and Z_i is an unobserved variable that denotes the genealogical class of individual i. There are only two different species or populations (A and B) that an individual's genes may come from.

The model can be described by imagining simulating data from it, conditional on π and the allele frequencies. For the ith individual: (1) A coloured ball is drawn from a barrel with balls in the proportions of π. The colour of the ball gives the genealogical class (Z_i) of the individual. (2) Given the genealogical class, the population of origin of the individual's first gene copy ($W_{i,\ell,1}$) is drawn from a barrel much like the Q-barrel described before. (3) The origin of the second gene copy ($W_{i,\ell,2}$) is drawn from a distribution that depends not only on the genealogical class, but also on the origin of the first gene copy. For example, if the genealogical class is F_1, and the first gene copy came from population A, then the second gene copy must come from population B. (4) The allelic type of each gene copy is drawn from the allele frequencies in their respective populations of origin.

Visible in the DAG are the inference problems that can be tackled with NEWHYBRIDS. The value of π can be estimated, and the genealogical class of each individual in the sample can be inferred. Also, the allele frequencies in populations A and B may be estimated.

The number of genealogical classes used in NEWHYBRIDS can be determined by the user. The default is six: two pure species categories, F_1, F_2, A-backcross, and B-backcross categories. A considerable amount of genetic data is required to distinguish genealogical classes, even with as few as six classes (Vähä and Primmer 2006). It is even more difficult to resolve other genealogical classes like second- or third-generation backcrosses. Hence, NEWHYBRIDS is particularly appropriate for the study of hybrid zones in which hybridization has started to occur only recently, or in which the degree of introgression and backcrossing is limited due to selection against hybrids. It is worth pointing out that if only the two pure categories (Pure A and Pure B) are used, the NEWHYBRIDS model reduces to the standard mixture model of Fig. 2.2b with $K = 2$.

The data required are the multilocus genotypes of the sampled individuals. Learning samples are not necessary, but they may be included. It is not necessary to have prior information about subpopulations or species.

The BAYESASS+ model

The model in BAYESASS+ is a natural extension of the STRUCTURE model with admixture and prior population information. Figure 2.4b gives a schematic of the migration model. Importantly, migration rates are not constrained to be the same between all pairs of populations. Further, with BAYESASS+ it is not necessary to assume a value of the migration rate. Rather, BAYESASS+ endeavours to estimate the (possibly nonsymmetrical) rates of migration between all subpopulations. The other two advances over STRUCTURE are the correct modelling of patterns of gene inheritance and the inclusion of an inbreeding parameter $F = (F_1, ..., F_K)$ that tries to account for possible departures from Hardy–Weinberg equilibrium within each subpopulation.

Comparing the DAG for the BAYESASS+ model (Fig. 2.5b) to that of STRUCTURE with admixture and prior population information (Fig. 2.3b) shows that the two are similar, differing only in a few variables, and a few extra arrows. Proceeding from top to bottom in the DAG, we first see that v has been replaced with a matrix v of individual migration rates between the populations (Fig. 2.4b). There is a new arrow connecting v to O_i because, since immigration rates are no longer symmetrical and equal, the origin of immigrants depends both on their destination S_i *and* on the migration matrix v. The two new arrows, from T_i and $W_{i,\ell,1}$ to $W_{i,\ell,2}$ are there as a consequence of the fact that BAYESASS+ models the inheritance of genes from migrants in the same way that NEWHYBRIDS does genes in F_1s and backcrosses. Finally, the arrows from $W_{i,\ell,1}$, $Y_{i,\ell,1}$, and F to $Y_{i,\ell,2}$ describe the interdependence of those variables induced by the possibility of inbreeding (departures from Hardy–Weinberg equilibrium). In words, the type of the second gene copy at a locus is no longer independent of the type of the first gene copy even if they both originate from the same population.

The primary goal of inference using this model is the estimation of the migration matrix. The data requirements for BAYESASS+ are the same as they are for STRUCTURE with admixture and prior population information – it requires multilocus genotypes sampled from K distinct subpopulations. The model provides a more faithful representation of the data than does STRUCTURE and it is appropriate for estimating recent migration between populations that are well differentiated genetically. However, it is apparent that if the populations are not greatly differentiated, then it may be difficult to estimate the migration rates between them. This could lead to misleading results if attempting to estimate migration rates between demes of a recently fragmented population. The various demes will be similar genetically due to recent common ancestry, and this might lead to inflated estimates of migration rates, even if no migration is presently occurring due to the recent fragmentation. Similarly, users should be suspicious of nonmigration rates close to $\frac{2}{3}$ as this is the minimum allowed by the

program and may indicate that populations are not genetically differentiated to the level required to get reliable results.

PRACTICAL ISSUES

Quite reasonably, an entirely separate chapter could be written dealing with
practical issues involved in running the programs described here; issues
from 'How large should my samples be?' and 'How many loci should I use?'
to issues like 'Why do I get different results in NewHybrids using different
priors for the allele frequencies?' and 'Can I trust the results from these
programs?' While some recent simulation studies (Evanno *et al.* 2005; Vähä
and Primmer 2006) have addressed these sorts of questions, and have
provided some general answers, the behaviour of these methods is affected
by many different features of the data, including the genetic differentiation
between the populations, the number of alleles at each locus, the degree of
admixture, etc. It is unlikely that any simulations that have been done will
correspond well to all such features in your own data set. Furthermore, you
may have different questions in mind than the ones that were addressed in
any particular simulation study. In such cases, it is valuable to compare your
results to the results obtained by analysing data simulated to look like your
own data set under different hypotheses of interest.

An excellent example of this type of effort appears in an analysis of
structure in cod (*Gadus morhua*) populations in the seas around Denmark
(Nielsen *et al.* 2003). The authors were interested in whether the patterns of
genotypes they observed in a contact zone were concordant with mechanical
mixing of pure members of two populations, or with a zone of admixture
between two populations. This is not a question that STRUCTURE automatically addresses, so the two different scenarios were simulated with a program called HybridLab (see Nielsen *et al.* 2003 for details of the program)
using allele frequencies from the two different pure populations. The
results from the simulated admixture scenario were more similar to the
results from the real data than were the results from the simulated mechanical mixing scenario, providing evidence that admixture between the populations may be occurring.

Simulating multilocus genotype data from specific allele frequencies
is not a difficult task, but is not a standard feature in many genetic simulation programs. In addition to HybridLab, the program spip (Anderson
and Dunham 2005) simulates multilocus genotypes from specified allele
frequencies, and the program simdata_nh (available from eric.anderson@
noaa.gov) simulates genotypes of individuals of different genealogical classes

under the NewHybrids model. These programs can be used to test the inferences from the three programs STRUCTURE, NewHybrids and BayesAss+.

The methods reviewed in this chapter are complex enough that it is difficult (even for the authors of the programs) to make specific predictions about how these methods will behave when confronted with specific data sets. For this reason, the most important practical advice I can give is that it is incumbent upon the careful user of these programs to simulate data that are similar to their own and then analyse them with the program they are using. In order to gain insight about the results of these programs, there really is no substitute for comparing your results to the results achieved using simulated data that look like your own, *but in which you know the truth* (i.e. you know which individuals are F_1s, and F_2s, or which ones are migrants).

ACKNOWLEDGEMENTS

I thank Devon Pearse, Ingrid Spies and Greg Wilson for their insightful comments on a draft of this chapter, and Robin Waples and Kristen Ruegg for helpful discussions on the topic.

REFERENCES

Anderson, E. C. and Dunham, K. K. (2005). SPIP 1.0: a program for simulating pedigrees and genetic data in age-structured populations. *Molecular Ecology Notes*, **5**, 459–461.

Anderson, E. C. and Thompson, E. A. (2002). A model-based method for identifying species hybrids using multilocus genetic data. *Genetics*, **160**, 1217–1229.

Bowcock, A. M., Ruiz-Linares, A., Tomfohrde, J. *et al.* (1994). High resolution of human evolutionary trees with polymorphic microsatellites. *Nature*, **368**, 455–457.

Corander, J., Waldmann, P., Marttinen, P., and Sillanpaa, M. J. (2004). BAPS 2: enhanced possibilities for the analysis of genetic population structure. *Bioinformatics*, **20**, 2363–2369.

Dawson, K. J. and Belkhir, K. (2001). A Bayesian approach to the identification of panmictic populations and the assignment of individuals. *Genetical Research*, **78**, 59–77.

Evanno, G., Regnaut S. and Goudet, J. (2005). Detecting the number of clusters of individuals using the software STRUCTURE: a simulation study. *Molecular Ecology*, **14**, 2611–2620.

Falush, D., Stephens, M. and Pritchard, J. K. (2003). Inference of population structure using multilocus genotype data: linked loci and correlated allele frequencies. *Genetics*, **164**, 1567–1587.

Manel, S., Gaggiotti, O. E. and Waples, R. S. (2005). Assignment methods: matching biological questions with appropriate techniques. *Trends in Ecology and Evolution*, **20**, 136–142.

Milner, G. B., Teel, D. J., Utter, F. M. and Burley, C. L. (1981). *Columbia River Stock Identification Study: Validation of Method.* Technical report, NOAA. Seattle: Northwest and Alaska Fisheries Center.

Nielsen, E. E., Hansen, M. M., Ruzzante D. E., Meldrup, D., and Gronkjaer, P. (2003). Evidence of a hybrid-zone in Atlantic cod (*Gadus morhua*) in the Baltic and the Danish Belt Sea revealed by individual admixture analysis. *Molecular Ecology*, 12, 1497–1508.

Paetkau, D., Calvert, W., Stirling, I. and Strobeck, C. (1995). Microsatellite analysis of population structure in Canadian polar bears. *Molecular Ecology*, 4, 347–354.

Pearse, D. E. and Crandall, K. A. (2004). Beyond F-ST: analysis of population genetic data for conservation. *Conservation Genetics*, 5, 585–602.

Pella, J. and Masuda, M. (2001). Bayesian methods for analysis of stock mixtures from genetic characters. *Fisheries Bulletin (Seattle)*, 99, 151–167.

Piry, S., Alapetite, A., Cornuet J.-M. *et al.* (2004). GENECLASS2: A software for genetic assignment and first-generation migrant detection. *Journal of Heredity* 95, 536–539.

Pritchard, J. K., Stephens, M., and Donnelly, P. (2000). Inference of population structure using multilocus genotype data. *Genetics*, 155, 945–959.

Rannala, B. and Mountain, J. L. (1997). Detecting immigration by using multilocus genotypes. *Proceedings of the National Academy of Sciences USA*, 94, 9197–9201.

Roques, S., Sevigny, J. M. and Bernatchez, L. (2001). Evidence for broadscale introgressive hybridization between two redfish (genus *Sebastes*) in the North-west Atlantic: a rare marine example. *Molecular Ecology*, 10, 149–165.

Smouse, P. E., Waples, R. S. and Tworek, J. A. (1990). A genetic mixture analysis for use with incomplete source population data. *Canadian Journal of Fisheries and Aquatic Science*, 47, 620–634.

Vähä, J.-P. and Primmer, C. R. (2006). Efficiency of model-based methods for detecting hybrid individuals under different hybridization scenarios and with different numbers of loci. *Molecular Ecology*, 15, 63–72.

Wilson, G. A. and Rannala, B. (2003). Bayesian inference of recent migration rates using multilocus genotypes. *Genetics*, 163, 1177–1191.

3

How to use MIGRATE or why are Markov chain Monte Carlo programs difficult to use?

PETER BEERLI

Population genetic analyses often require the estimation of parameters such as population size and migration rates. In the 1960s, enzyme electrophoresis was developed; it was the first method to gather co-dominant data from many individuals in many populations relatively easily. Summary statistics methods, such as allele-frequency based F-statistics (Wright 1951), were used to estimate population genetics parameters from these data sets. These methods matured and expanded into many variants that were enthusiastically accepted by many researchers. F-statistics are still a hallmark of any population genetic study, especially in conservation genetics, although over the years, limitations have become evident (Neigel 2002). Many of these methods use restrictive assumptions, for example, disallowing mutation. F-statistics, such as F_{ST} methods, are often employed on pairs of populations; this can lead to biased parameter estimates (see Beerli 2004; Slatkin 2005) and the reuse of data in these pairwise methods is undesirable from a statistical viewpoint.

In 1982, Sir John Kingman developed the coalescence theory (Kingman 1982a, b). His overview of the developments of this theory (Kingman 2000) gives an interesting insight into the development of new ideas. This new development opened the door to methods in population genetics that go beyond the F-statistics methods and have led to several theoretical breakthroughs (Hein et al. 2005; although inferences based on coalescence theory were not practicable until about 1995 because of computational constraints). In recent years, computer-intensive programs that can estimate parameters using genetic data under various coalescent models have been developed; for example, programs that estimate gene flow (Beerli and Felsenstein 1999, 2001; Bahlo and Griffiths 2000; Wilson et al. 2003; De Iorio and Griffiths 2004; Hey and Nielsen 2004; Beerli 2006; Ewing and Rodrigo 2006; Kuhner 2006). These programs use different

Population Genetics for Animal Conservation, eds. G. Bertorelle, M. W. Bruford, H. C. Hauffe, A. Rizzoli and C. Vernesi. Published by Cambridge University Press. © Cambridge University Press 2009.

models and different approaches, but in all of them, the quantities of interest are difficult to calculate. Very generally, the goal of these applications is to calculate the probability of the parameters of the chosen model given the data. Population genetics methods often use the relationship among the sampled individuals to get accurate estimates of population size, migration rate or other parameters. These relationships, called genealogies, are typically unknown. Therefore, an optimal approach is to look at all genealogies and weight them using the data. Such approaches can be expressed as integrals over all possible relationships. Unfortunately, there are too many possible genealogies and such an integral cannot be solved exactly. Several numerical integration methods have been developed over the centuries, but only recently Metropolis *et al.* (1953) developed a general approach allowing the integration of complicated multidimensional functions and named this approach the 'Markov chain Monte Carlo method'. Their original algorithm, the Metropolis algorithm, was extended by Hastings (1970) and Green (1995). Many coalescence-based programs use the Metropolis–Hastings or the Metropolis–Hastings–Green algorithm to approximate this integral over all possible genealogies. In the following explanations, I will focus on the program Migrate (Beerli and Felsenstein 1999, 2001; Beerli 2006) but all discussions of Markov chain Monte Carlo approximations and most, if not all, problems are shared with the other programs that use such an approximation.

WHAT IS 'MARKOV CHAIN MONTE CARLO'?

The Markov chain Monte Carlo (MCMC) method is an integration technique for problems that have no simple analytical solution. Instead of exploring the function to integrate in a systematic manner, as in standard numerical integration techniques, MCMC is an autocorrelated method, where each step or sample depends on the last one, but it also has no memory because no step prior to the last one is remembered and thus, cannot influence the choice of the next step. Requirements for the method to work are:

- It must be possible to calculate the integration-function up to a constant. We can often reduce the function of interest to two functions: one that we can calculate and another one that we cannot solve analytically but can hold constant throughout the analysis. Replacing this constant with 1 typically does not change the relationship among the steps or the steepness of the function but only the height of the function.

- Each point on the probability-landscape must be reachable from any other point, if necessary in multiple steps.
- Moves from an old point to a new point on this probability-landscape are reversible and equally likely; if not, this directional bias needs to be corrected.

An almost too simple example

Integration takes a central role for calculating the expectation of a probability distribution. It is standard procedure to calculate the integral analytically or to solve it piecewise, most often by discretizing the continuous distributions. The only requirement for such an approach is that we must be able to calculate the function at any point. With many discrete pieces this function can be integrated with high accuracy. Unfortunately, with many parameters (many dimensions) this approach does not work very well. Often, the function cannot be calculated on an absolute scale but only relative to an arbitrary quantity; therefore, all evaluations using this unscaled function will be off by a constant. When we compare function-values within the same analysis, the differences of these unscaled function-evaluations are the same as those using the correctly scaled function, which we typically cannot calculate easily. This new unscaled function can, however, be used in an MCMC context. The algorithm works like this

Step 1.1:	Start with a random assignment of parameters (for example migration rates, population sizes, genealogy)
Step 1.2:	Evaluate the function for this first step (L_{old})
Step 2.1:	Change the parameters (or a single parameter at a time)
Step 2.2:	Evaluate the function for this step (L_{new})
Step 3.1:	Evaluate the ratio R = L_{new} / L_{old}
Step 3.2:	Draw a random number r from a uniform distribution between 0 and 1
Step 3.3:	If $r < R$ then accept the parameter change and record the new state; otherwise stay at the old state, and record it
Step 4:	Go to 2.1 and repeat many, many times.

For a simple illustration of the steps above, I used a convolution of two normal distributions: in this case the absolute probability density function is known and can be calculated (smooth curve in Fig. 3.1). The histograms were built up using a very simple MCMC procedure that was optimized for this problem. Figure 3.1 shows an MCMC run for a single parameter after 3 steps, 300 steps, 300 000 steps, and 3 000 000 steps. Improvement of the approximation to the area under the curve of the function is obvious.

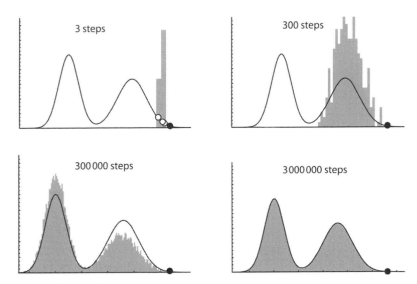

Figure 3.1. Approximation of the area under a curve using MCMC. The curve is the exact function, the grey area is the approximation using MCMC. The black dot marks the starting point of the run, the white dots in the top left panel show the three sampled states that make up the histogram.

Figure 3.1 clearly shows that without running many steps, the approximation is very crude. However, the problem is that there are no clear stopping rules; for example if we are only interested in the maxima of the function, a sample of 300 000 steps would be fine, but the area under the curve is still not approximated very well. If we do not know the function well enough, we would still not know whether there are more than two peaks. This example is very simple and it is important to remember that any integration in the context of multiple parameter estimation will almost certainly be more difficult and less accurate.

MIGRATE – A PROGRAM FOR INFERRING POPULATION GENETIC PARAMETERS

I will use my program MIGRATE to explain some general difficulties of using software that employs MCMC, and will also give some ideas on how to analyse data using such software.

MIGRATE uses two frameworks: (1) coalescence theory to model population genetics forces, such as population sizes and migration rates, and (2) mutation models that explain the change of alleles or nucleotides at sites over time. Both models are simplistic, but for many reasons, no better alternatives are available. It certainly is a strong assumption that Kingman's population genetic model fits all natural populations, but comparisons with other statistics, for example F_{ST} (Beerli 1998), have shown that coalescence theory recovers population scenarios at least as well as or better than some of the other methods. The mutation models are borrowed from phylogenetics (cf. Swofford *et al.* 1996; Felsenstein 2004) or 'old-fashioned' population genetics (Kimura and Crow 1964; Kimura and Ohta 1978a; Ohta and Kimura 1973). In phylogenetics, the distinction of the terms substitution and mutation is important, but, within this population genetics framework, we assume that mutations are neutral or nearly neutral, and therefore, substitution and mutation are equivalent.

Coalescence theory

Kingman (1982a, b) extended Sewall Wright's observation (1951) that it takes two randomly chosen chromosomes in a population of size N about $2N$ generations until they meet in their most recent common ancestor. Kingman showed that it is possible to calculate the probability of a genealogy of any number of individuals. His findings allowed the use of a random sample of individuals to infer parameters for the whole population. Hudson (1991) popularized Kingman's n-coalescent among biologists and today, many extensions of the basic n-coalescent exist; for example, models on recombination (Hudson and Kaplan 1988), gene flow (Notohara 1990; Hudson *et al.* 1992; Wilkinson-Herbots 1998), speciation (Nielsen 1998), selection (Kaplan *et al.* 1988; Neuhauser and Krone 1997; Felsenstein 2004) and many more. The coalescent was derived using a rather general population model, the Cannings model, which is a generalization of the Wright–Fisher population model. The Cannings model allows for variance in the offspring function, whereas the Wright–Fisher model fixes this variance at 1 (Ewens 2004). The coalescent fits simulated data that were generated using a time-forward process almost perfectly when the population model is the Wright–Fisher model. Although the coalescent is robust, caution is needed because it is a diffusion approximation and holds in principle only when the population size is much larger than the sample size, because with either large sample size or very small population size, we expect an increased probability of multiple coalescence per generation, which Kingman's n-coalescent ignores. The effects of multiple coalescences in a generation and effects of sample numbers were explored by several authors. Additions to the coalescence theory by Pitman (1999), Möhle (2000), Schweinsberg (2000), Möhle and Sagitov (2003) and Fu (2006) allow for situations in which more than two lineages merge in the same generation and therefore, for a less restrictive ratio of sample size and population size. Fu (2006) compared the standard coalescent with his multiple-merger coalescent and found that the standard coalescent works astonishingly well even with small populations and large sample sizes; this corroborates the finding of Wakeley and Takahashi (2003) that the standard coalescence is robust as long as the sample size is smaller than the effective population size. If the reproductive success is very uneven among individuals, the concept of effective population size could, in principle, become meaningless, for example if one individual produces all the offspring for the next generation (Eldon and Wakeley 2006). Such a 'neutral sweep' would be indistinguishable from a selective sweep. The risk for such a sweep decreases as the size of the population increases. It is perhaps most pronounced in species that can have small population sizes and produce millions of gametes per individual, as is the case for many fish species.

Mutation models

Readers familiar with phylogenetics know that many studies are preoccupied with using the best substitution model. In population genetics, the problem of misspecification of the mutation model is less severe because the gene trees (genealogies) typically occupy a much shorter time period than phylogenetic trees. MIGRATE accommodates only a few nucleotide mutation models; the default is the Felsenstein 84 model (F84: Hasegawa *et al.* 1985). This model is similar to the Hasegawa–Kishino–Yano (HKY) model: both allow for different nucleotide frequencies and uneven transition rates between purines and pyrimidines (see Swofford *et al.* 1996). Restricting the F84 model, for example by setting all base frequencies equal to 0.25, makes it equivalent to simpler models. This model is not very sophisticated, but it incorporates important features of sequence evolution without many additional parameters. Population genetic inference uses a much more recent time window than phylogenetics and more sophisticated models are warranted only for very rapidly evolving microbes. Researchers in population genetics often accept much simpler models for sequence data, such as the infinite sites model or no-mutation models. MIGRATE does not estimate mutation model parameters, such as transition–transversion ratio and site rate-variation parameters. To get good results, it is better to input specifics about the mutation model and whether rate variation among sites should be assumed. Such parameters can be derived using other programs such as PAUP* (Swofford 2003) or MODELTEST (Posada and Crandall 1998). Recently, single nucleotide polymorphism data were used to investigate population genetics features in humans (Wakeley *et al.* 2001). Programs like MIGRATE and LAMARC (Kuhner 2006) can adjust for the fact that only variable sites are used in the analysis. This is important because, without correction, population genetics parameters would be overestimated (Kuhner *et al.* 2000; Nielsen 2000; Nielsen and Signorovitch 2003; Clark *et al.* 2005).

The models for electrophoretic markers and microsatellite markers are even less sophisticated than the sequence models, although a large number of possible models is known (Calabrese and Sainudiin 2005). Most of these more sophisticated models are difficult to apply many millions of times during a single run: each might need a separate MCMC run to estimate a single branch length. MIGRATE allows the use of mutation models for allozyme data (Kimura and Crow 1964) and for microsatellites (single-step mutation model: Ohta and Kimura 1973; Kimura and Ohta 1978b) and a Brownian motion model that approximates the single-step mutation model (Beerli 1997; Blum *et al.* 2004). DNA or RNA sequence data often contain more information about the history of mutations in the sample and therefore, usually allow for better inferences than other types of data. Nevertheless, these other data types (allozymes, microsatellites) still contain useful information about the population genetics processes. The genealogies generated with such data may look uninformative but, as the example in this section shows, allow us to make inferences that go beyond F_{ST}-based analyses.

How are these pieces combined?

MIGRATE infers parameters either by (1) maximum likelihood or (2) Bayesian inference. A central probability in MIGRATE is the probability of the parameters for a specific data set and a specific genealogy. This probability is calculated as the product of the probability of the data given the parameter and the probability of a genealogy for a given parameter value. Finally, the likelihood is the sum over all genealogies (topologies and branch lengths) of this weight:

Likelihood of the parameters X

Sum of all different labelled histories

Integral over all different branch lengths

$$L(D|X) = \sum_T \int_B \mathrm{Prob}(T, B|X)\mathrm{Prob}(D|T, B)\mathrm{d}B$$

Likelihood of the genealogy

Probability of the genealogy given the parameters

Bayesian inference uses an arbitrary prior distribution for each parameter and the coalescent as a prior distribution for the genealogy, but it also needs the likelihood machinery to sum over all genealogies. Details were given by Beerli and Felsenstein (1999, 2001) and Beerli (2006). This sum over all genealogies is approximated using MCMC and the likelihood is scaled by an unknown constant: it is a relative likelihood. It is important to recognize that a specific log-likelihood value is uninformative, and that the likelihoods of different independent runs with MIGRATE typically cannot be compared. This topic is discussed in the section 'Likelihood ratio tests and related test statistics'.

Running in maximum likelihood mode

Maximum likelihood analysis (ML) and Bayesian inference (BI) use different schemes to estimate parameters. The likelihood method starts with arbitrary values for parameters and genealogy. A new set of genealogies is found with these arbitrary parameter settings using MCMC (these parameter are called the driving parameters because they drive the MCMC). Maximum likelihood estimates of the parameters are then found using this new set of genealogies. These maximum likelihood estimates are probably quite different from the driving parameter values because the data are pushing the likelihood function (and thus the parameter values) towards values that are compatible. A second MCMC chain uses these new parameter values as driving parameters and samples a new set of genealogies after which a new set of parameter values is estimated. This iterative procedure inches towards parameter values that are compatible with the data. By trial and error we (Mary Kuhner, Jon Yamato, Joseph Felsenstein and Peter Beerli, unpubl.) found that several chains that are relatively short allow the exploration of the parameter space. It typically takes about five to ten chains to find sufficiently good driving values, as marked by small changes of parameters between consecutive chains; then two or three very long chains are run and the last chain is used to report the maximum likelihood estimates. Approximate confidence intervals are calculated using profile likelihoods.

Running in Bayes inference mode

For Bayesian inference, it seems most profitable to run one single long chain with a prior distribution for each parameter or combination of parameters. Parameters and genealogy are updated randomly using a user-specified frequency of genealogy-changes. For likelihood, the driving values need adjusting, whereas in a Bayesian framework the prior distribution of the parameters provides a mechanism for exploring different parameter values to change the genealogy during the MCMC run. The parameter values recorded during the run of this single long chain are then used to generate a posterior probability density for each parameter. MIGRATE displays these posterior distributions as histograms and also tabulates quantiles, mode, median and mean. The most important features are the mode of the posterior distribution (i.e. the maximum posterior estimate), and the 2.5% and the 97.5% quantiles, the borders of the 95% credibility interval.

In ML, the success of a run depends on the length and number of short and long chains, whereas in BI the choice of the prior distribution is critical. This prior distribution is often a simple distribution that reflects our knowledge of the parameters before the analysis. Researchers often apply uninformative prior distributions, such as the uniform distribution, perhaps hoping not to bias the posterior distribution. However several Bayesian statisticians suggest using prior information and advocate the use of informative prior distributions. Informative data will overpower any reasonable prior distribution, but informative priors will influence the result when the data are weak. Effects of choices of prior boundaries are discussed using an example in a later section. In MIGRATE, several prior distributions are implemented: a uniform distribution with lower and upper bounds that need to be chosen more extreme than any parameter compatible with the data, and two types of exponential distribution that put more emphasis on small values dependent on the mean of the distribution.

A SHORT EXPLANATION OF WHAT MIGRATE DOES AND DOES NOT DO

MIGRATE, like other population genetic model-based methods, is based on several assumptions. It shares almost all of these assumptions with other programs that infer population sizes or magnitude of gene flow. These assumptions are:

- *Population sizes are constant through time or are randomly fluctuating around an average population size.* This assumption is very common for many population genetics analyses, especially F_{ST} -based analyses. Only a few programs that estimate gene flow relax this assumption, for example LAMARC (Kuhner 2006), and IM (Hey 2005). The program BEAST (Drummond *et al.* 2005) estimates varying population sizes through time for a single locus and a singe population. Additionally, some tests are now available for detecting whether a drastic decrease in population size occurred in the past (for example Cornuet and Luikart 1996); however, many loci are needed and the effects of the population bottleneck must be severe for it to be recognized. Such tests often ignore gene flow among populations or other population genetic forces.

- *Individuals within a population are randomly mating, and each individual has the same potential to have offspring.* Therefore, it is assumed that no selection is acting on the loci under study. The creation of programs for the inference of selection coefficients with a coalescence-based framework is underway.
- *Mutation rate is constant through time and is the same in all parts of the genealogy.* Although MIGRATE assumes rate constancy on the genealogy, it allows using of site rate variation among nucleotide sites and mutation rate differences among loci. Only phylogenetic methods, for example *r8s* (Sanderson 2002), and the program BEAST (Drummond *et al.* 2005) allow for different rates on different branches, but these programs either do not account for population parameters at all or only population sizes.
- *Immigration rate is constant through time, but can differ among populations.* All programs that allow for the estimation of migration rates force rate constancy through time or some segments of time (for example IM: Hey and Nielsen 2004); in addition, F_{ST} -based analyses also impose symmetric rates or symmetric numbers of migrants.
- *Populations exchange genetic material only through migrants, so no population divergence is allowed.* If the time of the most recent common ancestor is younger than the divergence time then MIGRATE is a perfect tool. If you have a data set with two populations that have split only very recently you might want to compare your MIGRATE results with the results from IM (Hey and Nielsen 2004). In contrast to IM, MIGRATE can analyse one, two, or more than two populations; using only population pairs can lead to overestimations of parameters (Beerli 2004; Slatkin 2005).

What happens when the population history violates the assumptions?
One of the most frequent comments from users MIGRATE is that it is not applicable because the population history of their species violates the assumptions of MIGRATE. However, it is important to remember that no program will be able to relax all assumptions, and practitioners need to assess whether an assumption violation will harm their conclusions. Figure 3.2 highlights the direction in which the program will err when assumptions are violated. Several population scenarios that deviate from the assumption that the population size is constant through time were simulated (see

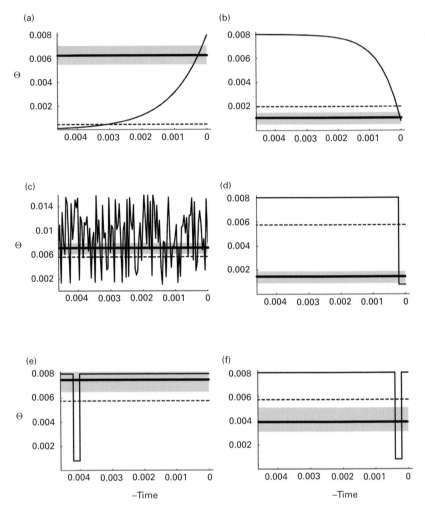

Figure 3.2. Estimation of population size under different population histories. The *x*-axis shows time scaled by mutation rate: past to the left, today is at 0. The *y*-axis shows the mutation scaled population size Θ that is 4 * effective population size * mutation rate per site. Thin lines show the true population size through time; the dashed line was calculated from the true population sizes using a harmonic mean to estimate the average long-term population size; the grey area is the 95% credibility interval and the thick line is the value at the mode of the posterior distribution evaluated by MIGRATE using simulated data sampled at time 0 (1 population with 50 individuals sampled; 10 loci each 10 000 base pairs long; details in Appendix).

Appendix for the simulation and run details). With growing or shrinking populations, MIGRATE will under- or overestimate the effective population size, respectively (Fig. 3.2a, b). The results show that the estimates are mainly influenced by the situation close to the sampling date. On a genealogy with concurrent tips, most lineages are present close to the tip date and will contribute more to the final estimate. With randomly fluctuating population sizes (Fig. 3.2c), the estimate will roughly track the average size. Interestingly, before this experiment, I had expected this estimate to be the harmonic mean, which is believed to track the long-term population size; however, the most recent fluctuations contribute more to the estimate and so many replicates might show an average at the harmonic mean. Short bottlenecks in the past have little effect on the estimate (Fig. 3.2e), whereas recent bottlenecks might mimic a smaller population size (Fig. 3.2f). If the population decline to moderate numbers is very sudden and very recent, MIGRATE is strongly influenced by the bottleneck (Fig. 3.2d). These outcomes need to be explored in more depth, and more simulations with different number of sampled individuals need to be done (Beerli, unpubl.). In any case, it is already possible to say that MIGRATE is influenced by recent changes in population size despite the fact that it delivers long-term estimates.

Example data set

As an example data set, I will use the one for water frogs from my Ph.D. thesis (Beerli 1994). The data are listed in the Appendix and include five populations and 31 electrophoretic marker loci; Beerli et al. (1996) and Beerli (1994) provide details about the different loci. Today, electrophoretic marker data may seem outdated, but it has only recently become easy to sample more than 30 anonymous sequence loci (Brumfield et al. 2003) or microsatellites for most species groups. A complete analysis is difficult because of uneven sampling, uneven distribution of alleles, and (perhaps even worse) lots of missing data. The localities are mapped in Fig. 3.3. This data set is interesting because additional information about the geological history of this area is available. After the last glaciation period (Würm period) ended, the water level rose about 120 m and so isolated the island Samos from the mainland around 10 000 years ago (R. A. Rohde at http:// globalwarmingart.com/wiki/Image:Post-Glacial_Sea_Level_png based on Fleming et al. 1998; Fleming 2000; Milne et al. 2005). The salt water barrier between Samos and Anatolia is shallow. However, the sea between Samos and Ikaria is rather deep and the two islands were probably only connected during the most severe of the more recent glaciation periods (Mindel period) about 200 000 years ago.

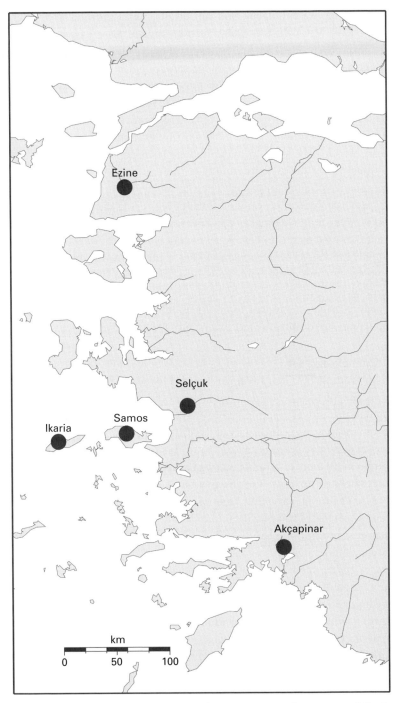

Figure 3.3. Map of water frog sampling locations on Anatolia, Samos and Ikaria.

ANALYSIS USING MIGRATE

I will now analyse the frog data set to estimate the gene flow pattern to and from the mainland (Selçuk) and islands Samos and Ikaria. We will assume that more gene flow occurs from the mainland to the islands than from the islands to the mainland, and in the following sections we will explore this hypothesis. The analysis in this chapter is incomplete, but reveals both difficulties and successes.

Basic analysis – getting familiar with MCMC-based software and data

MIGRATE version 2.0 and newer (Beerli 2006) has the capability of inferring the parameters using either maximum likelihood (ML) or Bayesian inference (BI). For a first analysis, BI is preferred over ML because simulations have shown that, with non-informative data, results using MCMC-based ML analyses are more error-prone (Beerli 2006). This chapter will give a sketch of a possible way to analyse any data and gain confidence that the results are correct. In a first encounter with the program and the data set, I suggest experimenting with the program using the default values for the run conditions. Once you are convinced that the data have been read correctly and the program runs to completion, run the program with the default values. Be aware that default values are chosen so that the program can finish in a reasonable time frame for small to moderate data sets. Depending on the number of parameters to explore, such defaults can be inappropriate and should only be considered as the roughest guide. The number of populations in the example data set is five, so there are 5 population-size and 20 migration parameters. The default values, and so the first default ML or BI run, will not be very trustworthy because these defaults were set for much smaller data sets. With 25 parameters, the MCMC runs will be 'too short'. The MCMC procedure adds variance to the variance introduced by the data, and only multiple runs of different lengths will help to evaluate the magnitude of this variance.

One of the common mistakes of such analyses is that researchers want to do it right on the first try; they will run all the data on very long chains and are disappointed when the program fails or the reported end of that single run is in the following month. A better practice is to use several trial runs to see how the software behaves (this is true for any program that uses MCMC). For BI, change the settings in the Strategy menu of MIGRATE and make sure to visit all submenus, especially the menu entries on the prior distributions. For a first run, choose one 'long' chain to explore around a million steps and save around 100 000 steps. On small data sets with few

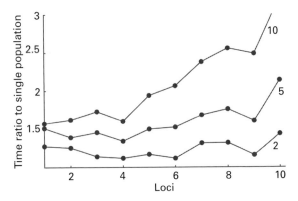

Figure 3.4. Comparison of the runtime of simulated data sets with 1, 2, 5 and 10 populations and 1 to 10 loci. The y-axis shows the runtime ratio of the multi-population parameter estimation compared with the single population. The effort for each run was the same: each run used a total of 100 sampled individuals with a total of 10 000 base pairs each. For example the last data point for the 5-population setting uses 20 individuals per population and 10 loci, each 1000 base pairs long.

loci and few populations this will take minutes, but it might take a couple of hours on data sets with more than four populations and a single locus. Figure 3.4 gives a rough comparison of runtime of different population scenarios and numbers of loci compared to a single-population run. With 10 populations and 10 loci, the runtime is about three times longer than with a single population when the amount of data is the same for all scenarios. In reality, researchers will have 10 times more data from 10 populations than from one population, therefore runtime will be probably about 30 times longer.

We can think of this first run with the default values as a baseline run. We expect that the resulting posterior distribution will not be smooth, and it is quite possible that some parameters will show strange posterior distributions (Fig. 3.5a). For example, if your data suggest a population size of 0.1, but your prior distribution is uniform on the interval 0 to 100, then most proposals will be rejected because most of the suggested population sizes are incompatible with the data. In such cases, we need to shrink the upper bounds of the uniform prior, increase the number of samples considerably, or use another prior, for example an exponential prior. Figure 3.5 gives examples of what could go wrong with prior specification. Once we get an idea how long to run the MCMC chains, set up an even longer chain and use this to report results. For ML analyses, a similar iterative approach is useful. The default settings will often work for two-population data sets that are moderately or highly variable. The example data set needs longer

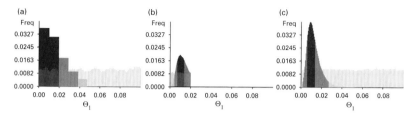

Figure 3.5. Effect of mis-specification of prior distribution on the posterior distribution. A sample of the prior distribution is shown with thin lines; histograms are posterior distributions: shading marks approximate 50% (black), 95% (dark grey) credibility sets. (a) A uniform prior in the range between 0.0 and 10.0, which is too diffuse combined with too few samples from the MCMC, does not lead to an informative posterior distribution. (b) A prior distribution that has too slow an upper limit (0.02) cuts off the posterior distribution at that upper limit. (c) Uniform prior distribution that facilitates fast convergence without truncation for this data set (upper limit 0.1, many more steps saved). Detailed run conditions in Appendix.

runs than the defaults and the sampled chains for the short and long chains should be large. ML uses an iterative scheme of several short and long chains because it does not change the parameter values that drive the MCMC. If these driving parameters are too small, convergence to good estimates is very slow. An iterative improvement of the driving values with several shorter chains moves these driving values towards the 'true' values (Wilson *et al.* 2000). When the driving values are sufficiently close to the 'true' values the ML approach delivers good estimates. ML estimates are very useful for establishing a likelihood ratio test framework (as discussed in the section 'Likelihood ratio test and related test statistics').

Comparison of effect of gene flow using the Bayesian framework

In contrast to a DNA sequence locus, an individual allozyme locus is not very informative because the history of the sampled mutations cannot be inferred; but with many loci there is a good chance that we can recover directionality in gene flow. Figure 3.6 shows such an analysis. MCMC run conditions are specified in the Appendix. The migration rates were calculated assuming that migration (gene flow) is only possible between nearest neighbours and geographic distance is also taken into account. A user can supply a geographic distance matrix between the localities and these distances will scale the migration rate. If migration rates are only a function of distance then all values should be similar. For frogs, salt water is a barrier; therefore, we expect lower migration rates than over land. Hence, I expected lower migration rates between Samos and Selçuk, and Samos and Ikaria, compared to migration rates between mainland locations. In fact, the

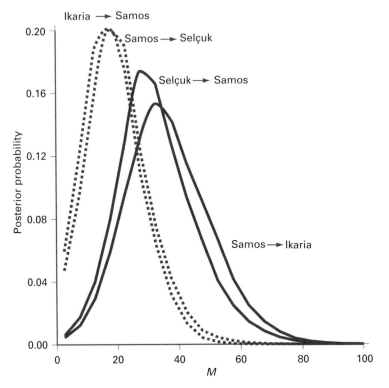

Figure 3.6. Posterior probability distributions of the mutation-scaled migration rate $M_{ji} = m_{ji}/\mu$ where m is the immigration rate per generation into a population i from j and μ is the mutation rate. All six pairwise migrations between the mainland (Selçuk) and the island of Samos (close to the mainland) and between Samos and Ikaria are shown.

migration rate between Samos and Ikaria should be the smallest because the sea strait separating Ikaria persisted for the longest time. The migration rate from the mainland (Selçuk) to the islands is much larger than from the islands to the mainland; for example the rate from Samos to Selçuk is about half of the rate from Selçuk to Samos (Fig. 3.6). The difference in geographic distance between Samos and Ikaria is larger than between Samos and the mainland, so we would expect a difference in gene flow; in this case, however, the difference seems smaller than expected.

Comparison of Bayesian inference and maximum likelihood

It is difficult to make a fair comparison between BI and ML, because each program use slightly different models and programs. Recently, the programs Migrate (Beerli 2006) and Lamarc (Kuhner 2006) were improved and can run both BI and ML. Only the portions of the program that constitute the individual statistics are different. ML works well with very variable data (Beerli 2006; Kuhner and Smith 2006), but has problems with low-variability data (Beerli 2006; Kuhner and Smith did not evaluate low-variability cases). When the data do not contain many variable sites the ML approach has difficulties in converging and needs very long MCMC chains. Often with such data, the ML approach does not give good guidance whether the data can support or reject a population model. In contrast, BI calculates posterior distributions that are similar to the prior distribution, thus alerting the user that the data may not support a complicated population model. In a Bayesian context, it is possible to use the distribution similar to that of the prior distribution to assess whether the data are overfitted with too complicated a model. When the posterior is identical to the prior then the data do not contribute to the result. In fact, programmers use this no-data case as one test to check whether the programs run correctly. In the ML analysis this is somewhat trickier: in current implementations, the MCMC algorithms describe a Brownian motion walk because the data have no influence. Running from the same starting point many times will produce results that are 'normally' distributed around the starting value.

Runs using BI and ML of the water frog data set reveal some differences, but the overall picture is about the same. A comparison of Figs. 3.6 and 3.7 shows that the two approaches agree that the gene flow to islands is higher than from the islands to the mainland.

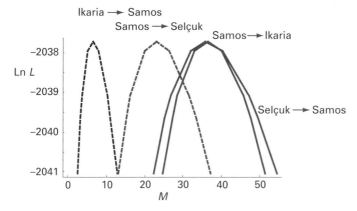

Figure 3.7. Log profile likelihood (Ln L) of mutation-scaled migration rates $M_{ji} = m_{ji}/\mu$ where m is the immigration rate per generation into a population i from j. The two curves closer to zero are for gene flow towards the mainland.

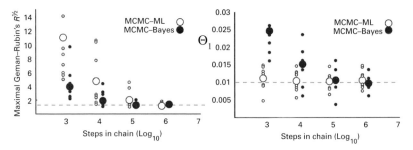

Figure 3.8. (a) Gelman–Rubin statistic of the Bayesian and ML schemes when run for different numbers of sampled steps in the last chain. Values below the dashed line show convergence. (b) Values of Θ estimates using the same runs as in (a). The dashed line in (b) is the population size used to simulate the data sets. Large dots are averages of 10 independent runs (small dots). The data were modelled using two populations; only the size of population 1 was reported.

How long to run

MCMC runs of complicated models need much longer to converge than simple models. The convergence rate is dependent on the data: when the true branching pattern and the mutation events are well distributed, convergence is fast; with low variability or very long terminal branches, the convergence is typically slow. The example data set needs longer chains than the default in MIGRATE. Although the program calculates the Gelman–Rubin convergence diagnostic (Brooks 1998), the best test is longer and longer trial runs. For example, increase the run-length by a factor of 10, until different runs return similar, consistent, results. This exercise is also useful because you become more familiar with the output file format and the program in general. Convergence diagnostics can show successful convergence, but the results may still be very different among runs when too few samples are taken. In a two-population scenario with simulated data from 10 loci (Fig. 3.8), BI seems to converge faster than ML when judged by the convergence diagnostic, but the estimates of ML converge faster to the true value than BI. This is only a single, very simple example, but still it needs to run for at least 10^5 steps. For most data sets, simple MCMC runs do not achieve good results because the chain does not explore the possible solutions very easily and improvements of the MCMC strategy are needed.

Replication and heating

Geyer (1991; Geyer and Thompson 1992) developed a replication scheme that allows combining different MCMC chains for ML estimation. This scheme calculates relative weights for each chain and so adjusts the contribution of each chain to the final ML. This replication scheme is used in MIGRATE and LAMARC (see Wilson *et al.* 2000).

Geyer and Thompson (1995) and others developed a method that uses several chains run with different acceptance ratios powered by the inverse of a 'temperature' (Metropolis-coupled MCMC or MCMCMC). With a temperature of 1.0, standard acceptance ratios are used; with a temperature of ∞, all changes in the MCMC are accepted. This powering up of the acceptance ratio essentially flattens the solution space and so makes it easier to cross deep valleys and descend from very steep peaks. After each chain has made a step, a random pair of temperatures is compared using a Metropolis algorithm-based acceptance ratio and, if the move is accepted, the chains running at different temperatures swap parameter states.

With more than two populations, I suggest exploring heating very early in the experimental runs because you do not know what the solution space looks like. It might be jagged and then you need chains that can jump between peaks. MCMCMC is a possible solution to such problems. MIGRATE allows to set arbitrary temperatures, and a static or an adaptive heating scheme. The adaptive heating scheme takes the start temperatures and decreases the temperature difference by 10% between chains that do not swap for a preset number of trials. If the chains swap more than once in the preset number of trials, the temperature difference increases by 10%. Adaptive heating with a fixed number of heated chains is not the cure-it-all for difficult mixing problems; a system that allows insertion or deletion of chains would be superior over simply increasing or shrinking the temperature difference of existing chains.

How long to wait
Runtime on a single CPU machine depends on the number of loci and the number of replicates. As a simple rule of thumb you can expect that time to increase linearly with the number of loci; for example, if one locus takes a couple of hours then with 31 loci, expect a run of several days on a single CPU machine. The run-length is highly dependent on the number of populations: the time to evaluate genealogies depends on the number of possible events on the genealogies. With n populations there are n different coalescent events, and with the default connection matrix among populations there are $n(n-1)$ possible migration events. Increasing the population number by 1 increases the possible number of events by a factor of $2n-1$ (Fig. 3.4). This increase is typically accompanied by an increase of the total number of individuals, which results in an additional slow-down.

For data sets with many populations, many loci are needed to get accurate estimates. Figures provided by Beerli and Felsenstein (1999) and Beerli (2006) show the reduction of the variance when using more than one locus. Estimates based on many loci take a long time and for such data sets, it is often more convenient to run them on a computer cluster. MIGRATE can run on a large number of computer systems. Difficulties arise when users have a large data set with many loci and want to run it on their laptop or desktop computer. Runs as outlined in this chapter will often take much too long and either the machines are needed for some other tasks or the power goes out.

The program can use symmetric multiprocessing (multiple threads) for running parallel chains with different temperatures. The use of a threaded program is not different from a non-threaded program. This is an efficient use of many high-end desktop machines with two CPUs or, very recently, with dual-core CPUs that can be found even in laptops. Typical gain in speed over non-thread runs is about 1.6 for Bayesian runs, and a little less than that for ML runs because the calculations for the approximate confidence intervals are not threaded.

The fastest way to run MIGRATE is to compile it for use on a computer cluster. The program can take advantage of large clusters running multiple loci and replicates on different CPUs. It uses the message passing interface (MPI: Gropp *et al.* 1999a, b). Several free programs, such as OPENMPI (Gabriel *et al.* 2004), LAM-MPI (Burns *et al.* 1994; Squyres and Lumsdaine 2003) and MPICH2 (http://www-unix.mcs.anl.gov/mpi/mpich/index.htm) are available to set up a virtual cluster on top of the real computer cluster. This real computer cluster can be a single machine or a network of idle lab computers, or a dedicated set of machines connected with a very fast network. Once the virtual cluster is functional, it is only a matter of compiling MIGRATE for such a cluster and running it. The MIGRATE manual gives details of installing and running MIGRATE on such machines. The speed gain depends on the number of loci, number of replicates, and how many real CPUs are available. I typically run MIGRATE on a small cluster of 15 computers with 30 single core 2 GHz AMD Opteron CPUs. The runtime difference is remarkable: the default run of the example data set took about 1 hour and 17 minutes whereas an Intel Core Duo (dual core) 2.16 Ghz machine took about 15 hours. For a researcher with some computer administration knowledge it is rather simple to establish an ad hoc cluster using desktop computers if they run some form of the UNIX operating system (for example LINUX or MacOS X); Windows might be trickier.

Can we trust the support intervals in a MCMC-assisted maximum likelihood analysis?

The support or approximate confidence interval of the maximum likelihood estimate is evaluated using profile likelihoods. In contrast to maximum likelihood, which finds the set of parameters with the highest likelihood, profile likelihood fixes one parameter at an arbitrary value and then finds the set of other parameters that maximize the likelihood. Often, we assume that the likelihood function approximates a χ^2 distribution. Significance levels of this χ^2 distribution then allow specifying quantiles and, thus, support intervals. With short MCMC runs the landscape of genealogies is not well explored and, therefore, the uncertainty of the parameters might be under-estimated. This is somewhat disturbing because it means we will be over-confident in our results. With informative data, very long runs often allow a good approximation of the support intervals. Recently, Abdo *et al.* (2004) claimed that the profile likelihood tables of MIGRATE are inadequate. Their simulation study used the program defaults and ignored guidelines in the manual about how long to run MIGRATE. They showed that the 95% support interval in MIGRATE is often too narrow. In simple scenarios, such as the one they tested, it should be possible to achieve appropriate confidence limits with informative data. Beerli (2006) showed in a much more complicated four-population scenario that, with certain parameter configurations, the data do not contain enough information to estimate migration rates with confidence. Such data sets typically do not produce consistent results when run several times using ML in MIGRATE, and therefore fail to deliver consistent support intervals. Using BI, we can recognize that the posterior distribution is similar to the prior distribution. The example data set does not contain much information per locus, but the 31 loci produce consistent results using BI. ML produces somewhat more variable results but the directionality and magnitude are the same (compare the modes of Figs. 3.6 and 3.7).

LIKELIHOOD RATIO TESTS AND RELATED TEST STATISTICS

Often, we might want to test one migration scenario against another. The MCMC approximations make this rather cumbersome because only relative likelihoods are calculated, and in normal (default) runs there is no control about the driving values that define the denominator of the relative likelihood. MIGRATE allows estimating an approximate likelihood ratio test (LRT) by using the sampled trees to test nested migration models. For example, using the ML scheme, many genealogies are sampled using the default connection matrix among populations: all can connect directly. By supplying an alternative to the most general model, we can test whether the power of the more restricted model to explain the migration scenario is similar to that of the full model. Accepting a parsimony criterion, we would choose the model with fewer parameters.

Comparison of two different migration models

Can we exclude migration from the islands to the mainland (Fig. 3.9)? Running MIGRATE using the likelihood ratio test allows us to make a comparison, but this comparison is only approximate because the full (or the more complete) model is used to sample genealogies. These are then used to evaluate the likelihood both of the model that was used to sample the genealogies and of the model with fewer parameters. Such a procedure seems likely to reject the null hypothesis that there is no difference between the two models too often. In a first application of the built-in LRT, Miura and Edwards (2001) successfully compared several scenarios and could exclude some but not all alternative models.

I describe a different approach that seems more appropriate but is much more time consuming and might be prohibitive without good computing resources. Carstens *et al.* (2005) described an even better, but even more expensive method to evaluate migration models. The reported likelihood in a single program run is a relative likelihood: it is relative to the likelihood of the last chain times an unknown constant. A procedure to make the runs for both models using the same unknown constant is outlined here:

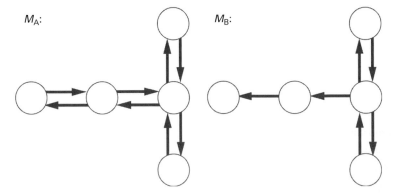

Figure 3.9. A possible, testable hypothesis: is gene flow between the islands and the mainland bidirectional (M_A) or unidirectional (M_B) resulting in the null hypothesis: $M_A = M_B$ and the alternative hypothesis $M_A \neq M_B$. Relative geographic position of the sample populations (circles) is simplified from Figure 3.3.

(1) Run data under model A; record parameters. This run needs to sample the MCMC chains appropriately and needs to be run for many steps (compare with section 'How long to run').

(2) Run data under model B; record parameters. This run needs to sample the MCMC chains appropriately and needs to be run for many steps (compare with section 'How long to run').

(3) Run data under model A for one very long single chain: no short chains, only one very long one, sampling, for example, the same number of genealogies as the total of run 1 or 2. Use the average parameter estimates from runs 1 and 2 for start parameters.

(4) Run data under model B for one very long single chain. Use the average parameter estimates from runs 1 and 2 for start parameters.

(5) Evaluate the likelihood ratio; calculate the degrees of freedom, which is the number of parameters that are different between the hypotheses; under some normality condition we can compare the LRT statistic with a χ^2 distribution with the same degree of freedom. MIGRATE calculates the probability of acceptance of the null hypothesis. Alternatively we can compare the LRT with tabulated χ^2 values for different significance levels typically printed in the Appendix of many introductory statistics texts.

The example data sets allow testing of whether there is only unidirectional migration from the mainland to Samos (the closest island) and from Samos to Ikaria. First, we set the model that allows for migration in both directions between mainland and the islands as the full model A (M_A) in which the unidirectional model B (M_B) is nested. Our null hypothesis specifies that there is no difference between the two models, and the alternative hypothesis is that the two models are different.

$$
\begin{aligned}
\text{LRT} &= -2\ln(L(D|B)/L(D|A)) \\
&= -2[\ln L(D|B) - \log L(D|A)] \\
&= -2(144.767 - 149.162) = 8.79, \quad (p = 0.012, \text{df} = 2)
\end{aligned}
$$

where LRT is the likelihood ratio test statistic for the two models. The probability that the improvement in likelihood for model B is caused by chance is small. Therefore, we reject the null hypothesis that assumes equality of model M_A and M_B. Therefore, we should use the full model (having a higher likelihood) and not the smaller model. We used a fair number of parameters and in these cases the likelihood ratio test may be conservative (Burnham and Anderson 2002). In addition, the LRT assumes nested hypotheses, whereas other model selection criteria, such as Akaike's information criterion (AIC: Akaike 1972) or Schwartz' Information Criterion (the Bayesian information criterion – BIC: Schwartz 1978) can be applied to nested and non-nested models. These information criteria use the number of parameters to penalize the likelihood ratio favoring models with fewer parameters. Burnham and Anderson (2002) gave an extensive discussion of LRT, AIC and other information criteria and suggested using a version of AIC that corrects for small sample size, the AIC_c (Hurvich and Tsai 1989). Applying AIC and AIC_c to the models A and B we get the following values:

$$AIC^{(A)} = -2 \ln L(D|A) + 2k_A = -2 \times 149.162 + 2 \times 13 = -272.32$$

$$AIC^{(B)} = -2 \ln L(D|B) + 2k_B = -2 \times 144.767 + 2 \times 11 = -267.53$$

$$AIC_c^{(A)} = -2 \ln L(D|A) + 2k_A c_A = -2 \times 149.162$$
$$+ 2 \times 13 \times 31/(31 - 13 - 1) = -250.91$$

$$AIC_c^{(B)} = -2 \ln L(D|B) + 2k_A c_B = -2 \times 144.767$$
$$+ 2 \times 11 \times 31/(31 - 13 - 1) = -253.639$$

where k_i is the number of parameters in the model i, n_L is the number of samples, and the small sample correction factors $c_A = n_L/(n_L - k_A - 1)$ and $c_B = n_L/(n_L - k_B - 1)$. For AIC_c, I chose the number of loci in the study as samples, ignoring the number of individuals in the study. It is not clear how to specify n_L when the samples are correlated. The different information criteria cannot be mixed for comparison. The model with the lowest score is the best model in the set. The example compares two models and using AIC we choose model M_A with a score of -272.32 over the model M_B with -267.32. Using AIC_c we choose model M_B with -253.639 over model M_B with -250.91. Burnham and Anderson (2002) suggested that for most cases we should use AIC_c because it corrects for small sample size and is equivalent to the original AIC with large sample sizes. For these data it might be a tough call to decide whether we should prefer the simpler model M_B as suggested by AIC_c or the full model M_A as suggested by the LRT. Given the large number of parameters in the models, the few informative loci, the quality of the data (many values missing), and the use of MCMC, it might be wise to explore both models further before concluding that there is no gene flow from the islands to the mainland.

The likelihoods are approximated by MCMC; it is important to show that the chains have converged and that one has sampled enough genealogies, by either replicated runs and/or convergence diagnostics. Replicated runs from random starting points (for example random genealogies and different parameter values) that arrive at similar estimates after long runs are most promising. Carstens et al. (2005) developed an even better method to estimate a more accurate likelihood ratio test than the procedure shown, but their method is very time intensive and requires bootstrapping the LRT because the commonly used assumption that the test-statistic is χ^2 distributed might be incorrect; as a consequence, the null hypothesis will be rejected too often. In MIGRATE, the described LRT comparisons and the built-in LRT approximation are used to justify the replacement of a more complicated model with a simpler model. In a worst-case scenario, we would use the test with too narrow confidence intervals and, therefore, inflated differences of the two likelihood values caused by insufficient MCMC runs or lack of congruence with a χ^2 distribution. The outcome would be conservative because we would reject the null hypothesis that the full model and the simpler model are equivalent, and we would stick with the more complicated (full) model.

Use of the coalescent in conservation genetics

In conservation genetics, most of the tools used with a single genetic sample in time are derivatives of the coalescence theory, and can be explained by summary statistics based on the coalescent, or are simply derived expectations of the coalescent, for example F_{ST} -based measures (Slatkin 1991; Neigel 2002). One of the biggest concerns in conservation biology is the long-term maintenance of variability in a population and, therefore, large effective population sizes, but changes in population size are difficult to estimate. With a single locus, positive growth in exponential growth models is often reported, but this result is strongly biased (Felsenstein *et al.* 1999). Populations that fluctuate randomly are often not distinguishable from estimates of populations with constant population sizes, and so an analysis using a model assuming constant population size will trace an average population size that is influenced by recent generations.

Programs such as MIGRATE that assume constant population sizes over time average the population size over time. Even programs such as LAMARC and IM, which allow for other models than constant population size through time average over time: LAMARC averages out fluctuations to fit an exponential growth model, and IM forces constant or linear growing population sizes before and after the population split. Only the program BEAST allows for changes of a set of time segments with different population sizes in the past for a single population and a single locus. It is very versatile in the treatment of past population size variability, but needs to allow the use of multiple loci to achieve precise results. With a constant population size model, the population size is averaged over the time interval between the date of the most recent common ancestor and the date of the sample. The expected time of the most recent common (diploid) ancestor is $4N_e$ generations in the past. In large populations the average is, therefore, over a longer time than in small populations. The coalescence-based population genetic parameter estimates are based on the number of mutation events, and also the frequencies of these alleles in the populations. Therefore, very recent changes in population size or migration rate are not necessarily visible using genetic data. Still, these long-term estimates deliver baselines for further management of these populations, for example protection or (moderate) harvest. For example, estimates of past population sizes of humpback whales estimated from mtDNA data (Roman and Palumbi 2003) are very different from current population sizes and from estimates using whaling logbooks. If the differences are real and not an artifact of the analysis, then management of whale populations should increase their protection. The whale study is based on a single locus, and further studies using multilocus data are urgently needed to corroborate Roman and Palumbi's findings. Using the probability distribution of the most recent common ancestor (Tavareé 1984) with the whaling logbook value as the true population size of humpback whales reveals a tiny probability ($p < 10^{-10}$) for a population size value at the 2.5% quantile of Roman and Palumbi's data. This result suggests that it will be difficult to justify the logbook values even with multiple loci. Still, studies based on a single locus are easy to criticize because different population genetic forces can deliver similar signatures; for example, a small population size estimate can be the result of a population bottleneck, a long-term small population size, or a recent selective sweep. Only studies with multiple unlinked loci will be able to distinguish the selective sweep from the small effective population size. Recently, the program BEAST (Drummond et al. 2005) working with single-locus sequence data from a single population was able to estimate population size changes over time using samples from different times.

Researchers often contrast results from census sizes (N_c) with effective population sizes (N_e) using the ratio of N_e/N_c. In some marine fishes these ratios are very small (for example Turner et al. 2002). We can interpret this result in a variety of ways, including the following:

- The population size today as measured with the census size could have increased strongly in the last generation or two, so that there are not enough new mutations to see this same increase in the effective population size measured by genetic variability. Given the dire situation for most species this is a rather unlikely scenario, and can be excluded rather easily with a historical observation that does not need to be based on genetics, although randomly fluctuating population size over genealogical time scale could well explain the difference.

- The effective population size and the census size are measured on a different population scale: census size is measured over a structured population and the genetic measurements came only from a single subpopulation. This is a highly unlikely scenario, even with unknown structure.
- Very few individuals have far more offspring than others. This will result in a small effective population size, and if the carrying capacity is large, large numbers of closely related individuals could be maintained. A comparison of multiple species with known life histories should reveal that when this sweepstakes scenario is correct, we would expect a correlation between number of eggs and ratio of N_e/N_c.

It will be important to explore these effects of high variance of reproduction success on the estimates of population sizes not only practically but also theoretically (Eldon and Wakeley 2006).

SUMMARY

Many powerful new methods for population genetic analysis have been developed in recent years. Almost all of them use heuristic techniques to calculate probabilities of model parameters given the observed data. Researchers that use such methods not only need to explore the variability in their data, but need to understand the variance introduced by the heuristic strategy. In this chapter, I have tried to point to ways that can help to minimize the error introduced by MCMC. The most important lesson is that such programs need to be run for a long time. If a convergence diagnostic is supplied, use it, but remember that convergence diagnostics only detect the grossest errors. Sometimes the diagnostic shows convergence, but the parameter estimates of interest still are not optimal. Run the program multiple times increasing run length. If you get different results, then you either need to run longer or resort to use MCMCMC. Replication is only useful when you have multiple computers to distribute the work. If you get different results using different prior distributions, try to understand why. Possible sources of the problem, ordered from the least likely to the most likely, are: (a) programming error; (b) in BI: bounds of priors are misspecified; in ML: driving values are not at equilibrium; (c) program has not been run long enough.

ACKNOWLEDGEMENTS

I thank Thomas Uzzell, two anonymous reviewers, Nathaniel K. Jue, Sonali Joshi and Michal Palczewski for many helpful comments and Heidi Hauffe for her patience. This work was supported by grant GM078985–01 of the National Institutes of Health through the joint NSF/NIGMS Mathematical Biology Program.

REFERENCES

Abdo, Z., Crandall, K. A. and Joyce, P. (2004). Evaluating the performance of likelihood methods for detecting population structure and migration. *Molecular Ecology*, **13**, 837–851.

Akaike, H. (1972). Information theory and an extension of the maximum likelihood principle. In *Second International Symposium on Information Theory*. Budapest: Akademiai Kaidó, pp. 267–281.

Bahlo, M. and Griffiths, R. C. (2000). Inference from gene trees in a subdivided population. *Theoretical Population Biology*, **57**, 79–95.

Beerli, P. (1994). Genetic isolation and calibration of an average protein clock in western Palearctic water frogs of the Aegean region. Ph.D. thesis, University of Zurich (http://www.people.sc.fsu.edu/~beerli/ownpapers/phd-thesis-beerli-1994.pdf).

Beerli, P. (1997). MIGRATE v. 0.3: a maximum likelihood program to estimate gene flow using the coalescent. http://popgen.sc.fsu.edu

Beerli, P. (1998). Estimation of migration rates and population sizes in geographically structured populations. In *Advances in Molecular Ecology*, NATO Science Series A: Life Sciences vol. 306, ed. G. R. Carvalho. Amsterdam: IOS Press, pp. 39–53.

Beerli, P. (2004). Effect of unsampled populations on the estimation of population sizes and migration rates between sampled populations. *Molecular Ecology*, **13**, 827–836.

Beerli, P. (2006). Comparison of Bayesian and maximum likelihood inference of population genetic parameters. *Bioinformatics*, **22**, 341–345.

Beerli, P. and Felsenstein, J. (1999). Maximum-likelihood estimation of migration rates and effective population numbers in two populations using a coalescent approach. *Genetics*, **152**, 763–773.

Beerli, P. and Felsenstein, J. (2001). Maximum likelihood estimation of a migration matrix and effective population sizes in n subpopulations by using a coalescent approach. *Proceedings of the National Academy of Sciences USA*, **98**, 4563–4568.

Beerli, P., Hotz, H. and Uzzell, T. (1996). Geologically dated sea barriers calibrate a protein clock for Aegean water frogs. *Evolution*, **50**, 1676–1687.

Blum, M. G. B., Damerval, C., Manel, S. and François, O. (2004). Brownian models and coalescent structures. *Theoretical Population Biology*, **65**, 249–261.

Brooks, S. P. (1998). Markov chain Monte Carlo method and its application. *Journal of the Royal Statistical Society Series D*, **47**, 69–100.

Brumfield, R. T., Beerli, P., Nickerson, D. A. and Edwards, S. V. (2003). The utility of single nucleotide polymorphisms in inferences of population history. *Trends in Ecology and Evolution*, **18**, 249–256.

Burnham, K. P., and Anderson, D. R. (2002). *Model Selection and Multimodel Inference: A Practical Information-theoretic Approach*, 2nd edn. New York: Springer Verlag.

Burns, G., Daoud, R. and Vaigl, J. (1994). LAM: an open cluster environment for MPI. In *Proceedings of Supercomputing Symposium*, pp. 379–386.

Calabrese, P. and Sainudiin, R. (2005). Models of microsatellite evolution. In *Statistical Methods in Molecular Evolution*, ed. R. Nielsen. New York: Springer-Verlag, pp. 289–305.

Carstens, B., Bankhead, A., Joyce, P. and Sullivan, J. (2005). Testing population genetic structure using parametric bootstrapping and migrate-n. *Genetica*, **124**, 71–75.

Clark, A. G., Hubisz, M. J., Bustamante, C. D., Williamson, S. H. and Nielsen, R. (2005). Ascertainment bias in studies of human genome-wide polymorphism. *Genome Research*, **15**, 1496–1502.

Cornuet, J. and Luikart, G. (1996). Description and power analysis of two tests for detecting recent population bottlenecks from allele frequency data. *Genetics*, **144**, 2001–2014.

De Iorio, M. and Griffiths, R. C. (2004). Importance sampling on coalescent histories. II: Subdivided population models. *Advances in Applied Probability*, **36**, 434–454.

Drummond, A., Rambaut, A., Shapiro, B. and Pybus, O. (2005). Bayesian coalescent inference of past population dynamics from molecular sequences. *Molecular Biology and Evolution*, **22**, 1185–1192.

Eldon, B. and Wakeley, J. (2006). Coalescent processes when the distribution of offspring number among individuals is highly skewed. *Genetics*, **172**, 2621–2633.

Ewens, W. J. (2004). *Mathematical Population Genetics*. New York: Springer-Verlag.

Ewing, G. and Rodrigo, A. (2006). Coalescent-based estimation of population parameters when the number of demes changes over time. *Molecular Biology and Evolution*, **23**, 988–996.

Felsenstein, J. (2004). *Phylogenetic Inference*. Sunderland: Sinauer Associates.

Felsenstein, J., Kuhner, M. K., Yamato, J. and Beerli, P. (1999). Likelihoods on coalescents: a Monte Carlo sampling approach to inferring parameters from population samples of molecular data. In *Statistics in Molecular Biology and Genetics*, ed. F. Seillier-Moiseiwitsch. Hayward, California: Institute of Mathematical Statistics and American Mathematical Society, pp. 163–185.

Fleming, K., Johnston, P., Zwartz, D., Yokoyama, Y., Lambeck, K. and Chappell, J. (1998). Refining the eustatic sea-level curve since the last glacial maximum using far- and intermediate-field sites. *Earth and Planetary Science Letters*, **163**, 327–342.

Fleming, K. M. (2000). Glacial rebound and sea-level change constraints on the Greenland ice sheet. Ph.D. thesis, Australian National University.

Fu, Y. X. (2006). Exact coalescent for the Wright–Fisher model. *Theoretical Population Biology*, **69**, 385–394.

Gabriel, E., Fagg, G. E., Bosilca, G., *et al.* (2004). Open MPI: Goals, concept, and design of a next generation MPI implementation. In *Proceedings, 11th European PVM/MPI Users' Group Meeting*, Budapest, Hungary, pp. 97–104.

Geyer, C. J. (1991). *Estimating Normalizing Constants and Reweighting Mixtures in Markov Chain Monte Carlo*, Technical Report No. 568. St Paul: School of Statistics, University of Minnesota.

Geyer, C. J. and Thompson, E. A. (1992). Constrained Monte-Carlo maximum-likelihood for dependent data. *Journal of the Royal Statistical Society Series B*, **54**, 657–699.

Geyer, C. J. and Thompson, E. A. (1995). Annealing Markov chain Monte Carlo with applications to ancestral inference. *Journal of the American Statistical Association*, **90**, 909–920.

Green, P. J. (1995). Reversible jump Markov chain Monte Carlo computation and Bayesian model determination. *Biometrika*, **82**, 711–732.

Gropp, W., Lusk, E. and Skjellum, A. (1999a). *Using MPI Portable Parallel Programming with the Message-Passing Interface: Scientific and Engineering Computation*, 2nd edn. Cambridge, Mass.: MIT Press.

Gropp, W., Lusk, E. and Thakur, R. (1999b). *Using MPI-2 Advanced Features of the Message-Passing Interface: Scientific and Engineering Computation*. Cambridge, Mass.: MIT Press.

Hasegawa, M., Kishino, K. and Yano, T. (1985). Dating the human–ape splitting by a molecular clock of mitochondrial DNA. *Journal of Molecular Evolution*, **22**, 160–174.

Hastings, W. K. (1970). Monte Carlo sampling methods using Markov chains and their applications. *Biometrika*, **57**, 97–109.

Hein, J., Schierup, M. H. and Wiuf, C. (2005). *Gene Genealogies, Variation and Evolution: A Primer in Coalescent Theory*. Oxford: Oxford University Press.

Hey, J. (2005). On the number of new world founders: a population genetic portrait of the peopling of the Americas. *PLoS Biology*, **3**, 965–974.

Hey, J. and Nielsen, R. (2004). Multilocus methods for estimating population sizes, migration rates and divergence time, with applications to the divergence of *Drosophila pseudoobscura* and *D. persimilis*. *Genetics*, **167**, 747–760.

Hudson, R. R. (1991). Gene genealogies and the coalescent process. *Oxford Surveys in Evolutionary Biology*, **7**, 1–44.

Hudson, R. R. and Kaplan, N. L. (1988). The coalescent process in models with selection and recombination. *Genetics*, **120**, 831–840.

Hudson, R. R., Slatkin, M. and Maddison, W. P. (1992). Estimation of levels of gene flow from DNA sequence data. *Genetics*, **132**, 583–589.

Hurvich, C. M. and Tsai, C.-L. (1989). Regression and time series model selection in small samples. *Biometrika*, **76**, 297–307.

Kaplan, N. L., Darden, T. and Hudson, R. R. (1988). The coalescent process in models with selection. *Genetics*, **120**, 819–829.

Kimura, M. and Crow, J. (1964). The number of alleles that can be maintained in a finite population. *Genetics*, **49**, 725–738.

Kimura, M. and Ohta, T. (1978). Stepwise mutation model and distribution of allelic frequencies in a finite population. *Proceedings of the National Academy of Sciences USA*, **75**, 2868–2872.

Kingman, J. (1982a). The coalescent. *Stochastic Processes and Their Applications*, **13**, 235–248.

Kingman, J. (1982b). On the genealogy of large populations. In *Essays in Statistical Science*, ed. J. Gani and E. Hannan. London: Applied Probability Trust, pp. 27–43.

Kingman, J. F. (2000). Origins of the coalescent: 1974–1982. *Genetics*, **156**, 1461–1463.

Kuhner, M. (2006). LAMARC 2.0: maximum likelihood and Bayesian estimation of population parameters. *Bioinformatics*, **22**, 768–70.

Kuhner, M. and Smith, L. (2006). Comparing likelihood and Bayesian coalescent estimators of population parameters. *Genetics*, **75**, 155–165.

Kuhner, M. K., Beerli, P., Yamato, J. and Felsenstein, J. (2000). Usefulness of single nucleotide polymorphism data for estimating population parameters. *Genetics*, **156**, 439–447.

Metropolis, N., Rosenbluth, A. W., Rosenbluth, N., Teller, A. H. and Teller, E. (1953). Equation of state calculation by fast computing machines. *Journal of Chemical Physics*, **21**, 1087–1092.

Milne, G. A., Long, A. L. and Basset, S. E. (2005). Modelling Holocene relative sea-level observations from the Caribbean and South America. *Quaternary Science Reviews*, **24**, 1183– 1202.

Miura, G. I. and Edwards, S. V. (2001). Cryptic differentiation and geographic variation in genetic diversity of Hall's babbler *Pomatostomus halli*. *Journal of Avian Biology*, **32**, 102–110.

Möhle, M. (2000). Ancestral processes in population genetics: the coalescent. *Journal of Theoretical Biology*, **204**, 629–638.

Möhle, M. and Sagitov, S. (2003). Coalescent patterns in diploid exchangeable population models. *Journal of Mathematical Biology*, **47**, 337–352.

Neigel, J. E. (2002). Is F_{ST} obsolete? *Conservation Genetics*, **3**, 167–173.

Neuhauser, C. and Krone, S. M. (1997). The genealogy of samples in models with selection. *Genetics*, **145**, 519–534.

Nielsen, R. (1998). Maximum likelihood estimation of population divergence times and population phylogenies under the infinite sites model. *Theoretical Population Biology*, **53**, 143–151.

Nielsen, R. (2000). Estimation of population parameters and recombination rates from single nucleotide polymorphisms. *Genetics*, **154**, 931–942.

Nielsen, R. and Signorovitch, J. (2003). Correcting for ascertainment biases when analyzing SNP data: applications to the estimation of linkage disequilibrium. *Theoretical Population Biology*, **63**, 245–255.

Notohara, M. (1990). The coalescent and the genealogical process in geographically structured populations. *Journal of Mathematical Biology*, **29**, 59–75.

Ohta, T. and Kimura, M. (1973). A model of mutation appropriate to estimate the number of electrophoretically detectable alleles in a finite population. *Genetical Research*, **22**, 201–204.

Pitman, J. (1999). Coalescents with multiple collisions. *Annals of Probability*, **27**, 1870–1902.

Posada, D. and Crandall, K. A. (1998). MODELTEST: testing the model of DNA substitution. *Bioinformatics*, **14**, 817–818.

Roman, J. and Palumbi, S. (2003). Whales before whaling in the North Atlantic. *Science*, **301**, 508–510.

Sanderson, M. (2002). Estimating absolute rates of molecular evolution and divergence times: a penalized likelihood approach. *Molecular Biology and Evolution*, **19**, 101–109.

Schwartz, G. (1978). Estimating the dimension of a model. *Annals of Statistics*, **6**, 461–464.

Schweinsberg, J. (2000). Coalescents with simultaneous multiple collisions. *Electronic Journal of Probability*, **5**, 1–50.

Slatkin, M. (1991). Inbreeding coefficients and coalescence times. *Genetical Research*, **58**, 167–175.

Slatkin, M. (2005). Seeing ghosts: the effect of unsampled populations on migration rates estimated for sampled populations. *Molecular Ecology*, **14**, 67–73.

Squyres, J. M. and Lumsdaine, A. (2003). A Component Architecture for LAM/MPI. In *Proceedings, 10th European PVM/MPI Users' Group Meeting, number 2840 in Lecture Notes in Computer Science*. Venice, Italy, pp. 379–387.

Swofford, D. (2003). PAUP*: phylogenetic analysis using parsimony (*and other methods), v. 4. Sunderland: Sinauer Associates.

Swofford, D., Olsen, G., Waddell, P. and Hillis, D. (1996). Phylogenetic inference. In *Molecular Systematics*, ed. D. Hillis, C. Moritz and B. Mable. Sunderland: Sinauer Associates, pp. 407–514.

Tavareé, S. (1984). Line-of-descent and genealogical processes, and their applications in population genetics models. *Theoretical Population Biology*, **26**, 119–164.

Turner, T. F., Wares, J. P. and Gold, J. R. (2002). Genetic effective size is three orders of magnitude smaller than adult census size in an abundant, estuarine-dependent marine fish (*Sciaenops ocellatus*). *Genetics*, **162**, 1329–1339.

Wakeley, J. and Takahashi, T. (2003). Gene genealogies when the sample size exceeds the effective population size of the population. *Molecular Biology and Evolution*, **20**, 208–213.

Wakeley, J., Nielsen, R., Liu-Cordero, S. N. and Ardlie, K. (2001). The discovery of single-nucleotide polymorphisms – and inferences about human demographic history. *American Journal of Human Genetics*, **69**, 1332–1347.

Wilkinson-Herbots, H. M. (1998). Genealogy and subpopulation differentiation under various models of population structure. *Journal of Mathematical Biology*, **37**, 535–585.

Wilson, I. J., Weale, M. E. and Balding, D. J. (2003). Inferences from DNA data: population histories, evolutionary processes and forensic match probabilities. *Journal of the Royal Statistical Society Series A*, **166**, 155–188.

Wilson, L., Stephens, D. A., Harding, R. M., *et al.* (2000). Inference in molecular population genetics: discussion. *Journal of the Royal Statistical Society Series B*, **62**, 636–655.

Wright, S. (1951). The genetical structure of populations. *Annals of Eugenics*, **15**, 323–354.

Data set

This chapter used a data set from my thesis (Beerli 1994) as an example of how I would analyse such a data set. The data set is imperfect, probably like most real data sets, and simple but complex enough to highlight difficulties in its analysis.

5 31 Electrophoretic loci data: Anatolian water frogs
16 SELCUK

SELC ?1072 EE BB BB BB AA AA BC BB ?? ?? CD AA ?? AA CC DD DD BB BB ?? CC BB AA AA ?? EE BD ?? BC BB
SELC ?1071 EE BB BB BB AB AA AA CJ BB ?? ?? CC AA ?? AC CC DD DD BB BB ?? CC BB AA AA ?? EE BB ?? BB BB
SELC ?1074 EE BB BB BB AB AA AA BC BB ?? BB CC AA ?? BC CC DD DD BB BB ?? CC BB AA AA BB EE BD ?? BC BB
SELC ?1073 EE BB BB BB AA AA AA CC BB ?? ?? CC AA ?? CC CC DD DD BB BB ?? AC BB AA AA ?? EE BB ?? BC BB
SELC ?1068 EE BB BB BB AA AA AA CC BB ?? ?? CC AA ?? ?? CC DD DD BB BB ?? CC BB AA AA ?? EE BB ?? BC BB
SELC ?1067 EG BB BB BB AB AA AA CJ BB ?? ?? CC AA BB BB CC CD DD BB BB ?? CC BB AA AA ?? EE DD ?? BB BB
SELC ?1070 EE BB BB BB AB AA AA BC BB ?? BB CC AA ?? CC CC DD DD BB BB ?? CC BB AA AA BB EE DD ?? BC BB
SELC ?1069 EE BB BB BB AB AA AA CC BB ?? ?? CC AA ?? BC CC CD DD BB BB ?? AC BB AA AA ?? EE BB ?? BB BB
SELC ?1080 EE BB BB BB AA AA ?? BB ?? ?? CC AA ?? CC CC DD DD BB BB ?? AC BB AA AA ?? EE DD ?? BB BB
SELC ?1079 EE BB BB BB AA AA AA CC BB ?? ?? DD AA ?? CC CC DD DD BB BB ?? CC BB AA AA ?? EE DD ?? BC BB
SELC 17164 EE BB BB BB AA AA ?? CC BB ?? ?? CC AA ?? BB CC ?? ?? BB AA ?? ?? EE BB ?? BB BB
SELC 17163 EE BB ?? BB BB AA ?? CC BB ?? ?? CC AA ?? CC CC ?? ?? BB BB ?? ?? BB AA ?? ?? EE BB ?? BB BB
SELC ?1076 EE BB BB BB AA AA AA CC BB ?? ?? CD AA ?? CC CC DD DD BB BB ?? CC BB AA AA ?? EE BD ?? BC BB
SELC ?1075 EE BB BB BB AA AA AA CC BB ?? ?? CC AA ?? CC CC DD DD BB BB ?? AC BB AA AA ?? EE BD ?? BB BB
SELC ?1078 EE BB BB BB AB AA AA BC BB ?? ?? CC AA ?? CC CC DD DD BB BB ?? AC BB AA AA ?? EE DD ?? BB BB
SELC ?1077 EE BB BB BB AA AA AA CC BB ?? BB CC AA ?? AC CC DD DD BB BB ?? AA BB AA AA BB EE BD ?? BB BB

15 AKCAPINAR

AKCA 16804 EE BB AB BB ?? AA AA CC BB ??? CC AA BB BB ?? DD DD BB BB AA AC BB AA AC BB EE DD AA CC BB
AKCA ?1065 EE BB BB BB EE AA AA CC BB ?? ?? CC AA ?? DD CC DD DD ?? BC AA AC BB AA AA BB EE DD ?? CC BB
AKCA ?1064 EE BB BB BB AB AA AA BB BB ?? ?? CC AA ?? DD CC DD DD ?? CC AA CC BB AA AA BB EE DD ?? CC BB
AKCA 16805 ?? ?? BB ?? ?? AA AA BC BB ?? CC AA BB BB ?? DE DD BB CC AA AC BB AA AA BB EE DD AA CC BB
AKCA 16808 ?? ?? AB ?? BE AA ?? ?? BB ?? ?? ?? BB ?? ?? ?? DD DD BB BB AA CC BB AA AA ?? ?? DD ?? CC BB
AKCA 16807 ?? ?? ?? BB AB AA ?? ?? BB ?? ?? ?? BB ?? ?? ?? DD DD BB BB BC AA AC BB AA AA ?? ?? DD ?? CC BB
AKCA 16806 EE BB BB BB ?? AA AA BJ BB ?? ?? CC AA BB BC ?? DD DD BB BB AA CC BB AA AA BB EE DD AA CC BB
AKCA ?1063 EE BB BB BB AB AA AA BC BB ?? ?? CC AA BB DD CC DD DD BB BC AA CC BB AA AA BB EE DD ?? CC BB
AKCA ?1058 EE BB BB BB AB AA AA ?? BB ?? BB CC AA BB CC AA BB CD CC ?? DD BB BC AA AA BD AA AC BB ?? DD ?? CC BB
AKCA ?1057 EE BB AB BB BB AA AA ?? BB ?? BB ?? BB ?? BB CC AA BB DD CC ?? DD BB BE AA AA BB AA AJ BB ?? DD ?? CC BB
AKCA ?1066 EE BB AB BB BB AA AA ?? BB ?? ?? CC AA ?? BB ?? ?? CC AA ?? CD CC DD DD BB BB AA AC BB AA AJ BB ?? DD ?? CC BB
AKCA ?1059 EE BB AA BB AA AA AA ?? BB ?? ?? BB BC CC ?? DD BB BB AA AC BB AA AJ BB ?? DD ?? CC BB
AKCA ?1062 EE BB AA BB AB AA AA AA BC BB ?? ?? CC AA BB DD CC DD DD BB BC AA CC BB AA AA ?? EE DH ?? CC BB
AKCA ?1061 EE BB BB BB AB AA AA BC BB ?? ?? CC AA BB BB AA AA BB EE DH ?? CC BB
AKCA ?1060 EE BB BB BB AB BE AA AA ?? BB ?? ?? CC ?? BB DD ?? ?? DD BB BB AA AC BB AA AC BB ?? DD ?? CC BB

11 EZINE

EZIN?1081 EE BB BB BB ?? AA AA CC BB ?? ?? CC AA ?? BC CC DD DD BB ?? AA CC BB AA AA BB EE BB AA BB BB
EZIN 16782 EE BB ?? BB BB AA ?? CC BB ?? ?? BB CC ?? ?? BB AA AA BB EE BB AA ?? ?? ??
EZIN 16781 EE BB BB BB AA AA ?? CC BB ?? ?? BB CC ?? BB BB CC CD DD BB AA AC BB AA CC BB CE BD AA BB BB
EZIN 16783 EE BB BB BB ?? BB AA ?? CC BB ?? ?? CD ?? ?? BB CC BB ?? CC DD DD BB BC AA CC BB AA AA ?? EE BD AA BB BB
EZIN 16785 ?? BB BB ?? AB AA ?? CC BB ?? ?? CC AA ?? BB CC DD DD BB BB AA AC BB AA CC BB AA AA BB BB
EZIN 16784 EE BB BB BB ?? ?? AA ?? CC BB ?? ?? CC AA ?? CC AA AC ?? CE BD AA BB BB

EZIN ?1083 EE BB BB BB AB AB AA AA CC BB ?? ?? CC AA ?? BB CC CD DD BB BB AA CC BB AA AA BB EE BB AA BB BB
EZIN ?1082 EE BB AB BB AA AA AA CC BB ?? ?? CC AA BB BB CC DD DD BB BB AA CC BB AA AA BB EE BB AA BB BB
EZIN ?1084 EE BB AB BB AA AA AA CC BB ?? ?? CC AA ?? BC CC CD DD BB BB AA AC BB AA AC AB EE BB AA BB BB
EZIN 16780 ?? BB AB ?? AB AA ?? CC BB ?? ?? CC AA BB BB CC DD BB BB AA AC BB AA AA BB EE BD AA BB BB
EZIN ?1085 EE BB AB BB BB AA AA CC BB ?? ?? CC AA BB BB CC BB DD BB ?? CC BB AA AC BB EE BB AA BB BB

11 IKARIA

IKAR 17331 EE BB BB BB AA AA ?? CC BB ?? ?? CD AA BB DD ?? DD DD BB BB AA CC BB AA AJ ?? EE BB AA CC BB
IKAR 17330 EE BB BB BB AA AA ?? CC BB ?? ?? CD AA BB DD ?? DD DD BB BB AA CC BB AA AA ?? EE BB AA CC BB
IKAR 17332 EE BB BB BB AA AA ?? CC BB BB BB CD AA BB DD ?? ?? DD BB BB AA CC BB AA AA ?? EE DD AA CC BB
IKAR 17379 EE BB BB ?? AA AA ?? CC ?? ?? ?? CC AA ?? DD CC DD ?? ?? BB AA CC BB AA AA BB EE BD AA CC BB
IKAR 17378 EE BB BB BB AA AA ?? CC ?? ?? BB DD AA ?? DD CC DD ?? BB BB AA CC BB AA AA BB EE BB AA CC BB
IKAR 17329 EE BB BB BB AA AA ?? CC BB ?? ?? CC BB BB DD ?? DD AD BB BB AA CC BB AA AA ?? EE DD AA CC BB
IKAR 17325 EE BB BB BB AA AA ?? CC ?? BB ?? DD AA ?? CC ?? BB ?? ?? BB AA CC BB AA AA ?? EE BB AA CC BB
IKAR 17324 EE BB BB BB BB AA ?? CC ?? ?? DD CC DD DD BB BB AA AA BB EE BB AA CC BB
IKAR 17326 EE BB BB BB BB AA ?? CC ?? BB BB DD AA ?? DD ?? BB ?? CC ?? BB ?? CC BB AA CC BB AA AA ?? EE BB AA CC BB
IKAR 17328 EE BB BB BB BB AA ?? BC BB BB ?? DD AA BB DD ?? DD DD BB ?? BB AA CC BB AA AA ?? EE BB AA BC BB
IKAR 17327 EE BB BB BB AA AA ?? CC ?? BB BB CD AA ?? CD CC DD DD ?? BB ?? CC BB AA AA BB EE BD AA CC BB

4 SAMOS

SAMO 17320 EE BB AB BB AA AA ?? CC BB ?? ?? DD AA BB BB ?? DD DD ?? BB AA AC BB AA AA BB EE DD AA BC BB
SAMO 17321 EE BB BB BB AA AA ?? CC BB BD BB DD AA BB DD CC DD ?? ?? BB AA CC BB AA AA BB EE DD AA BC BB
SAMO 17323 EE BB AB BB AA AA ?? CC BB ?? ?? AA BB BB CC DD BB BB AA AC BB AA AA BB AA AA CE BB
SAMO 17322 EE BB AB BB AA AA ?? ?? BB BD BB CD AA BB BB DD ?? DD DD ?? BB AA AA AA EE DD AA CC BB

Run conditions for specific examples in this chapter

Figure 3.2: For each of the six panels a data set for a single population with 50 sampled individuals, each with 10 unlinked loci, each 10 000 bp long, was generated.

MIGRATE was run using the Bayesian inference mode. The runs were done on a computer cluster with one master and 10 compute nodes. Four parallel heated chains using an adaptive heating scheme were run for each locus. Each chain sampled 10 000 MCMC updates of parameters and genealogies every 200 steps, after discarding the first 100 000 updates. Only the values of the cold chains were used for the posterior distributions. Each run took about 10 minutes.

Figure 3.5: For each of the three panels MIGRATE was run twice, first with an mtDNA data set from 10 individuals of *Rana lessonae* (Plötner *et al.*, unpubl.) to generate the posterior distribution, and then with no data (all nucleotides were replaced by '?') to generate a sample from the prior distribution. Run condition: the runs were done on a computer cluster with one master and four compute nodes and combinations (replicates) of four parallel long chains, each chain sampled 10 000 MCMC-updates of parameters and genealogies every 200 steps, after discarding the first 10 000 updates. The optimal strategy to run this on a single computer would have been different: one long chain, sampling 40 000 every 200, and discarding 10 000. This would have run about four times longer. The prior distribution for the scaled population size Θ was uniform with bounds for (A) at 0 and 10, (B) 0 and 0.02, and (C) 0 and 0.1. The histograms were copied from the PDF result file and combined with the program Adobe Illustrator.

Figure 3.6: Allozyme data set was run on a parallel computer cluster with a total of 72 compute nodes for about 2.5 hours. The run used a customized migration matrix that allowed gene flow only between geographic neighbours, the distance between neighbours was adjusted using a geographic distance file. One cold chain and three heated chains were used during the run: temperatures were 1.0, 1.2, 3.0, and 6.0. Ten replicates of one long chain were used to visit 10 000 000 steps per locus and saving 50 000 steps (50% genealogy change trials, 50% parameter change trials). The recorded parameters were then used to generate the posterior distributions.

Figure 3.7: Allozyme data set was run on a parallel computer cluster with a total of 72 compute nodes for about 1.5 hours. The run used a customized migration matrix that allowed gene flow only between geographic neighbours, the distance between neighbours was adjusted using a geographic distance file. For each locus a total of 10 short chains each visiting 10 000 genealogies and using 500 to improve the driving values for the next chain. Finally, 3 long chains each visiting 100 000 samplings are used. The last chain delivers the MLE and profile likelihood curves shown in Fig. 3.7. To improve mixing, I used a heating scheme with four chains with temperatures of 1.0, 1.2, 3.0 and 6.0.

4

Nested clade phylogeographic analysis for conservation genetics

JENNIFER E. BUHAY, KEITH A. CRANDALL AND
DAVID POSADA

INTRODUCTION

Genetic sequence data have become widely used in evaluating the unique relationship between geography and evolutionary history for conservation of species. Traditional methods, such as bifurcating trees and Wright's *F*-statistics, often fall short in detailing past and contemporary events and contribute little intraspecific information (Posada and Crandall 2001; Pearse and Crandall 2004). Phylogenetic techniques, when applied in lower level systematic studies, show poor resolution, often resulting in polytomies and ambiguous connections (Crandall *et al.* 1994). This is particularly the case when species have recently diverged or have complicated metapopulation structure, in which case, bifurcating trees do not have the ability to accurately depict their evolutionary history (Posada and Crandall 2001). Despite this lack of resolution, broad geographic patterns can still be elucidated for older taxa using phylogenetic approaches. The field of phylogeography began by overlaying phylogenies onto geography and making broad inferences about evolutionary histories of species and populations (Avise 1989). This approach, however, does not provide the opportunity to (1) statistically test the null hypothesis of no geographic association between populations, (2) test whether samples (number of individuals and collection localities) are sufficient, or (3) infer historical and contemporary processes and patterns that dictate current genetic variation (Carbone and Kohn 2004). However, approaches such as Nested Clade Analysis (NCA: Templeton *et al.* 1995), also known as Nested Clade Phylogeographic Analysis or NCPA (Templeton 2004), provide a statistical framework in which to test hypotheses about historical events and current population structure within species.

Indeed, conservation of a species is highly dependent on understanding the processes and the patterns that gave rise to the current phylogeographic

Population Genetics for Animal Conservation, eds. G. Bertorelle, M. W. Bruford, H. C. Hauffe, A. Rizzoli and C. Vernesi. Published by Cambridge University Press. © Cambridge University Press 2009.

composition of each unique taxon. The NCPA approach also has important applications to species delimitation and diagnosis, as it can be used to test for exchangeability and genealogical 'exclusivity' (Crandall *et al.* 2000). In this chapter, we detail the methodology of the NCPA of haplotype trees in phylogeographic studies and its application to a wide range of issues in conservation biology. Using examples from some published NCPA studies, we will discuss the method and its applications to conservation and to the study of population history within species.

NETWORK APPROACHES REQUIRE THOROUGH SAMPLING

There are particular cases where species are severely endangered and there are not enough populations or individuals to sample for an in-depth phylogeographic analysis. It is these species that are most in need of protection, yet it is very difficult to gather enough samples to detail biogeographic patterns and metapopulation dynamics for management purposes. One such example is the United States federally threatened freshwater bivalve (*Potamilus inflatus*), which once ranged across the entire southeastern United States but is now limited to a few rivers, including the Black Warrior and Amite Rivers (Roe and Lydeard 1998). Due to the conservation status and rarity of these freshwater clams, thorough sampling (both numbers of individuals and sampling localities within the distribution) seems impossible, and therefore, direct comparison of sequence data coupled with a multi-species phylogeny was used to assess geographic variation. Twelve nucleotide sites of a 600 base pair portion of cytochrome oxidase I showed variation between the two rivers in a sample of eight individuals. A phylogenetic tree revealed distinct differences between the two rivers as well as between other *Potamilus* species. Based on these results, the authors recommended that *P. inflatus* be recognized as two separate species rather than as two disjunct populations based on the presence of genetically diagnosable characters and a 2% sequence divergence between the Amite form and the Black Warrior form of *P. inflatus*. Although the phylogeny and unique nucleotide differences were sufficient for species' diagnosis, there is still no information about the evolutionary history (i.e. dynamics within species) of the imperilled clams. Furthermore, Crandall *et al.* (2000) have argued that mere genetic distinctiveness at neutral genetic loci is not necessarily the sole criterion for diagnosing species or evolutionarily significant units (ESUs) for conservation. Additional information on the ecological exchangeability would be desirable to further substantiate the diagnosis of distinct species.

The highly endangered Tasmanian freshwater crayfish species *Astacopsis gouldi* also presented geographic sampling difficulties due to its endangered status. This species was historically found throughout all drainages in northern Tasmania, but overhunting has lead to local extirpations and an imperilled conservation assessment. Sinclair *et al.* (2005) sampled several drainage basins across its range, including only a few individuals per site (fewer than ten) as permitted by authorities. Despite the restricted sampling, it was still possible to construct a haplotype network to help determine genetic structure across the rivers that currently harbour isolated populations of the species. A phylogenetic tree of the haplotypes was uninformative for evolutionary processes within the species because there were unresolved polytomies, but the network suggested extensive gene flow and migration of the crustacean species across many drainages. Despite the inability to conduct statistical tests for significant associations between sampled sites and genetic variation (as would be provided by NCPA), conservation management plans could effectively use the haplotype network information for reintroduction and augmentation efforts across various watersheds.

For cases where the species of focus is widespread and common, NCPA can be used to understand contemporary and historical evolutionary processes and patterns. It is critical that populations across the entire distribution are sampled. Phylogeographic approaches, particularly NCPA, are dependent on both geographic sampling and the numbers of individuals at each site. Thorough sampling allows researchers to detect historical events, such as range expansions and fragmentation, as well as contemporary processes, such as ongoing gene flow and isolation. 'Thorough' sampling is becoming a contentious issue for metapopulation studies, particularly how to 'best' sample a taxon for a phylogeographic study or for cases where species' taxonomic status is in question (i.e. species' complexes or hybrid zones). Indeed, the sampling effort should be as homogeneous as possible, so for example, the errors in the allele frequencies are similar across the sampled range.

Issues of genetic sampling include the choice of gene (how much variation within and between species), the numbers of genes sequenced (total number of base pairs), and mitochondrial versus nuclear gene regions. The gene of choice differs between taxonomic groups, but should be variable enough to detect differences at the population level for your study species. For example, the mitochondrial 16S gene is appropriate for studies of freshwater crayfish phylogeography (Buhay and Crandall 2005; Finlay *et al.* 2006) while the COi gene has been used for spiders (Paquin and Hedin 2004) and NDi has been used for toads (Masta *et al.* 2003).

Typically, most studies use one gene region for NCPA and we recommend that other gene regions and analytical methods be used in support of the phylogeographic inferences provided by the NCPA method (see Carstens *et al.* 2004 for comparisons of analytical methods). Importantly, using several gene regions should greatly enhance the NCPA and provides cross-validation of the resulting inferences (Templeton 2002, 2004).

Issues of geographic sampling include the numbers of individuals sampled per locality and the numbers of sampled localities across the distribution of the species. Geographic sampling seems to be the most common question about the NCPA approach. Geographic sampling was recently addressed by Morando *et al.* (2003) in a phylogeographic study of a South American lizard species complex (*Liolaemus elongatus-kriegi*). They found that inadequate geographic sampling resulted in false patterns of regional genealogical exclusivity, and therefore recommended a sample size of five to ten individuals per site for as many sites as possible. Sampling density (number of localities across the distributional range of the focal group) should be determined based on biologically realistic dispersal ability of the species.

A recent example highlighting important issues with geographic and genetic sampling involved a meadow jumping mouse species *Zapus hudsonius* and the United States federally threatened subspecies *Z. h. preblei* contained within the taxon. King *et al.* (2006) sampled large numbers of individuals at few localities (348 individuals from 14 sites) and analysed many loci (21 nuclear microsatellites, the mtDNA control region, and the mtDNA cytochrome *b* gene). Their conclusion was that the subspecies in question is a valid taxon, and is genetically distinct from neighbouring subspecies. In contrast, Ramey *et al.* (2005) sampled extensively across the distribution of the mouse species, favouring more localities over large numbers of individuals. Ramey *et al.* (2005) gathered genetic data for the mitochondrial control region and five microsatellite regions, in addition to morphological measurements, from 195 individuals for over 80 localities. Their conclusion was that the subspecies in question is not a valid taxon because of evidence of recent gene flow with a neighbouring subspecies. NCPA was not conducted for either study, but it would have been possible if the datasets for the mitochondrial control region were combined. The use of NCPA would have been a beneficial and statistically based approach for examining the taxonomic status of the mouse subspecies. The inference procedure for the analysis asks explicitly 'is the species present between the sampled localities?' and if the species *is* present, then the inference would be 'inadequate geographic sampling' for clades showing regional genealogical exclusivity due to poor sampling design. The researchers would then be provided with areas that need to be sampled (geographic gaps) by using the inference procedure (Hedin and Wood 2002; Paquin and Hedin 2004). We provided this example of the contradictory mouse conclusions to illustrate that project design (and hence, gene sampling and locality sampling) along with subsequent adjustments to project design are critical in elucidating evolutionary history, contemporary processes, and species' boundaries for conservation.

HOW TO CONDUCT A NESTED CLADE PHYLOGEOGRAPHIC ANALYSIS

Network construction

In theory, any phylogenetic reconstruction of the history of the sampled haplotypes can be used for the NCPA. However, as we have argued above, at the population level, network approaches are often more useful. There are many different ways to construct a haplotype network, including clustering, hierarchy, distance, and least squares methods (reviewed in Posada and Crandall 2001), but in the case of NCPA, statistical parsimony is most often employed (although any other method could be used as well). A recent study by Cassens *et al.* (2005) found that the minimum spanning networks resulted in poor genealogical estimates, while parsimony and median-joining methods performed well, particularly in cases with extinct or unsampled interior haplotypes. The program TCS (Clement *et al.* 2000; freely available at http://darwin.uvigo.es) constructs haplotype networks using the method of statistical parsimony (Templeton *et al.* 1992). The input format is a simple nexus file with aligned DNA sequences from every individual. Sequences of closely related outgroups should be included in the input file to root the network. The output is the genealogical network depicting the number of mutational steps between haplotypes.

Network diagrams illustrate different types of information

The phylogeography of obligate cave crayfish in the genus *Orconectes* was examined using 485 base pairs of sequence data from the mitochondrial 16S gene (Buhay and Crandall 2005). These sequences were used to construct a statistical parsimony network (Fig. 4.1) resulting in 69 unique haplotypes identified from 421 individuals sampled at 67 cave localities, thoroughly covering the entire distribution along the western escarpment of the Cumberland Plateau in the Southern Appalachians. This network shows the mutational steps between each haplotype (haplotypes are

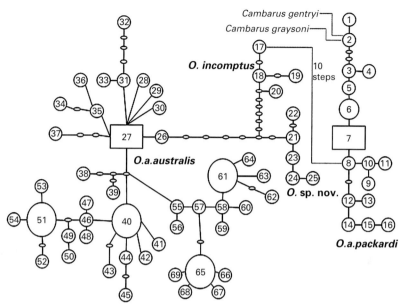

Figure 4.1. *Orconectes a. packardi* (haplotypes 1–16) was outside the 95% confidence limit (nine steps) while *O. incomptus* (haplotypes 17–20), *O. sp. nov.* (haplotypes 21–25), and *O. a. australis* (haplotypes 26–69) were connected within the 95% confidence level. Empty circles in the network represent unsampled, possibly extinct haplotypes. The outgroups *Cambarus gentryi* and *C. graysoni* were outside the 95% limit and connected to haplotype 2 of *O. a. packardi*. (from Buhay and Crandall 2005).

represented as circles with different numbers), including missing haplotypes (marked as small empty circles). Missing haplotypes may be extinct or unsampled haplotypes. A 95% confidence level is first calculated to decide whether we should connect two haplotypes. The 95% confidence level is the maximum number of mutational steps between two haplotypes under which we are 95% sure that no multiple mutations at the same site (overimposed changes) have occurred. The idea is that because we cannot see overimposed changes, we do not want to make those connections in which we can easily underestimate the actual number of differences between two haplotypes. The 95% confidence level for this network is nine steps, which means that there must be fewer than nine mutational differences for the method to directly connect two haplotypes. If the number of mutational steps between sampled haplotypes is greater than the 95% confidence level, multiple separate networks will result. In Fig. 4.1, this was the case with haplotypes 1 through 16 (*O. australis packardi*), which form a distinct network separated from the rest of the haplotypes (17 through 69 which included *O. incomptus, O.* sp. nov. *1*, and *O. a. australis*) by ten steps. In each network, the putative ancestral haplotype is the one with the highest outgroup probability (Castelloe and Templeton 1994) and is depicted as a rectangle, while the other haplotypes are drawn as circles. The *O. a. packardi* haplotype determined to be ancestral was 7, while the ancestral haplotype for the other species was 27 and they are both depicted as rectangles. Frequency (number of individuals) is indicated by the size of the circles. The outgroup probability for each haplotype is also provided by the TCS program and can be found by clicking on the haplotype in the network. The haplotypes with the greatest numbers of individuals in Fig. 4.1 are haplotypes 7, 27, 40, 51, 61 and 65, which are represented by the largest symbols.

A rough estimation of the relative age of haplotypes is determined by both the frequency of the haplotypes in the sample (which is why it is important to include all the sequence data in the analysis and not just the unique sequences) and the number of connections (Castelloe and Templeton 1994). Neutral coalescent theory suggests that high frequency haplotypes are usually older than low frequency haplotypes and are typically found in more internal locations in the network. Rarer haplotypes are thought to have arisen recently, occupy tips, and often have fewer connections to other haplotypes (Crandall and Templeton 1993).

How to resolve loops and ambiguous connections

When there is homoplasy (due to parallel changes, reversals or recombination) in the data, some haplotypes may be connected to several other haplotypes forming loops or reticulations, resulting in unresolved networks. Using predictions from coalescent theory and information about the sampling, loops can be broken to facilitate nesting through the higher nesting levels, although rules exist for nesting with very simple ambiguous connections (Templeton and Sing 1993). Three different criteria can be used to resolve loops: frequency, network location, and geography (Crandall and Templeton 1993). First, haplotypes are most likely connected to higher frequency haplotypes, rather than to haplotypes representing a single individual. Second, haplotypes are most likely connected to interior haplotypes than to haplotypes on the tips of the network. And third, haplotypes are most likely connected to haplotypes from the same geographic area than to haplotypes found in distant areas.

An example of an unresolved network and how to resolve the connections can be found in Pfenninger and Posada (2002). In this study, 16S sequences were sampled in 204 land snails (*Candidula unifasciata*) from 37 localities. The initial network included 46 haplotypes and four major loops (Fig. 4.2).

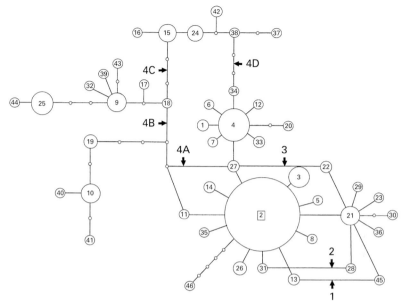

Figure 4.2. Unresolved network constructed using TCS showing different break possibilities (marked by numbers and arrows) for resolution. Numbered arrows 1, 2 and 3 were broken between singletons, while loop 4 had four different break options (4A through 4D) but was broken at arrow 4C (along with the unsampled haplotype between 4A and haplotype 11) based on geographical criteria (from Pfenninger and Posada 2002).

Three of the loops (marked 1, 2 and 3) included connections between single-tons (haplotypes represented by only a single sequence in the sample) and therefore were broken based on the frequency criterion (see arrows). The fourth loop was more complex (labelled 4A to 4D in Fig. 4.2). The authors in fact explored all solutions of loop 4, and despite the differences in the resulting nesting designs, the NCPA inferences were essentially identical.

Building the nesting design

Once the haplotype network is resolved, the next step in the NCPA is to build the nesting design. Beginning at the tips, clades will include haplo-types connected by one mutational step while working toward the interior of the network (Templeton and Sing 1993; Crandall 1996). Using the example of Pfenninger and Posada (2002) (with loop 4 cut at 4C), haplotype 44 and 25 are joined into clade 1-5 (Fig. 4.3). Missing haplotypes, represented by open circles, are considered when making the nesting decisions in the same terms as sampled haplotypes. Haplotype 19 is connected to a missing

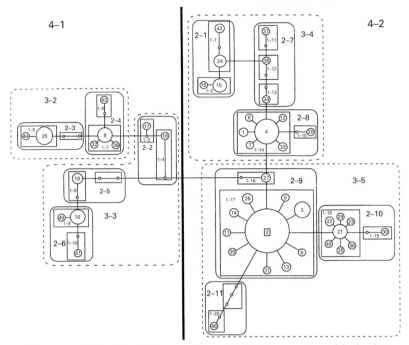

Figure 4.3. Nesting levels shown as hierarchical clade groupings of the haplotype network. The total cladogram comprises two four-step clades: 4-1 and 4-2 (Pfenninger and Posada 2002).

haplotype to build clade 1-8. Two missing haplotypes can be grouped together in a clade. All haplotypes, sampled or unsampled, must be grouped. There will be cases where more than one haplotype is connected to another haplotype by one step. Haplotypes 9, 32 and 39 are joined into clade 1-7 and haplotypes 1, 4, 6, 7, 12 and 33 are joined to form clade 1-14. Each of the nesting groups is called a step clade. Haplotypes are the 0-step clades. The next step is to hierarchically join the 1-step clades into 2-step clades based on one mutational step and so on, using the same rules, until the entire network is grouped at the highest level, working from the tips to the interior of the network. Clade 1-1 and Clade 1-2 are joined to form Clade 2-1 (Fig. 4.3). The nesting process is completed when all clades are nested together at the highest nesting level, which is the total cladogram. Clade 4-1 and Clade 4-2 comprise the total cladogram in Fig. 4.3.

From haplotype network to geographical analysis using GeoDis

GeoDis is a program that statistically tests the associations between the genetic and geographical distances (Posada *et al.* 2000). GeoDis is freely

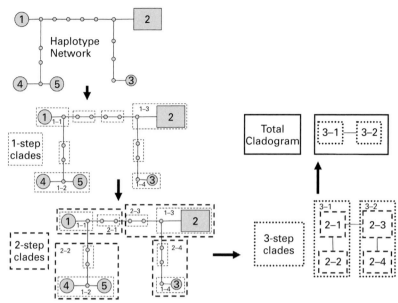

Figure 4.4. Example network using five haplotypes and the nesting levels. This network was then used to write the input file for GEODIS shown in Table 4.1.

available (at http://darwin.uvigo.es) and can be run on a PC or Mac platform. The input file is a written description of the nested cladogram and the corresponding geographic information. The process of writing the input file can be exceptionally tedious for large datasets, and must be done carefully to prevent mistakes.

The first step in constructing this input file for GEODIS is to make a list of the haplotypes obtained from the TCS network (Fig. 4.4) and all the sampled individuals that are represented by each haplotype. Using a simple example of five haplotypes and 18 salamanders for illustration, Haplotype 1 = 4 individuals (2 from Arkansas, 2 from Missouri); Haplotype 2 = 7 individuals (1 from Illinois, 2 from Iowa and 4 from Michigan); Haplotype 3 = 1 individual from Iowa, Haplotype 4 = 3 individuals from Alabama, and Haplotype 5 = 3 individuals (1 from Georgia, 2 from Tennessee).

The next step is to make a geographic description of the sampled localities. If the studied organism can move between localities through the shortest possible path (i.e. a straight line), this is done by specifying latitude–longitude coordinates (in Degree Minute Second: DMS or decimal degrees: DD format). Pairwise distances (km) in table format can also be specified by the user, particularly for cases where species are limited by habitat barriers, such as in the case of aquatic species and linear river systems (see discussion below on Fetzner and Crandall 2003).

A sample input file using the example salamander dataset above is given in Table 4.1. The input file only includes clades that have *both* genetic and

Table 4.1 *Sample input file for GEODIS and line-by-line explanation of the input. Data correspond to Fig. 4.4.*

Input file	Line-by-line explanation of the input
Salamander mtDNA CO1	project name
8	number of sampled localities
1 Missouri	first locality
2 38 25 33 N 93 17 08 W	number individuals and geog. coords.
2 Michigan	second locality
4 42 16 00 N 83 38 36 W	number individuals and geog. coords.
3 Illinois	third locality
1 37 58 36 N 88 30 11 W	number individuals and geog. coords.
4 Iowa	fourth locality
3 43 03 04 N 91 46 54 W	number individuals and geog. coords.
5 Arkansas	fifth locality
2 36 24 12 N 92 58 30 W	number individuals and geog. coords.
6 Georgia	sixth locality
1 32 31 12 N 84 38 56 W	number individuals and geog. coords.
7 Alabama	seventh locality
3 34 00 53 N 85 52 27 W	number individuals and geog. coords.
8 Tennessee	eighth locality
2 36 12 32 N 85 30 23 W	number individuals and geog. coords.
4	number nested clades with genetic and geog. info
Clade 1-2	name of first nested clade
2	number of haps in clade 1-2
4 5	name of haps in clade 1-2
1 1	position of haps: both are tips (1)
3	number of localities within clade 1-2
7 6 8	locality numbers from list (AL, GA, TN)
3 0 0	Hap 4 = 3 AL, 0 GA, 0 TN
0 1 2	Hap 5 = 0 AL, 1 GA, 2 TN
Clade 3-1	name of second nested clade
2	number of 2-step subclades in clade 3-1
1-1 1-2	names of subclades in clade 3-1
0 1	position: 1-1 is interior (0), 1-2 = tip
5	number of localities within 3-1
5 1 7 6 8	locality numbers (Ark, MO, AL, GA, TN)
2 2 0 0 0	2 Ark, 2 MO, 0 AL, 0 GA, 0 TN in 1-1
0 0 3 1 2	0 Ark, 0 MO, 3 AL, 1 GA, 2 TN in 1-2
Clade 3-2	name of third nested clade
2	number of 2-step subclades in clade 3-2
1-3 1-4	names of subclades in clade 3-2

Table 4.1 (*cont.*)

Input file	Line-by-line explanation of the input
0 1	position: 1-3 is interior (0), 1-4 = tip
3	number of localities within 3-2
3 4 2	localities numbers (IL, Iowa, Mich)
1 2 4	1 IL, 2 Iowa, 4 Mich in 1-3
0 1 0	0 IL, 1 Iowa, 0 Mich in 1-4
Total Cladogram	Final grouping of highest clades
2	number of 3-step clades within Total
3-1 3-2	names of subclades within Total
1 1	position: both are tips
8	number of localities in Total
5 1 7 6 8 3 4 2	Ark, MO, AL, GA, TN, IL, Iowa, Mich
2 2 3 1 5 0 0 0	
0 0 0 0 0 1 3 4	
END	

geographic variation, such as in Clade 1-2 on Fig. 4.4. Clade 1-2 includes two haplotypes (haplotype 4 and 5) AND includes individuals from three locales: Alabama, Georgia and Tennessee. Table 4.1 provides the line-by-line input on the left, with the explanation on the right for clarification. Out of the possible eleven nested clades (four 1-step clades, four 2-step clades, two 3-step clades and 1 total cladogram), only four clades (Clades 1-2, 3-1, 3-2 and Total) contained both genetic and geographic variation for analyses. Once the program GEoDis is opened, input the data file, choose what format your geographic information is in (DMS or DD or pairwise table) and select 'run'. The program will output the statistical relationships between the genetic and geographic distances for each clade based on the number of permutations chosen (default = 1000 resamples for a 5% level of significance). For each clade, the observed χ^2 is given along with its probability of being observed under the null hypothesis of no association between geography and genetic variation (Table 4.2).

For each clade, the NCPA statistics are reported as 'within clade distance' (D_c) and 'nested clade distance' (D_n) (Table 4.3). When both interior and tip subclades exist, there is also a test for interior vs. tip clades, reported as $I–T$ distance. The D_c is calculated as the average distance of the individuals from the geographical centre of the clade. The nested clade distance (D_n) is calculated as the average distance of the clade individuals from the next higher-level clade's geographical centre. Significantly small (reported as 'S') and large (reported as 'L') deviations at the 0.05 confidence level are

Table 4.2 *Nested contingency results based on 9999 permutations for clades with genetic and geographic associations. Probability (P) is the probability of obtaining a χ^2 statistic larger than or equal to the observed statistic. Clades with P values less than 0.05 suggest significant geographic structure*

Clades nested with:	Permutational χ^2 statistic	P
Clade 1-1	7.00	0.046
Clade 1-1	466.93	0.044
Clade 1-17	459.69	0.000
Clade 2-1	7.47	0.024
Clade 2-9	64.65	0.020
Clade 3-1	17.00	0.016
Clade 3-2	21.43	0.000
Clade 3-4	22.42	0.022
Clade 3-5	124.90	0.044
Clade 4-1	97.99	0.000
Clade 4-2	83.70	0.000
Entire cladogram	187.65	0.000

Source: Pfenninger and Posada (2002).

key measures for making inferences with the key of Templeton (2005) found at http://darwin.uvigo.es/software/geodis.html.

Using the inference key to uncover evolutionary processes and patterns

Researchers often present the geographic results from GeoDis as a flowchart between step levels, as shown by Table 4.3 from Tarjuelo *et al.* (2004). Once the NCPA results are organized, the next step is to examine the results of each clade with significant genetic–geographic variation (significantly small and significantly large values) using the dichotomous inference key provided by Templeton (2005). This key is primarily used to translate the statistical output of GeoDis into biological inferences. Some of the inferences include 'restricted gene flow/isolation by distance', 'contiguous range expansion', 'allopatric fragmentation' and 'long-distance colonization'. The key is only used for the clades that show significantly large or small values for D_c, D_n, or I–T. When there are no significant distances within a clade, the null hypothesis of no geographical association of haplotypes cannot be rejected (Templeton 1998, 2001). The chain of inference is also usually reported for each outcome, such as the inference chain (1-2-3-4-9-NO) for the Total Cladogram in Table 4.3.

Table 4.3 *Results of the nested geographical analysis for* Pseudodistoma crucigaster. *Column 'Name' is the name of the clade,* D_c *is the clade distance and* D_n *is the nested clade distance at each one of the levels of the analysis (haplotype, 1-step and 2-step levels). The row I–T indicates the average difference between interior and tip clades. Superscript S means that the statistic was significantly small and superscript L that the statistic was significantly large (both at the 5% level). The lines in bold describe the steps followed in the inference key and the conclusion reached by this method: NS (not significant), Past Frag (past fragmentation), RE (range expansion), RGF (restricted gene flow) and CRE (contiguous range expansion)*

Haplotype			1-step clade			2-step clade		
Name	D_c	D_n	Name	D_c	D_n	Name	D_c	D_n
IV	183.15	176.46						
XI	0	102.31	**NS**			(Yellow-Grey		
I–T	183.15	74.14	1-1	176.57	201.04^L	**Morphotypes)**		
X	0	0	1-2	0	181.51			
V	0	0						
VI	0	0						
VII	0	0						
VIII	0	0						
IX	0	0	1-3	0^S	176.44	**1-2-3-4-9-No: Past Frag**		
			I–T	176.57^L	24.21	2-1	191.75^L	172.26^L
I	80.24^S	135.56^L				(Orange Morphotype)		
II	22.90^S	142.97^L	**1-2-3-5-6-'too few clades'** **RE/RGF**					
I–T	57.33^L	−7.408^S	1-4	140.08^S	139.44			
III	0	0	1-5	0	130.25	**1-2-11-12-No: CRE**		
			I–T	140.08	9.19^S	2-2	140.78^S	149.47^S
			Total cladogram			**1-2-3-4-9-No: Past Frag**		

Source: Tarjuelo *et al.* (2004).

NCPA also provides information about evolutionary time, with lower-level nesting processes occurring more recently than the processes that are significant at higher nesting levels. Based on the inferences of the older clades, researchers can gather information about past geographic and environmental events, such as the effects of Pleistocene glaciations or the uplift of mountain ranges on distribution patterns (Templeton 1998). Using younger clade groupings, we can elucidate the impacts of human activities on species' ranges, or show recent expansions by invasive species.

The NCPA inference key has recently come under debate because of the 'subjective' interpretations that are made with respect to the data being examined (Knowles and Maddison 2002). Although the NCPA approach is indeed a statistical framework, the inference key of processes and patterns is largely flexible, lacking a 'standard' method to test the inferences themselves. Knowles and Maddison (2002) argue that the inferences made are outside the realm of confidence, meaning that there is no way to test statistically between inferred events of long-distance colonization, isolation by distance, migration or past fragmentation. Indeed, model-based approaches (and model selection) for phylogeography are desirable, but if one has no a priori hypotheses to test, there are infinite models (through time and space) that one could use. Obviously, the NCPA approach is not a stand-alone answer to questions of evolutionary history of a species, and although NCPA provides many insights and generates hypotheses when none exist, the use of other analytical methods implemented by programs such as MESQUITE (Maddison and Maddison 2004), IM (Hey and Nielsen 2004), ARLEQUIN (Schneider *et al.* 2000) and MISMATCH DISTRIBUTIONS (Rogers and Harpending 1992) can be a tremendous asset in complementing and validating the inferences from the NCPA (Carstens *et al.* 2004). By incorporating results from multiple sources, a stronger case can be made for the phylogeographic patterns of the species that were elucidated by the NCPA.

A special case: terrestrial versus riparian species
As was previously stated, straight-line distances may not be always the best method of representing physical distances between populations. Aquatic species are restricted to the current paths of the waterways they inhabit, whereas terrestrial species are not limited to a linear habitat. Therefore, geographic distances reflected by latitude and longitude coordinates may not be appropriate for riparian species (Fig. 4.5). This idea was empirically tested using mitochondrial data from the 16S gene of a widespread freshwater crayfish species in the Ozarks (Fetzner and Crandall 2003). *Orconectes luteus* were collected from 35 stream sites mostly across Missouri. Geographic coordinates were recorded for comparison to river distances, which were measured using topographic maps. A distance matrix (in kilometres) between pairwise comparisons for every site was used as input for GEODIS, and a separate input file was assembled using latitude–longitude coordinates for the collection sites. The objective of the study was to compare the two distance methods and their outcomes in inferring phylogeographic structure. It should be noted that the NCPA statistics are not exactly the same in both cases, and differences are expected because of the variation

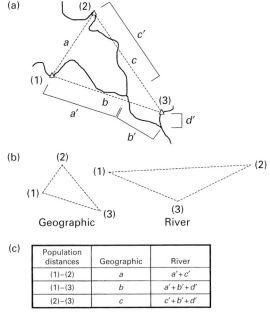

Figure 4.5. Diagram of the distance differences between geographic and river approaches for three sites labelled (1), (2) and (3). (a) Great circle geographic distances are labelled *a*, *b* and *c*, while river/linear distances are labelled *a'*, *b'*, *c'* and *d'*. (b) Illustration of the differences in the geographic and river distances between sites. (c) How the calculations are done between sites (from Fetzner and Crandall 2003).

in calculated geographic distances between locales and their genetic diversity. Results from the NCPA showed distinct differences between the significant clades, often inferring different processes from the two distance methods, particularly for lower nested clades (Table 4.4). Results from Fetzner and Crandall (2003) highlight the importance of using the appropriate (biologically relevant) geographical distances when implementing the NCPA.

Inferring biogeographic patterns

Regional biogeographic patterns can be elucidated by examining the population structure of species, and the inferences provided by NCPA reflect both historical and contemporary patterns of genetic variation. In the land snail example (Pfenninger and Posada 2002), isolation by distance with long-range dispersal was the inferred pattern for clade 4-2 (Fig. 4.6). Northward contiguous range expansion was inferred for clades nested within clade 4-1, and included areas of secondary contact with clade 4-2. Because these inferred patterns relate to the highest nesting levels (which are the oldest groupings), they are possibly responses to historical

Table 4.4 *Comparison of inferences drawn from the geographic and linear river distance methods for geographically significant clades. At lower nesting levels, the use of linear river distances made a drastic difference in the inferences made about contemporary patterns*

Clade	Geographic distances	Linear distances
1-1	restricted gene flow with isolation by distance (rgf/ibd)	past fragmentation (pf)
1-18	panmixia	inconclusive outcome
2-1	panmixia	restricted gene flow with isolation by distance (rgf/ibd)
2-5	restricted gene flow with some long-distance dispersal (rgf/ldd)	past fragmentation
2-6	past fragmentation (pf)	panmixia
2-8	contiguous range expansion (cre)	panmixia
2-13	restricted gene flow with isolation by distance (rgf/ibd)	restricted gene flow/dispersal with some long-distance dispersal (rgf/ dispersal ldd)
3-1	restricted gene flow with isolation by distance (rgf/ibd)	long-distance colonization (ldc)
3-2	past fragmentation (pf)	past fragmentation (pf)
3-4	past fragmentation (pf)	more sampling needed
4-1	past fragmentation (pf)	past fragmentation (pf)
4-2	past fragmentation (pf)	past fragmentation (pf)
4-3	allopatric fragmentation (af)	allopatric fragmentation (af)
5-1	long-distance colonization (ldc)	long-distance colonization (ldc)

Source: Fetzner and Crandall (2003).

environmental changes, such as glaciation events. Many phylogeographic studies have investigated responses of species to glacial advance and retreat cycles (Cooper *et al.* 1995; Comes and Abbott 1998; Turgeon and Bernatchez 2001; Branco *et al.* 2002; Hoffman and Blouin 2004). No other phylogeographic method incorporates both temporal and spatial structure in a statistically testable framework, and many of these aforementioned studies also validated the inferences using other metapopulation analyses.

NCPA CAN BE USED TO DELIMIT SPECIES' BOUNDARIES

A contentious issue in conservation biology is the diagnosis of species and the methods employed to delimit species' boundaries (Sites and Crandall 1997; Sites and Marshall 2003, 2004; Agapow *et al.* 2004). There seems to be little agreement on the definition of 'species' even though 'species' are

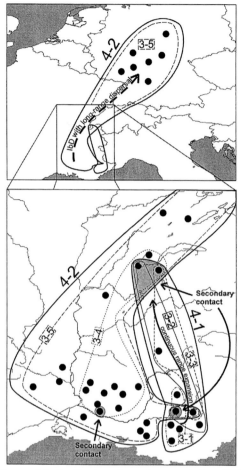

Figure 4.6. Geographic distribution of 3- and 4-step clades with inferred events determined with the inference key of Templeton (2005). Areas of secondary contact between clades are shaded (from Pfenninger and Posada 2002).

deemed by many to be the fundamental units of conservation biology. Species are certainly important entities regardless of one's concept, but conservation biology is also deeply concerned with intraspecific variation within and among populations (Sites and Crandall 1997; Crandall *et al.* 2000). It is at the population level that evolutionary forces operate to drive speciation processes, and in this regard, it is necessary to simultaneously recognize the importance of species and populations for conservation measures. NCPA can be used to diagnose species under the Genealogical Concordance Species Concept (Avise and Ball 1990) and the Cohesion

Species Concept (Templeton 1989). Both approaches use concordance as criteria for species delimitation as well as 'exclusivity'.

In the case of genealogical concordance, multiple types of markers (such as mitochondrial and nuclear DNA, morphological, ecological, behavioural, habitat, etc.) must show consistent patterns. There is, however, no set 'level' of concordance. In other words, the number of markers and degree of concordance among the markers is subjective (Hudson and Coyne 2002). This concept is largely based on coalescent theory (Hudson 1990), and that concordance is the result of long evolutionary separation, which will be reflected in the concordant gene genealogies (Baum and Shaw 1995). A *genealogical species* is a group of organisms whose members are more closely related to each other ('exclusivity') than to any other organisms outside the group (Baum and Shaw 1995). The boundaries of genealogical species can be defined using the testable null hypotheses of Templeton (1989) in the NCPA framework. The first null hypothesis is that the sampled group represents a single evolutionary lineage. If the first null hypothesis is rejected, for example through the inference of fragmentation, then a second null hypothesis is tested. The second null hypothesis is that there is no significant difference across lineages with respect to genetic and/or demographic adaptations. Genealogical concordant species are recognized after both null hypotheses are rejected.

The cohesion species concept is largely based on the ability to rigorously test a set of null hypotheses concerned with the association of geography and genotypes/phenotypes (Templeton *et al.* 1995). The first null hypothesis is the same as that listed above for genealogical species diagnosis. If the organisms represent multiple lineages with 95% confidence, then the first null hypothesis is rejected. The second null hypothesis is that populations of different lineages are genetically and/or ecologically interchangeable among each other. The second null hypothesis is rejected when there is a significant association between geography and genetic and/or ecological variables, determined by NCPA. The ecological basis is that individuals can be 'exchanged' or moved between populations because they occupy the same niche (see Rader *et al.* 2005 for a variety of approaches to test ecological exchangeability and see Finlay *et al.* 2006 for an application of the ecological exchangeability approach). The genetic basis is that individuals are 'exchangeable' if there is extensive gene flow among populations.

NCPA was recently applied to delimit species boundaries of a South American lizard complex *(Liolaemus elongatus-kriegi*: Morando *et al.* 2003) using the combined approach of Wiens and Penkrot (2002) and Templeton (2001). The Wiens–Penkrot protocol complements species' delimitation studies that combine haplotype phylogenies and NCPA, but does not require 'exclusivity'. Morando *et al.* (2003) followed the protocol but with modifications: (1) multiple gene regions were used to test for genealogical concordance and (2) exclusivity was a criterion for species' boundaries. Because the authors did not have ecological data to address exchangeability, they only tested the first null hypothesis of Templeton's cohesion species criterion, with independent lineages arbitrarily defined as those outside the 95% confidence limit determined from their most variable mitochondrial gene region. The combined approach, using a priori defined criteria, supported the same clades and identified many more independent lineages than previously recognized under existing taxonomic names. Lineages that were supported by multiple lines of evidence were interpreted as 'candidate species' because they met the criteria for genetic concordance, geographic concordance, exclusivity and/or fragmentation/isolation by distance determined by NCPA.

SUMMARY

NCPA provides a statistical framework to elucidate historical and contemporary evolutionary processes that have contributed to the present-day genetic variation of a species. Some practical applications of NCPA include inferences about species' responses to past environmental events, current routes of gene flow and expansion, and the diagnosis of species under the Cohesion Species Concept and Genealogical Concordance Species Concept. It is a powerful tool for understanding population-level and species-level patterns of variability both temporally and spatially. Indeed, one of the primary goals of conservation biology is the protection of the evolutionary forces that naturally drive speciation and biodiversity.

ACKNOWLEDGEMENTS

We thank the editors for inviting us to participate in this publication and for their patience. Two anonymous reviewers and Jack Sites provided valuable comments, which greatly improved this manuscript. JEB was supported by a graduate fellowship from the Graduate Studies Office at Brigham Young University and an NSF Dissertation Improvement Grant (DEB 0508580).

REFERENCES

Agapow, P.-M., Bininda-Emonds, O. R. P., Crandall, K. A. *et al.* (2004). The impact of species concept on biodiversity studies. *Quarterly Review of Biology*, **79**, 161–179.

Avise, J. C. (1989). Gene trees and organismal histories: a phylogenetic approach to population biology. *Evolution*, **43**, 1192–1208.

Avise, J. C. and Ball, R. M. Jr (1990). Principles of genealogical concordance in species concepts and biological taxonomy. *Oxford Surveys in Evolutionary Biology*, **7**, 45–67.

Baum, D. A. and Shaw, K. L. (1995). Genealogical perspectives on the species problem. In *Experimental and Molecular Approaches to Plant Biosystematics*, ed. P. C. Hoch and A. G. Stephenson. St. Louis: Missouri Botanical Garden, pp. 289–303.

Branco, M., Monnerot, M., Ferrand, N. and Templeton, A.R. (2002). Postglacial dispersal of the European rabbit (*Oryctolagus cuniculus*) on the Iberian Peninsula reconstructed from nested clade and mismatch analysis of mitochondrial DNA genetic variation. *Evolution*, **56**, 792–803.

Buhay, J. E. and Crandall, K. A. (2005). Subterranean phylogeography of freshwater crayfishes shows extensive gene flow and surprisingly large population sizes. *Molecular Ecology*, **14**, 4259–4273.

Carbone, I. and Kohn, L. (2004). Inferring process from patterns in fungal population genetics. *Applied Mycology and Biotechnology*, **4**, 1–30.

Carstens, B. C., Stevenson, A. L., Degenhardt, J. D. and Sullivan, J. (2004). Tested nested phylogenetic and phylogeographic hypotheses in the *Plethodon vandykei* species group. *Systematic Biology*, **53**, 781–792.

Cassens, I., Mardulyn, P. and Milinkovitch, M. C. (2005). Evaluating intraspecific network construction methods using simulated sequence data: do existing algorithms outperform the global maximum parsimony approach? *Systematic Biology*, **54**, 363–372.

Castelloe, J. and Templeton, A. R. (1994). Root probabilities for intra-specific gene trees under neutral coalescent theory. *Molecular Phylogenetics and Evolution*, **3**, 102–113.

Clement, M., Posada, D. and Crandall, K. A. (2000). TCS: a computer program to estimate gene genealogies. *Molecular Ecology*, **9**, 1657–1659.

Comes, H. P. and Abbott, R. J. (1998). The relative importance of historical events and gene flow on the population structure of a Mediterranean ragwort, *Senecio gallicus* (Asteraceae). *Evolution*, **52**, 355–367.

Cooper, S. J. B., Ibrahim, K. M. and Hewitt, G. M. (1995). Postglacial expansion and genome subdivision in the European grasshopper *Chorthippus parallelus*. *Molecular Ecology*, **4**, 49–60.

Crandall, K. A. (1996). Multiple interspecies transmissions of human and simian T-cell leukemia/lymphoma virus type I sequences. *Molecular Biology and Evolution*, **13**, 115–131.

Crandall, K. A. and Templeton, A. R. (1993). Empirical tests of some predictions from coalescent theory with applications to intraspecific phylogeny reconstruction. *Genetics*, **134**, 959–969.

Crandall, K. A., Templeton, A. R. and Sing, C. F. (1994). Intraspecific phylogenetics: problems and solutions. In *Models in Phylogeny Reconstruction*, ed. R. W. Scotland, D. J. Siebert and D. M. Williams. Oxford: Clarendon Press, pp. 273–297.

Crandall, K. A., Bininda-Emonds, O. R. P., Mace, G. M. and Wayne, R. K. (2000). Considering evolutionary processes in conservation biology. *Trends in Ecology and Evolution*, **15**, 290–295.

Fetzner, J. W. Jr and Crandall, K. A. (2003). Linear habitats and the nested clade analysis: an empirical evaluation of geographic versus river distances using an Ozark crayfish (Decapoda: Cambaridae). *Evolution*, **57**, 2101–2118.

Finlay, J. B., Buhay, J. E. and Crandall, K. A. (2006). Surface to subsurface freshwater connections: phylogeographic and habitat analyses of *Cambarus tenebrosus*, a facultative cave-dwelling crayfish. *Animal Conservation*, **9**, 375–387.

Hedin, M. C. and Wood, D. (2002). Genealogical exclusivity in geographically proximate populations of *Hypochilus thorelli* Marx (Araneae, Hypochilidae) on the Cumberland Plateau of eastern North America. *Molecular Ecology*, **11**, 1975–1988.

Hey, J. and Nielsen, R. (2004). Multilocus methods for estimating population sizes, migration rates and divergence time, with applications to the divergence of *Drosophila pseudoobscura* and *D. persimilis*. *Genetics*, **167**, 747–760.

Hoffman, E. A. and Blouin, M. S. (2004). Evolutionary history of the Northern Leopard Frog: reconstruction of phylogeny, phylogeography, and historical changes in population demography from mitochondrial DNA. *Evolution*, **58**, 145–159.

Hudson, R. R. (1990). Gene genealogies and the coalescent process. *Oxford Surveys in Evolutionary Biology*, **7**, 1–44.

Hudson, R. R. and Coyne, J. A. (2002). Mathematical consequences of the genealogical species concept. *Evolution*, **56**, 1557–1565.

King, T. L., Switzer, J. F., Morrison, C. L. *et al.* (2006). Comprehensive genetic analyses reveal evolutionary distinction of a mouse (*Zapus hudsonius preblei*) proposed for delisting from the US Endangered Species Act. *Molecular Ecology*, **15**, 4331–4359.

Knowles, L. L. and Maddison, W. P. (2002). Statistical phylogeography. *Molecular Ecology*, **11**, 2623–2635.

Maddison, W. P. and Maddison, D. R. (2004). Mesquite: a modular system for evolutionary analysis. Version 1.05 http://mesquiteproject.org

Masta, S. E., Laurent, N. M. and Routman, E. J. (2003). Population genetic structure of the toad *Bufo woodhousii*: an empirical assessment of the effects of haplotype extinction on nested cladistic analysis. *Molecular Ecology*, **12**, 1541–1554.

Morando, M., Avila, L. J. and Sites, J. W. Jr (2003). Sampling strategies for delimiting species: genes, individuals, and populations in the *Liolaemus elongatus-kriegi* complex (Squamata: Liolaemidae) in Andean–Patagonian South America. *Systematic Biology*, **52**, 159–185.

Paquin, P. and Hedin, M. (2004). The power and perils of 'molecular taxonomy': a case study of eyeless and endangered *Cicurina* (Araneae: Dictynidae) from Texas caves. *Molecular Ecology*, **13**, 3239–3255.

Pearse, D. E. and Crandall, K. A. (2004). Beyond F_{ST}: Analysis of population genetic data for conservation. *Conservation Genetics*, **5**, 585–602.

Pfenninger, M. and Posada, D. (2002). Phylogeographic history of the land snail *Candidula unifasciata* (Helicellinae, Stylommatophora): fragmentation, corridor migration, and secondary contact. *Evolution*, **56**, 1776–1788.

Posada, D. and Crandall, K. A. (2001). Intraspecific gene genealogies: trees grafting into network. *Trends in Ecology and Evolution*, **16**, 37–45.

Posada, D., Crandall, K. A. and Templeton, A. R. (2000). GeoDis: a program for the cladistic nested analysis of the geographical distribution of genetic haplotypes. *Molecular Ecology*, **9**, 487–488.

Rader, R. B., Belk, M. C., Shiozawa, D. K. and Crandall, K. A. (2005). Empirical tests for ecological exchangeability. *Animal Conservation*, **8**, 1–9.

Ramey, R. R., Liu, H. P., Epps, C. W., Carpenter, L. M. and Wehausen, J. D. (2005). Genetic relatedness of the Preble's meadow jumping mouse (*Zapus hudsonius preblei*) to nearby subspecies of *Z. hudsonius* as inferred from variation in cranial morphology, mitochondrial DNA, and microsatellite DNA: implications for taxonomy and conservation. *Animal Conservation*, **8**, 329–346.

Roe, K. J. and Lydeard, C. (1998). Species delineation and the identification of evolutionarily significant units: lessons from the freshwater mussel genus *Potamilus* (Bivalvia: Unionidae). *Journal of Shellfish Research*, **17**, 1359–1363.

Rogers, A. R. and Harpending, H. (1992). Population growth makes waves in the distribution of pairwise genetic differences. *Molecular Biology and Evolution*, **9**, 552–569.

Schneider, S. D., Roessli, D. and Excoffier, L. (2000). ARLEQUIN version 2.0: a software for population genetic data analysis. Geneva: Genetics and Biometry Laboratory, University of Geneva.

Sinclair, E. A., Losada, M. P. and Crandall, K. A. (2005). Molecular phylogenetics for conservation biology. In *Phylogenies and Conservation*, ed. A. Purvis, T. Brooks and J. Gittleman. Cambridge: Cambridge University Press, pp. 19–56.

Sites, J. W. Jr and Crandall, K. A. (1997). Testing species boundaries in biodiversity studies. *Conservation Biology*, **11**, 1289–1297.

Sites, J. W. Jr and Marshall, J. C. (2003). Delimiting species: a Renaissance issue in systematic biology. *Trends in Ecology and Evolution*, **18**, 462–470.

Sites, J. W. Jr and Marshall, J. C. (2004). Operational criteria for delimiting species. *Annual Review of Ecology, Evolution, and Systematics*, **35**, 199–227.

Tarjuelo, I., Posada, D., Crandall, K. A., Pascuel, M. and Turon, X. (2004). Phylogeography and speciation of colour morphs in the colonial ascidian *Pseudodistoma crucigaster*. *Molecular Ecology*, **13**, 3125–36.

Templeton, A. R. (1989). The meaning of species and speciation. In *Speciation and Its Consequences*, ed. D. Otte and J. A. Endler. Sunderland: Sinauer Associates, pp. 3–27.

Templeton, A. R. (1998). Nested clade analyses of phylogeographic data: testing hypotheses about gene flow and population history. *Molecular Ecology*, **7**, 381–397.

Templeton, A. R. (2001). Using phylogeographic analyses of gene trees to test species status and processes. *Molecular Ecology*, **10**, 779–791.

Templeton, A. R. (2002). Out of Africa again and again. *Nature*, **416**, 45–51.

Templeton, A. R. (2004). Statistical phylogeography: methods of evaluating and minimizing inference errors. *Molecular Ecology*, **13**, 789–809.

Templeton, A. R. (2005). Inference key. http://darwin.uvigo.es/software/geodis.html.

Templeton, A. R. and Sing, C. F. (1993). A cladistic analysis of phenotypic associations with haplotypes inferred from restriction endonuclease mapping. IV. Nested analysis with cladogram uncertainty and recombination. *Genetics*, **134**, 659–669.

Templeton, A. R., Crandall, K. A. and Sing, C. F. (1992). A cladistic analysis of phenotypic associations with haplotypes inferred from restriction endonuclease mapping. III. Cladogram estimation. *Genetics*, **132**, 619–633.

Templeton, A. R., Routman, E. and Phillips, C. A. (1995). Separating population structure from population history: a cladistic analysis of the geographical distribution of mitochondrial DNA haplotypes in the tiger salamander, *Ambystoma tigrinum*. *Genetics*, **140**, 767–782.

Turgeon, J. and Bernatchez, L. (2001). Mitochondrial DNA phylogeography of lake cisco (*Coregonus artedi*): evidence supporting extensive secondary contacts between two glacial areas. *Molecular Ecology*, **10**, 987–1001.

Wiens, J. J. and Penkrot, T. A. (2002). Delimiting species using DNA and morphological variation and discordant species limits in Spiny Lizards (*Sceloporus*). *Systematic Biology*, **51**, 69–91.

5

A comparison of methods for constructing evolutionary networks from intraspecific DNA sequences

PATRICK MARDULYN, INSA CASSENS AND MICHEL
C. MILINKOVITCH

In phylogeography or population genetic studies, evolutionary relationships among DNA haplotypes can be depicted either as a graph, called a 'network', with cycles (or 'loops'), or as a set of phylogenetic trees (i.e. connected graphs with no circuits), possibly with multifurcation(s) and/or ancestral haplotype(s) (both represented by collapsing zero-length branches). For example, several equally optimal trees inferred under the maximum parsimony (MP) criterion display alternative relationships among haplotypes (Fig. 5.1a, b). A strict consensus tree can be used to summarize this set of trees (Fig. 5.1c), but this approach discards much of the historical information. Indeed, a strict consensus tree is typically compatible with many more alternative trees than those used to build it: e.g. the consensus in Fig. 5.1c is compatible with 105 different strictly bifurcating topologies although only two haplotypic trees have been used to build it. Furthermore, the consensus tree cannot easily summarize branch length information (e.g. in Fig. 5.1, taxon 4 is at the tip of a 0 step-long or a 1 step-long branch in trees (a) and (b), respectively). On the contrary, a network graph allows display much of the information contained in the data in a single figure (Fig. 5.1d). Therefore, the major advantage of such graphs over traditional phylogenetic trees is the possibility of using cycles (loops) to represent either ambiguities in the data or genuine reticulate evolution (due to e.g. recombination or horizontal gene transfer). In parsimony networks, sampled and unsampled haplotypes (white circles and black dots, respectively, in Fig. 5.1d) are symbolized by nodes (vertices) that are connected by edges, where each edge represents a single nucleotide substitution. Unsampled haplotypes are inferred to connect sampled haplotypes when

Population Genetics for Animal Conservation, eds. G. Bertorelle, M. W. Bruford, H. C. Hauffe, A. Rizzoli and C. Vernesi. Published by Cambridge University Press. © Cambridge University Press 2009.

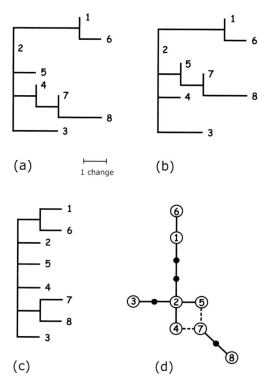

Figure 5.1. (a) and (b) Two different phylograms resulting from a parsimony analysis of a set of DNA sequences. (c) The strict consensus cladogram between trees (a) and (b) (note that much of the information contained in the source trees was discarded in the process of building the consensus). (d) Network built by combining the two maximum parsimony (MP) trees. Each edge in the graph represents one mutation step; black dots represent missing haplotypes (i.e. absent from the data set) that must be inferred to connect a sampled haplotype to the rest of the network. By discarding one of the two dashed edges, it is possible to recover each source MP tree.

the latter are separated by more than a single substitution. The so-called 'degree' of a node corresponds to the number of edges to which it is connected (e.g. in Fig. 5.1d, haplotype 2 is a node of degree 4).

In a previous empirical study on the phylogeography of dusky dolphins, Cassens *et al.* (2003) showed that the different algorithms available to build networks can produce significantly different graphs, and that these differences had an important impact on the interpretation of the evolutionary history of the species under investigation. In conservation genetic studies, accurate genealogies are required for the purpose of defining management units. Therefore, it is crucial to understand how the different algorithms work, and to identify under which circumstances each can be used appropriately. Recently, we have performed simulations of the evolution of sequences along known genealogies, in order to compare the efficiencies of different methods available for inferring haplotypic networks (Cassens *et al.* 2005). Here, (1) we briefly describe the principle and weaknesses of some of the most commonly used and/or promising methods (see Posada and Crandall (2001) for a more complete review of network construction algorithms); (2) we extend our simulation comparative work by analysing empirical data sets taken from the literature; and (3) we discuss, based on simulation and empirical data, how to choose an appropriate method for analysing population sequence data.

NETWORK CONSTRUCTION METHODS

Minimum spanning network (MSN)

A minimum spanning tree (MST) connects all sampled DNA sequences in a single tree without cycles. Excoffier and Smouse (1994) have modified an algorithm for constructing MSTs from a matrix of pairwise distances among haplotypes (Prim 1957; Rohlf 1973) in order to include all possible MSTs within a single graph, the MSN (implemented in the software ARLEQUIN, v. 2.000: Schneider *et al.* 2000). The minimum spanning network (MSN) is simply built by connecting, first, DNA sequences that are separated by one mutation, then those separated by two mutations, and so forth until the graph includes all sampled sequences. Although widely used in the literature for phylogeography studies, a MSN suffers from a serious problem: the connections are only formed among sampled haplotypes, i.e. the inference of a missing haplotype node of degree three or above is not possible. For this reason, the MSN usually does not contain the most parsimonious trees, unless all sampled sequences are separated from at least one other sampled sequence by a single mutation.

Reduced median network (RMN)

Bandelt *et al.* (1995) have described a method that generates a graph, called a 'median network', from strictly binary data (two character states). Each haplotype is modified as a vector of binary characters such that each character defines a unique split of haplotypes (i.e. characters that define the same split are pooled into one character). Starting from the initial set of vectors, the algorithm creates majority consensus vectors (or 'median vectors') of each triplet of vectors, thereby enlarging the set of vectors at each step. The process stops with a set of vectors that contains all consensus vectors of its triplets. The generated median network is guaranteed to include all most parsimonious trees. However, if the amount of homoplasy in the data is high, the number of intermediate nodes constructed to join the sampled haplotypes will become extremely large. Therefore, the authors (Bandelt *et al.* 1995) have described a method that reduces the number of connections of the median network, using other criteria, thereby generating a RMN. The criteria used to reduce the median network are (1) the compatibility criterion (Meacham and Estabrook 1985), and (2) the fact that mutations occur with a higher probability from more frequent haplotypes to less frequent haplotypes (e.g. Casteloe and Templeton 1994). The RMN is not guaranteed to include all MP trees, although the authors suggest this is often the case in practice. Although DNA sequences are not binary data, it is possible to transform them into binary data if not more than two states are present at each site. If more than two states are present, it is still possible to pre-process the data to produce a binary data set (Bandelt *et al.* 1995), although some loss of information will be associated with this process. The RMN approach is implemented in both the programs NETWORK, v. 4 (available at http://www.fluxus-engineering.com/sharenet.htm) and SPECTRONET (available at http://imbs.massey.ac.nz/download/spectronet/).

Median-joining network (MJN)

Bandelt *et al.* (1999) have proposed this approach as an alternative to the RMN approach, for dealing with larger data sets and with multistate characters. All MSTs are first combined within a single network (MSN) following an algorithm analogous to that proposed by Excoffier and Smouse (1994). Then, some consensus sequences (median vectors) of three mutually close sequences at a time are added to the graph, thereby generating missing haplotype nodes of degree three or above, in order to reduce its overall length. The triplets of sequences used for generating median vectors are selected according to specific rules, designed to increase the probability that the MP trees are included in the final graph. Although the MJN is not guaranteed to include all, or even one, MP tree, the MJN algorithm can improve considerably the original MSN graph (especially when analysing distant sequences) by reducing its overall length. This method is implemented in the program NETWORK, v. 4 (available at http://www.fluxus-engineering.com/sharenet.htm).

Statistical parsimony network

Statistical parsimony was first introduced by Templeton *et al.* (1992) and is implemented in the program TCS (Clement *et al.* 2000). In the initial description of the algorithm by Templeton *et al.* (1992), all connections are iteratively established among haplotypes starting with the smallest distances and ending when all haplotypes are connected. This approach, at least in its initial description by Templeton *et al.* (1992), is therefore very similar to the MSN algorithm mentioned above. Its main originality lies in the a priori definition of a 'parsimony limit', i.e. an estimate of the number of mutations separating any two sequences, above which the probability of multiple substitutions at a single site is more than 5%. Connections above this limit are considered non-parsimonious, and building the graph ends either when all haplotypes are connected or when the distance corresponding to the parsimony limit has been reached. Thus, in certain cases, the TCS graph will not include all the sampled sequences. Rather, two or more unconnected subgraphs will be produced. Importantly, missing haplotype nodes of degree three or above can be inferred by TCS, thereby allowing the program to generate graphs with reduced lengths compared to the MSN. To the best of our knowledge, however, the exact procedure of how these missing haplotype nodes are inferred is not described in the literature yet.

Union of maximum parsimonious trees (UMP)

In the literature, networks are usually built using one of the algorithms described above, without the help of an optimality criterion to compare different possible networks. Most network construction methods are therefore purely algorithmic, sensu Swofford *et al.* (1996; i.e. a method defined solely on the basis of an algorithm). This is in contrast with many phylogeny inference methods that use an optimality criterion (e.g. parsimony or likelihood) for exploring the space of all possible trees, and selecting the best one(s) (i.e. the trees that best explain the observed data given the optimality criterion). This feature could confer an advantage to phylogeny inference methods over network building methods. In an attempt to combine the advantages of using an optimality criterion and displaying all ambiguous relationships in a single graph, we have recently proposed combining all MP trees into a single network graph (Cassens *et al.* 2005). Thus, this approach requires two consecutive steps. First, an MP analysis is performed and all most parsimonious trees are saved with their respective branch lengths. This step is not algorithmic, but clearly based on the use of an optimality criterion, parsimony. Second, all the saved MP trees are combined into a single figure (Fig. 5.2). While transforming a single MP phylogram into a network-type graph without cycle, and vice versa, is trivial, combining the information contained in several trees into a single network-type graph can be a more difficult task. We have proposed an algorithm that

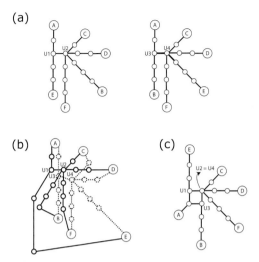

Figure 5.2. Principle of the union of maximum parsimonious trees (UMP) algorithm combining the MP trees in a single graph. (a) Two example MP trees. Each missing haplotype node of degree ≥3 is given a unique label (U1–U4). (b) All connections from both graphs are combined into a single figure. Tree number 1 is drawn with plain lines, whereas tree number 2 is drawn with dashed lines. (c) Identical edges and nodes from different trees are merged (an edge/node from one tree can only be merged with an edge/node from another tree). Only the edges and nodes that are not present in all trees are kept separate, generating cycle(s) in the network.

(1) combines all connections from all MP trees into a single graph (Fig. 5.2b), and (2) merges edges and missing haplotype nodes that are identical among different trees (Fig. 5.2c) (see Cassens *et al.* (2005), for a detailed description of the algorithm). Hence, during step (2), some cycles are maintained (i.e. some edges and missing haplotype nodes from different trees are not merged) where unique genealogical pathways are suggested in one or several (but not all) MP trees. Each MP tree, including branch lengths, can be reconstructed by removing some edges from the final UMP graph. This second step (merging edges and nodes) is clearly algorithmic, and builds one of possibly different graphs that include all the saved MP trees. We have indeed found that the placement and number of loops constructed by this algorithm can depend on the order with which connections are compared among trees. An optimality criterion could also be used to select the best graph containing all MP trees; for example, the minimum number of cycles included in the graph. It then remains to be investigated how to guarantee that all possible networks with the minimum

number of loops are found. It is important to note that a UMP graph may be compatible with more alternative trees (possibly less parsimonious) than those used to build it, i.e. with additional trees that are not supported by the data, although this problem is much less important than in strict consensus trees. This impossibility of guaranteeing the presence in the network of only the MP trees was already noted by Fitch (1997).

The UMP approach shares the goal of attempting to generate a graph that includes all MP trees with the RMN approach. Like the RMN approach, it has some practical problems. First, when the number of haplotypes is relatively high, an exact search for the MP trees is not possible, and a heuristic search strategy must be used instead, which does not guarantee

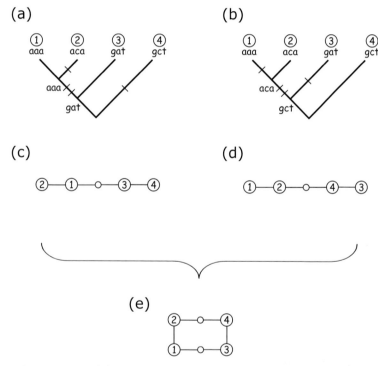

Figure 5.3. (a) and (b) Cladogram representing the most parsimonious (MP) tree among four sequences (3 nucleotides), with two alternative MP reconstructions of ancestral states. Mutations are represented by bars perpendicular to the branch along which they have occurred. (c) and (d) Network graphs (zero-branch lengths are collapsed) corresponding to the MP reconstruction of ancestral states in (a) and (b), respectively. (e) Combination of networks (c) and (d) in a single graph. By default, PAUP* infers only one MP reconstruction of ancestral states, the one shown in (a). In contrast, NETWORK and TCS reconstruct the graph shown in (e).

that all MP trees will be found. Second, when evaluating a tree, phylogeny inference programs (e.g. PAUP*: Swofford 2003) generally consider a single MP reconstruction of ancestral states. Construction of a network does not only require inferring the branching pattern, but also the number of mutations that have occurred along each branch. Therefore, it is important to consider all MP assignments of ancestral states, even if this would result in generating more trees, and thus more cycles in the final network. Figure 5.3 shows one simple example where considering two possible MP reconstructions of ancestral states results in one additional connection. Analyzing the four sequences of Fig. 5.3 by PAUP* (parsimony criterion) results in a single MP tree (Fig. 5.3a), while NETWORK (MJN) and TCS correctly reconstruct the network of Fig. 5.3e. Considering all MP reconstructions of ancestral sequences would therefore most probably improve the UMP approach. To our knowledge, however, no phylogenetic inference program offers this possibility at the moment. PAUP* allows users to select between 'accelerated transformation', 'delayed transformation' and 'minimization of the fCommands value' (i.e. three different methods for optimizing character states on a tree), potentially yielding different distributions of changes across the tree, but it does not consider all MP reconstructions. Note that including all MP reconstructions of ancestral sequences is likely to increase the number of cycles in the graph, and it remains to be investigated whether doing so will not produce a graph as complex as a median network graph (see above).

COMPARISON OF METHODS: SIMULATED SEQUENCE DATA

Following our empirical study on the phylogeography of dusky dolphins (Cassens *et al.* 2003) indicating that different methods of network construction can generate substantially different networks, we conducted a comparison analysis of the three most widely used methods in the literature (MSN, MJN, TCS), as well as of the UMP approach (Cassens *et al.* 2005). In that study, the evolution of DNA sequences was simulated along four different genealogies, some typically observed in empirical studies (Fig. 5.4). The inferred graphs were compared to the source genealogy. From this comparison, we computed the number of errors generated by each method. An error was defined as a branch that must be removed from the source genealogy (making it partly unconnected, i.e. each removal of a branch generates two subtrees) in order to make it totally compatible with the constructed graph (i.e. all the connections that remain in the subtrees are

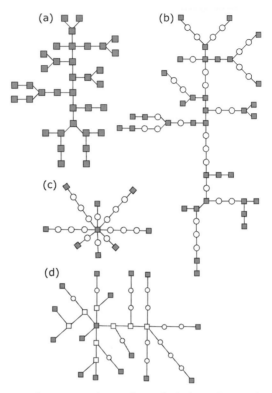

Figure 5.4. Template tree topologies along which the evolution of DNA sequences has been simulated in Cassens *et al.* (2005). Along each of the 1-step connections between any two haplotypes, sampled or missing, the probability of occurrence of one mutation is 0.01 per site. (a) The topology with single-step branches and all haplotypes sampled should be easy to reconstruct by most methods. The more complex topologies in (b), (c) and (d) were chosen as they allow testing of the impact of (b) the presence of long branches, (c) a star-like pattern, and (d) missing haplotype nodes of degree ≥3 on graph construction. Grey squares, sampled haplotypes; open circles and squares, missing haplotype nodes of degree 2 and ≥3, respectively (from Cassens *et al.* 2005).

included in the inferred graph). We also computed the number of cycles in each inferred network. When comparing networks generated by different methods, both measures (number of errors and number of loops) were taken into account. Indeed, a graph with a maximum number of cycles (in graph theory, this is called a 'fully connected graph', i.e. each node is connected to all other nodes) is, by definition, error-free (it is necessarily compatible with the source genealogy), yet is of little interest in this context because it conveys no genealogical information.

The most important result of our simulation study is the clear confirmation that different algorithms can generate different networks, at least with moderately distant sequences. Our analyses also demonstrate that the MSN algorithm yields the graphs with the highest number of errors in the case of sequences simulated along genealogies for which some nodes of degree ≥3 are not sampled (e.g. tree d in Fig. 5.4). This is not surprising, given its inability to infer missing haplotype nodes of degree ≥3. In the case of the three other genealogies that we used for simulations (Fig. 5.4a, b, c), the MSN algorithm produces slightly (but significantly) fewer errors than the other three methods. However, this is only because these three other methods sometimes wrongly infer missing haplotype nodes of degree ≥3, a mistake the MSN algorithm is not capable of doing. Furthermore, the MSN algorithm produces the highest number of cycles in all cases studied.

The three other methods (TCS, MJN and UMP) are less strikingly different in their respective abilities to minimize the number of cycles and the number of errors. Nonetheless, small but statistically significant differences were observed: although TCS builds networks that include a significantly smaller number of loops in most cases, it was also shown to produce slightly, but significantly, more errors than MJN and UMP, in the case of sequences simulated along genealogy (d) (Fig. 5.4). While building networks with fewer cycles may be appealing, it is not necessarily an advantage if this feature is associated with a higher probability of error. Finally, the numbers of errors in the MJN and UMP networks were not statistically different, but the number of cycles produced by UMP was significantly higher than the number of cycles generated by MJN.

COMPARISON OF METHODS: EMPIRICAL DATA SETS

Simulation studies allowed us to test network construction algorithms for a few case genealogies that we thought to be typical of real intraspecific data sets. However, many additional types of genealogies probably need to be explored before one can conclude on the strengths and weaknesses of each method.

Empirical data should allow investigation of whether the conditions that make different methods generate different networks are effectively met in real data sets. Indeed, our previous study (Cassens *et al.* 2003) is, to the best of our knowledge, the only one that explicitly highlights important differences among haplotype networks generated by different algorithms. For this reason, we chose 10 suitable empirical intraspecific data sets in the literature (see Table 5.1 for references and accession numbers), all

Table 5.1 *Empirical data sets analysed here; the method used in the original data set is indicated*

Data set number	Reference	Species name	GenBank accession number	Method used in original article[a]	Number of haplotypes
1	Balakrishnan et al. (2003)	*Cervus eldi*	AY137080-AY137125	MP	46
2	Caicedo and Schaal (2004)	*Solanum pimpinellifolium*	AY274227-AY294248	MP, TCS	22
3	Worheide et al. (2002)	*Leucetta 'chagosensis'*	AF45852-AF458870	TCS	19
4	Colgan et al. (2002)	*Lycosa sp. 'Wirra'*	AF42485I-AF424878	MP	28
5	Olsen (2002)	*Manihot esculenta*	AF136119-AF136149	MP	26
6	Neiman and Lively (2004)	*Potamopyrgus estuarinus*	AY570182-AY570227	MP, ML, NJ	41
7	Wilke and Pfenninger (2002)	*Hydrobia glyca* and *H. acuta*	AF467556-AF467653	TCS	37
8	Pitra et al. (2002)	*Hippotragus niger*	AF364780-AF364809	MP, ML, NJ, MJN, TCS	9
9	Dejong et al. (2003)	*Biomphalaria pfeifferi*	AY197982-AY198013	MP, ML, ME, TCS	27
10	Rawson et al. (2003)	*Chelonibia testudinaria*	AY174289-AY174367	NJ, MSN	54

[a] MP, maximum parsimony; ML, maximum likelihood; NJ; neighbor joining; ME, minimum evolution; TCS, statistical parsimony network; MJN, median-joining network.

published in issues of the journal *Molecular Ecology* between 2002 and 2004, and analysed these with three methods: TCS, MJN and UMP. We deliberately ignored data sets with highly similar sequences (i.e. whose evolution can be depicted with networks including only a small number of missing haplotype nodes), as these are the least likely to produce different results with different approaches. The MSN algorithm was not used because it would have predictably performed poorly given the many missing haplotype nodes it would have failed to reconstruct. Like in the simulation study, MJN networks were inferred with the program NETWORK (with parameter $\varepsilon = 0$), while statistical parsimony networks were constructed using the software TCS (gaps considered as missing character states, and increasing the parsimony limit until the program connected all haplotypes). To build the UMP graph, we first used the TBR branch swapping heuristic search option of the program PAUP* 4.0b10 (Swofford 2003) to find the MP trees (100 replicates with random sequence addition). These MP trees were then combined using the algorithm described in Cassens *et al.* (2005; program available at www.ulb.ac.be/sciences/ueg/html_files/combine-tree.html). As we found that the placement and number of loops constructed by this algorithm can depend on the order with which connections are compared among trees, the result with the lowest number of cycles in our analyses was selected among the 10 graphs produced with 10 different orders of connection comparisons.

Obviously, with empirical data, we do not have access to the true genealogy, which is the reference we used in computer simulations to identify the number of errors in the reconstructed networks (Cassens *et al.* 2005). Instead, we try to highlight here the differences among the results given by all three methods. More specifically, for each empirical data set, we made pairwise comparisons of networks as follows. First, we identify the subgraphs of the two networks that are identical (e.g. in Fig. 5.5, black nodes and edges are identical in TCS and MJN networks, and also happen to be included in the UMP graph) and compute D_a, the number of connections that need to be added to yield a single network (i.e. the number of subgraphs minus 1). For example, in Fig. 5.5, $D_a = 3$. Obviously, when two networks include at least one same tree, $D_a = 0$. Second, we compute D_b as simply the sum of connections that are different between two networks. Connections are defined iteratively. Each connection is a chain of edges and nodes that connects (1) a node belonging to one of the subgraphs identified above and (2) either a node belonging to one of the subgraphs, or to a node belonging to another connection already built in a previous iteration, with the restriction that edges cannot belong to more

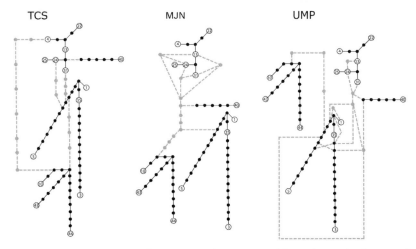

Figure 5.5. TCS, MJN and UMP networks constructed by analysing the data set of Balakrishnan *et al.* (2003). Subnetworks that are identical among all three networks are represented by black nodes and edges; parts that are different are drawn in grey.

than one connection. The TCS and NETWORK graphs of Fig. 5.5 include five and eight such connections (displayed in grey), respectively. We believe the two measures, D_a and D_b, are important. Indeed, two networks may have an entire genealogy in common ($D_a = 0$), yet they may be characterized by many different connections (because one network includes many cycles), hence exhibit a high D_b value. On the other hand, two networks each with no cycle but with very different topologies would exhibit both a high D_a and a high D_b.

D_a and D_b values are presented in Table 5.2 for each of the 10 selected data sets. Only two of the 30 pairwise comparisons yield both a D_a and a $D_b = 0$. All other comparisons reveal differences between networks, at the very least because some networks have more cycles than others, i.e. $D_b \neq 0$. More importantly, the networks are often incompatible with one another ($D_a > 0$), in the sense that it is not possible to reconstruct one identical genealogy from each of the two networks. Figure 5.5 shows, as an example, the networks constructed from data set number one (Table 5.1) that includes among the most divergent sequences of the test.

Clearly, empirical intraspecific sequence data sets result in different haplotype networks depending on the method used, such that the important differences among network construction methods identified by Cassens *et al.* (2003) on dusky dolphin sequences are not exceptional.

Table 5.2 *Pairwise comparisons of the networks constructed by the analysis of ten empirical data sets with TCS, MJN and UMP. The values of D_a and D_b are shown (see text for details). The number of each data set refers to Table 5.1*

Data set no.	D_a / D_b TCS–MJN	D_a / D_b TCS–UMP	D_a / D_b MJN–UMP
1	3/13	3/16	3/15
2	2/4	1/5	0/6
3	1/4	1/4	0/0
4	0/1	2/6	2/7
5	1/6	0/2	1/5
6	1/3	2/5	2/7
7	0/2	0/11	0/9
8	0/0	0/1	0/1
9	0/3	1/4	1/7
10	1/4	2/10	3/9

WHICH METHOD(S) SHOULD BE USED TO INFER GENEALOGICAL RELATIONSHIPS AMONG INTRASPECIFIC SEQUENCES?

Different networks can yield very different interpretations of the evolution-ary history of the taxa under scrutiny. Therefore, it is worth investigating the causes of these differences and identifying the best method for the problem at hand. At this stage, it is however not possible to recommend a single best method for building a network from intraspecific DNA sequences. At the very least, we strongly recommend researchers to compare the results obtained with different methods. Clearly, the problem increases as sequen-ces become more divergent. If all sampled sequences are connected to at least one other sequence by a single mutation, inferring the haplotype net-work is trivial and all the methods tested (MSN, MJN, TCS and UMP) will result in the same graph. As sequences become more divergent, genealog-ical relationships are more difficult to infer, and slight differences appear among different methods of network construction. Once missing haplotype nodes of degree ≥3 need to be inferred, important differences of inference arise among methods. The MSN algorithm is particularly problematic in this latter case as it is not capable of inferring such nodes. We recommend avoiding the use of this method, as it is not possible to know in advance whether a missing node haplotype of degree ≥3 must be inferred.

In our computer simulation studies (Cassens *et al.* 2005), the numbers of errors recorded for the TCS, MJN and UMP methods are reasonably low (on average between one and two errors per network), and the differences of performance among the three methods are quite small (although statistically significant). Nonetheless, several of the empirical data sets analysed here reveal large numbers of different connections among the TCS, MJN and UMP networks. A single connection difference between two networks can have a major impact on the interpretation of the population evolution under study.

The statistical parsimony approach (Templeton *et al.* 1992) implemented in TCS deals with the problem of very divergent sequences by taking into account a 'parsimony limit' (see above) and treating in isolation the subnetworks that are separated by a number of mutations above this limit. In other words, we first identify the maximum number of steps separating two sequences for which a connection is still parsimonious, then we give up the idea of resolving the entire genealogy of the sample, as the parts of the network that are too divergent are not connected to one another. In the case of DNA sequences, the estimation of the parsimony limit depends entirely upon the length of the sequences. However, the current model used for the estimation of this limit assumes that all nucleotides (including invariant sites) evolve at the same rate. In real data sets, this assumption is unrealistic, such that we recommend estimating first the proportion of invariable sites (using a model of DNA sequence evolution), then to remove the corresponding proportion of invariant sites from the data set before estimating the parsimony limit (i.e. before analysing the data with TCS). Nonetheless, many researchers wish to reconstruct a complete graph (i.e. a graph including all sequences), even if this causes an increase of uncertainty in the reconstructed graph.

Our computer simulations (Cassens *et al.* 2005) have shown that UMP is an appropriate method to infer genealogical relationships at the intraspecific level. Although it did produce slightly more cycles than MJN and TCS, it is also (with MJN) the method that generated the smallest number of errors. One could argue that the good results of MJN indicate that the use of an optimality criterion (maximum parsimony) to search tree space does not confer an advantage to UMP over MJN. However, it should be investigated whether the UMP approach can be improved further if all MP reconstructions of ancestral states are considered during the maximum parsimony search (first step; see above).

ACKNOWLEDGEMENTS

We thank two anonymous reviewers for their critical and constructive comments on a previous version of the manuscript. This work was supported by the Belgian Fund for Scientific Research (FNRS), the

'Communauté Française de Belgique' (ARC 11649/20022770), and the 'Région Wallonne'.

REFERENCES

Balakrishnan, C. N., Monfort, S. L., Gaur, A., Sing, L. and Sorenson, M. D. (2003). Phylogeography and conservation genetics of Eld's deer (*Cervus eldi*). *Molecular Ecology*, 12, 1–10.

Bandelt, H. J., Forster, P., Sykes, B. C. and Richards, M. B. (1995). Mitochondrial portraits of human populations using median networks. *Genetics*, 141, 743–753.

Bandelt, H.-J., Forster, P. and Röhl, A. (1999). Median-joining networks for inferring intraspecific phylogenies. *Molecular Biology and Evolution*, 16, 37–48.

Caicedo, A. L. and Schaal, B. A. (2004). Population structure and phylogeography of *Solanum pimpinellifolium* inferred from a nuclear gene. *Molecular Ecology*, 13, 1871–1882.

Cassens, I., Van Waerebeek, K., Best, P. B. *et al.* (2003). The phylogeography of dusky dolphins (*Lagenorhynchus obscurus*): a critical examination of network methods and rooting procedures. *Molecular Ecology*, 12, 1781–1792.

Cassens, I., Mardulyn, P. and Milinkovitch, M. (2005). Evaluating intraspecific 'network' construction methods using simulated sequence data: do existing algorithms outperform the global maximum parsimony approach? *Systematic Biology*, 54, 363–372.

Castelloe, J. and Templeton, A. R. (1994). Root probabilities for intraspecific gene trees under neutral coalescent theory. *Molecular Phylogenetics and Evolution*, 3, 102–113.

Clement, M., Posada, D. and Crandall, K. A. (2000). TCS: a computer program to estimate gene genealogies. *Molecular Ecology*, 9, 1657–1659.

Colgan, D. J., Brown, S., Major, R. E., Christie, F., Gray, M. R. and Cassis, G. (2002). Population genetics of wolf spiders of fragmented habitat in the wheat belt of New South Wales. *Molecular Ecology*, 11, 2295–2305.

Dejong, R. J., Morgan, J. A. T., Wilson, W. D., *et al.* (2003). Phylogeography of *Biomphalaria glabrata* and *B. pfeifferi*, important intermediate hosts of *Schistosoma mansoni* in the New and Old World tropics. *Molecular Ecology*, 12, 3041–3056.

Excoffier, L. and Smouse, P. E. (1994). Using allele frequencies and geographic subdivision to reconstruct gene trees within a species: molecular variance parsimony. *Genetics*, 136, 343–359.

Fitch, W. M. (1997). Networks and viral evolution. *Journal of Molecular Evolution*, 44, S65–S75.

Meacham, C. A. and Estabrook, G. F. (1985). Compatibility methods in systematics. *Annual Review of Ecology and Systematics*, 16, 431–446.

Neiman, M. and Lively, C. M. (2004). Pleistocene glaciation is implicated in the phylogeographical structure of *Potamopyrgus antipodarum*, a New Zealand snail. *Molecular Ecology*, 13, 3085–3098.

Olsen, K. M. (2002). Population history of *Manihot esculenta* (Euphorbiaceae) inferred from nuclear DNA sequences. *Molecular Ecology*, 11, 901–911.

Pitra, C., Hansen, A. J., Lieckfeldt, D. and Arctander, P. (2002). An exceptional case of historical outbreeding in African sable antelope populations. *Molecular Ecology*, 11, 1197–1208.

Posada, D. and Crandall, K. A. (2001). Intraspecific gene genealogies: trees grafting into networks. *Trends in Ecology and Evolution*, **16**, 37–45.

Prim, R. C. (1957). Shortest connection networks and some generalizations. *Bell Systems Technical Journal*, **36**, 1389–1401.

Rawson, P. D., Macnamee, R., Frick, M. G. and Williams, K. L. (2003). Phylogeography of the coronulid barnacle, *Chelonibia testudinaria*, from loggerhead sea turtles, *Caretta caretta*. *Molecular Ecology*, **12**, 2697–2706.

Rohlf, F. J. (1973). Hierarchical clustering using the minimum spanning tree. *Computer Journal*, **16**, 93–95.

Schneider, S., Roessli, D. and Excoffier, L. (2000). ARLEQUIN: a software for population genetics data analysis. v. 2.000. Genetics and Biometry Laboratory, University of Geneva.

Swofford, D. L. (2003). PAUP*, phylogenetic analysis using parsimony (* and other methods), v. 4b10. Sunderland: Sinauer Associates.

Swofford, D. L., Olsen, G. J., Waddel, P. J. and Hillis, D. M. (1996). Phylogenetic inference. In *Molecular Systematics*, ed. D. Hillis, C. Moritz and K. M. Mable. Sunderland: Sinauer Associates, pp. 407–514.

Templeton, A. R., Crandall, K. A. and Sing, C. F. (1992). A cladistic analysis of phenotypic associations with haplotypes inferred from restriction endonuclease mapping and DNA sequence data. III. Cladogram estimation. *Genetics*, **132**, 619–633.

Wilke, T. and Pfenninger, M. (2002). Separating historic events from recurrent processes in cryptic species: phylogeography of mud snails (*Hydrobia* spp.). *Molecular Ecology*, **11**, 1439–1451.

Worheide, G., Hooper, J. N. A and Degnan, B. M. (2002). Phylogeography of western Pacific *Leucetta 'chagosensis'* (Porifera: Calcarea) from ribosomal DNA sequences: implications for population history and conservation of the Great Barrier Reef World Heritage Area (Australia). *Molecular Ecology*, **11**, 1753–1768.

Molecular approaches and applications

Challenges in assessing adaptive genetic diversity: overview of methods and empirical illustrations

AURÉLIE BONIN AND LOUIS BERNATCHEZ

INTRODUCTION

Since its inception, the concept of 'evolutionarily significant unit' (ESU) has had several theoretical definitions differing mainly on the emphasis given to neutral versus adaptive genetic diversity (Ryder 1986; Waples 1991; Moritz 1994; Crandall *et al.* 2000; reviewed in Fraser and Bernatchez 2001). The 'neutral' definition highlights a genetic background shaped over a long-term evolutionary time scale by evolutionary forces such as genetic drift and migration (Moritz 2002). In contrast, the 'adaptive' definition underlines the existing, adaptively significant phenotypic variation resulting from the ongoing action of natural selection (McKay and Latta 2002; van Tienderen *et al.* 2002). Thus, neutral and adaptive diversities depict distinct temporal realms (Fig. 6.1) and should be assessed separately and differently (Bowen 1999; Moritz 2002).

The neutral component of genetic diversity was the first to be investigated, coinciding with the breakthrough in the development of molecular markers and new tools and concepts in phylogeography and population genetics. A wide range of methods is currently available to measure neutral genetic variability, which has resulted in crucial answers to several key conservation issues (see examples in Frankham *et al.* 2002). In contrast, characterizing adaptive genetic variation remains challenging (van Tienderen *et al.* 2002; Vasemägi and Primmer 2005), requiring the identification of a small number of adaptive loci scattered throughout the genome (Black *et al.* 2001). Moreover, molecular markers are generally assumed to be neutral and they have been shown to be poor indicators of adaptive genetic diversity (Merilä and Crnokrak 2001; Reed and Frankham 2001; McKay and Latta 2002). Given this background, we review here different strategies that can be adopted to reveal genetic polymorphisms with an adaptive role.

Population Genetics for Animal Conservation, eds. G. Bertorelle, M. W. Bruford, H. C. Hauffe, A. Rizzoli and C. Vernesi. Published by Cambridge University Press. © Cambridge University Press 2009.

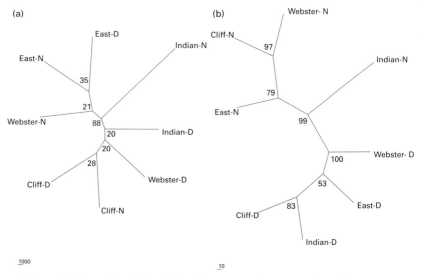

Figure 6.1. Neutral and adaptive loci do not carry the same biological message. In northeastern North America, two different and sympatric ecotypes of the lake whitefish (*Coregonus clupeaformis*) can be found in the St John River Basin. Here, the two ecotypes, referred to as N for normal (benthic) and D for dwarf (limnetic), were sampled in four localities (Cliff Lake, East Lake, Indian Pond and Webster Lake) along the St John River. Neighbour-joining trees were constructed using (a) markers expected to be neutral and (b) markers assumed to be adaptive. Numbers at nodes represent bootstrap values (in percent). In (a), the tree topology reflects the neutral history of populations that tend to gather according to the sampled locality, whereas in (b), populations group according to their ecotype, which mirrors their adaptive divergence (adapted from Campbell and Bernatchez 2004).

The goal of this chapter is not to present an exhaustive and detailed list of existing methods, but rather to highlight the underlying principles of those we believe to be the most useful and/or readily applicable, as well as their advantages and drawbacks. Several examples relevant to conservation are also presented.

The available methods for identifying loci that are under selection and therefore, potentially adaptive, differ according to the analytical and screening strategies they rely on (Table 6.1). Choice of analytical strategy is directly influenced by the type of information that is easily available for the studied populations or species. For example, the quantitative trait locus (QTL) analysis requires phenotypic data and results from experimental crosses in addition to the genotyping of hundreds of markers (Falconer and Mackay 1996). As for the marker screening strategy, this is chiefly determined by the trade-off between the resolution power of the method and the extent

Table 6.1 *The main methods currently available for searching for adaptive loci. A given method corresponds to a specific combination of either a single or multilocus screening strategy with a particular analytical strategy*

	Screening strategy	
Analytical strategy	Single locus	Multilocus
Association between genotype and phenotype or environment	Candidate gene/mutation analysis[a]	Linkage study QTL analysis
Inter-population differentiation versus neutrality	Study of sequence variation Allozyme study[b]	Genome scan based on at least several hundreds molecular markers (AFLPs, SNPs, microsatellites)[a] Q_{ST}–F_{ST} comparison[a]
Intra-population diversity versus neutrality	Study of sequence variation	Genome scan based on microsatellites
Differential gene expression	Quantitative RT-PCR analysis	Transcription profiling using a cDNA microarray[a]

[a] Methods that are described further in this chapter.
[b] Allozymes are entered under the single-locus strategy when searching for adaptive loci, but they can also be used in multilocus studies in population genetics.

of genome coverage. In general, the more molecular markers are examined, the more accurately the global adaptive diversity is represented at the genome scale (Bonin *et al.* 2006); however, such multilocus surveys usually fail to identify the precise locus (loci) or mutation(s) controlling the expression of an adaptive trait (Phillips 2005; Slate 2005).

The following sections will address in more detail four approaches that are commonly used to evaluate adaptive genetic diversity and, in some cases, to reveal the underlying loci (Vasemägi and Primmer 2005): (1) Q_{ST}–F_{ST} comparison; (2) analysis of a candidate gene or mutation; (3) genome scan; and (4) transcription profiling based on a cDNA microarray.

Q_{ST}–F_{ST} COMPARISON

In a rigorous assessment of adaptive diversity, it is necessary to ensure from the outset that the phenotypic divergence observed between populations has a genuine adaptive basis, i.e. that it is heritable and associated with a differential fitness in contrasting environments, or that it is driven by divergent natural selection (Kawecki and Ebert 2004). Therefore, it is important to rule out other potential causes of phenotypic variations, such as non-adaptive phenotypic plasticity (Pigliucci 2005) or maternal effects (Mousseau and Fox 1998), by conducting common-garden or reciprocal transplant experiments, direct measurements of selection coefficients, and estimates of adaptive landscapes using the distribution of spatial and trophic resources (Schluter 2000).

The most readily applicable method for identifying 'true' adaptive phenotypic divergence is to compare the phenotypic differentiation between populations with neutral expectations (i.e. the amount of differentiation expected from the effects of only mutation and genetic drift). To this end, the Q_{ST}–F_{ST} comparison (Spitze 1993) is the most frequently applied test (Podolsky and Holtsford 1995; Lynch et al. 1999), and is based on the theoretical principle that divergence at quantitative traits should be similar to that of allele frequencies at nuclear loci, if these traits are evolving neutrally and have a quasi-pure additive genetic basis (Wright 1951). Moreover, under the influence of migration, mutation and genetic drift, the among-population proportion of total genetic variance in phenotypic traits is expected to equal that of neutral molecular loci (Lande 1992). Therefore, as an indirect method for detection of natural selection, the extent of population differentiation at quantitative traits (Q_{ST}) can be compared with that quantified at neutral molecular markers (F_{ST}). The prediction is that divergent selection will cause Q_{ST} to be larger than that expected for neutral loci. Q_{ST} is quantified as the proportion of among-population genetic variance at quantitative traits:

$$Q_{ST} = \sigma^2_{GB}/(\sigma^2_{GB} + 2\,\sigma^2_{GW})$$

where σ^2_{GB} and σ^2_{GW} are the additive components of genetic variance between and within populations, respectively (Spitze 1993).

The strict use of the Q_{ST}–F_{ST} comparison implies that knowledge of genetic variation for quantitative traits is available. However, this method has also been applied using phenotypic variance (e.g. Bernatchez 2004; Merilä 1997) under the assumption that this estimate is an accurate surrogate for additive genetic variance. Indeed, previous studies have emphasized the similarity between genetic and phenotypic variances (e.g. Roff and Mousseau 1987; Cheverud 1988; Roff 1995, 1996), and measures of phenotypic covariance have been shown to be nearly as successful as those of genetic covariance in predicting the direction of divergence between species (e.g. Schluter 1996). Consequently, results of Q_{ST} studies based on phenotypic variance (especially if performed in a controlled environment) may be useful for identifying phenotypic traits that have evolved under directional selection and, therefore, are likely to be adaptive.

Unfortunately, the theoretical basis for rigorously comparing F_{ST} and Q_{ST} is not well understood, and Q_{ST} estimates may suffer from bias and imprecision which are context-dependent. For instance, the expectation that Q_{ST} = F_{ST} under neutrality depends on the assumption that mutation rates are considerably lower than migration rates. This assumption may be violated for some systems (e.g. where there is low gene flow) and markers with high mutation rates (e.g. microsatellites). Thus, Q_{ST}–F_{ST} comparisons should be considered more as exploratory tools for identifying traits potentially under the effect of selection, rather than actual or absolute measures of adaptive genetic variation (Hendry 2002; O'Hara and Merilä 2005).

Case study: the Q_{ST}–F_{ST} comparison applied to two different ecotypes of the lake whitefish (*Coregonus clupeaformis*)

Rogers *et al.* (2002) and Rogers and Bernatchez (2005) tested the null hypothesis of the absence of selective effects on the phenotypic divergence between lake whitefish ecotypes (*Coregonus* sp.) by adopting the Q_{ST}–F_{ST} strategy for various morphological, behavioural and physiological (growth) traits (Fig. 6.2). Dwarf and normal whitefish ecotypes co-inhabit several lakes of the St John River Basin in northeastern North America. These forms are adapted for the differential use of the limnetic or benthic trophic niches, and

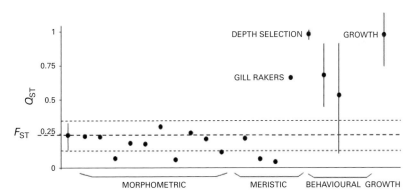

Figure 6.2. Q_{ST} estimates between a dwarf and normal whitefish populations for ten morphological traits (in order: preorbital length, orbital length, trunk length, dorsal fin length, caudal peduncle length, maxillary width, maxillary length, body depth, head depth, interorbital width), four meristic traits (suprapelvic scales, dorsal ray count, pectoral ray count, gill raker count), three behavioural traits (depth selection, directional changes, burst swimming) and one physiological trait (growth rate) with their 95% confidence intervals. The first dot represents the extent of genetic divergence between the two populations (θ value of 0.24), and the 95% confidence interval of this neutral expectation is delineated by dotted lines (adapted from Bernatchez 2004).

increased efficiency for the capture of small pelagic prey appears to be more important for the fitness of the dwarf than of the normal ecotype. The level of genetic differentiation between dwarf and normal ecotypes estimated from θ at microsatellite loci was 0.24 (95% CI = 0.13 – 0.33). For behavioural traits, the Q_{ST} estimates for depth selection (0.98, 95% CI = 0.96 – 1.00) and directional changes (0.68, 95% CI = 0.45 – 0.91) are high and significantly different from neutral expectations. The Q_{ST} estimates for morphological traits range from 0.06 (maxillary length) to 0.66 (gill raker counts), and gill rakers are the only morphological trait with a Q_{ST} significantly higher than neutral expectations. For growth rate, highly significant differences are observed between dwarf and normal whitefish when reared under identical conditions (normal 0.58 g/day, and dwarf fish 0.22 g/day). This translates into a high Q_{ST} value (0.91, 95% CI = 0.62 – 1.00), which is significantly higher than F_{ST}.

Overall, the whitefish example illustrates how the Q_{ST}–F_{ST} approach allows the identification of a few phenotypic traits (depth selection, directional changes, growth rate and gill rakers), among the many that were analysed, as those most likely to be adaptive in this system. In contrast, the remaining traits are more likely to have evolved neutrally and, thus, are less important from a conservation perspective.

ANALYSIS OF A CANDIDATE GENE OR MUTATION

The analysis of a candidate gene or mutation can be considered a 'bottom-up' strategy, because it starts with a particular DNA polymorphism and attempts to establish the role of this polymorphism in adaptive processes (Phillips 2005). Two different approaches can be adopted for such an analysis: first, we can test if the mutation under consideration is statistically associated with an adaptive phenotype or selective pressure (see for example Kohn et al. 2000; Nair et al. 2003; Rosenblum et al. 2004). If the test is significant, it does not establish a causal link between the studied phenotypic or environmental factor and the mutation; however, it does indicate that the particular mutation may be adaptive, or at the very least, that it may be in linkage disequilibrium with an adaptive allele. Second, when a gene sequence is known, sequence variability can be used to assess if the gene evolves under the influence of natural selection by monitoring patterns of intra-population diversity or inter-population differentiation (Ford 2002; Nielsen 2005; Wright and Gaut 2005).

Focusing on a candidate gene or mutation has been shown to be particularly efficient in unravelling the genetic architecture of adaptive traits when only one or a few major genes are involved, such as in heavy metal tolerance or insecticide resistance (Kawecki and Ebert 2004; Storz 2005). This approach has even led to the identification of important adaptive mutations in control regions or introns (Phillips 2005). However, the analysis of a candidate gene requires preliminary information about the gene function and/or sequence that may not be easily accessible for some species or adaptive phenotypes (Ford 2002). In addition, as the emphasis is only on one or a few genes, this approach does not estimate adaptive diversity over the entire genome, which may mean that some alleles with potentially important conservation values are overlooked.

Case study: melanism in the rock pocket mouse

The rock pocket mouse (*Chaetodipus intermedius*) is a small rodent living in the rocky habitats of southern Arizona, New Mexico and adjacent areas of northern Mexico. This species usually inhabits light-coloured rocks, and individuals have a typical sandy dorsal pelage. However, pocket mice can also be found on dark-coloured volcanic lava flows where they display a dark melanic coat (Fig. 6.3a). A strong correlation has been shown to exist between the dorsal pelage and habitat colour throughout the species' distribution range (Hoekstra *et al.* 2005). Hence, variation in pelage coloration in the pocket mouse has a putative adaptive significance, probably by providing protection from predators by improving camouflage in the surrounding habitat.

Using an association study, the genetic basis of the melanism has been identified in a population of pocket mice living on basaltic lava in the Pinacate site in Arizona (Nachman *et al.* 2003). There, the dark colour of the pelage has been found to be tightly associated with the substitution of four amino acids in the protein sequence of the melanocortin-1-receptor gene, *Mc1r*. Interestingly, these four mutations are not found in several other melanic populations originating from New Mexico (Hoekstra and Nachman 2003). This indicates that (1) melanism does not have the same genetic basis in the population from Arizona and that from New Mexico, and (2) melanism appeared and evolved independently in the two cases (Hoekstra and Nachman 2003).

Analyses of neutral diversity in this species have conveyed interesting additional information about the biology of the rock pocket mouse. Phylogeographic studies based on two mitochondrial genes revealed that populations are highly differentiated throughout the species distribution (Nachman *et al.* 2003; Hoekstra *et al.* 2005). Nonetheless, there is no correspondence between the phylogeographic structure and the variation in the pelage colour (Fig. 6.3b), which suggests a rapid and separate

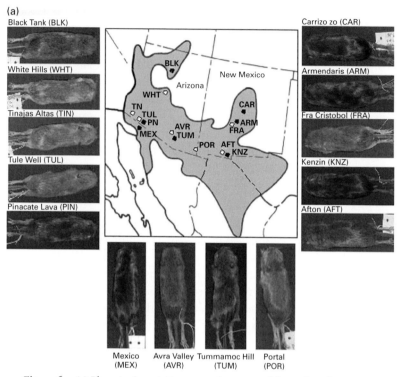

Figure 6.3. (a) Phenotypic variation across the range of the rock pocket mouse (*Chaetodipus intermedius*) in southwestern USA. Photographs represent the typical dorsal coloration of individuals from each of 14 collecting localities indicated by circles. Filled circles represent lava flows and open circles non-volcanic rocky regions (adapted from Hoekstra *et al.* 2005). (b) Neighbour-joining phylogeography of the rock pocket mouse (*Chaetodipus intermedius*) across its distribution range. Topology is rooted with *C. penicillatus* and *C. baileyi*. Asterisks indicate individuals inhabiting lava. Geographic regions are mentioned on the right. Bootstrap support is displayed under the internal branches (See also colour plate.) (adapted from Hoekstra *et al.* 2005).

evolution for this trait. On a local scale, dark- and light-coloured populations are occasionally connected by pronounced gene flow (Hoekstra *et al.* 2004, 2005). Therefore, the clear geographical genetic structure observed for the allele responsible for melanism most likely results from strong selective pressures counterbalancing the homogenizing effects of gene flow between phenotypically divergent populations.

The rock pocket mouse example is particularly enlightening from a conservation viewpoint because it illustrates two major tenets to be respected in a conservation study. First, it shows that examining the adaptive structure

(b)

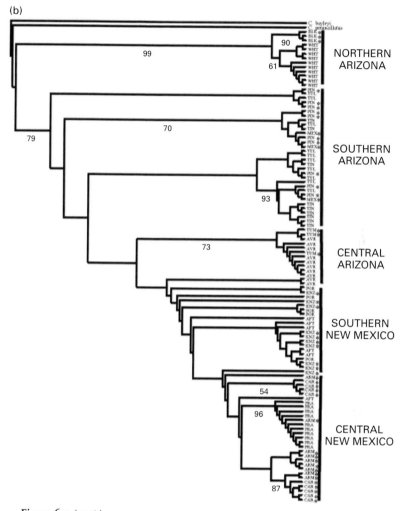

Figure 6.3. (cont.)

of populations may lead to inaccurate inferences regarding their evolution-ary history or levels of connectivity, which may be better depicted by neutral markers. As such, this case study strikingly illustrates the independence between neutral and adaptive components of genetic diversity and the need to assess them separately. Second, this particular example of melanism demonstrates that elucidating the genetic basis of adaptation is a complex matter, even in cases where the genotype–phenotype relationship is quite simple. Indeed, the same adaptive phenotype may arise from different adaptive pathways affecting different genes. This underlines the fact that a priori information regarding adaptive processes must be considered with caution in conservation plans and should not be transferred blindly between different case studies.

GENOME SCAN

Genome scans involve the analysis of many (several hundred to several thousand) molecular markers, that may or may not be genetically mapped, in order to reveal genome-wide patterns of genetic diversity, differentiation or linkage disequilibrium. Such multilocus surveys have recently been made possible by the improvement of genotyping throughputs (Schlötterer 2004) combined with the increased use of molecular markers such as single nucleotide polymorphisms (SNPs) or amplified fragment length polymorphisms (AFLPs), which permit a thorough screening of the genome (Vos *et al.* 1995; Morin *et al.* 2004). In this respect, the AFLP procedure is a particularly remarkable innovation, as it is theoretically applicable to any species, without any a priori knowledge of the genome, and provides hundreds and even thousands of low-cost random markers (Vos *et al.* 1995; Bensch and Akesson 2005).

Genome scans represent the practical facet of the emerging field of population genomics that seeks to understand the respective roles of the different evolutionary forces in shaping genetic variability across genomes and populations (Black *et al.* 2001; Luikart *et al.* 2003). The underlying principle of population genomics is that the dominant evolutionary forces are of two types (Black *et al.* 2001; Luikart *et al.* 2003): some affect all loci across the genome, without exception (e.g. genetic drift and gene flow), while others act locally on specific genes or chromosomal regions (e.g. selection). As a result, there is a 'background' genetic variability produced by genome-wide forces, whereas a few loci will display an atypical pattern of variation caused by the influence of locus-specific forces (Black *et al.* 2001; Storz 2005). It is possible to identify these 'outlier' loci by comparison with the rest of the genome (Fig. 6.4). In studies of adaptation, such loci deserve particular attention as they are likely to be under selection, or at least linked to a selected gene (Schlötterer 2002).

In practice, outlier loci can be detected by applying one of two types of statistical analysis, which differ mainly according to the level at which genetic diversity is measured (Schlötterer 2002; Luikart *et al.* 2003; Storz 2005). One category monitors reductions of genetic diversity within populations, possibly due to the spread of an advantageous allele (Payseur *et al.* 2002; Vigouroux *et al.* 2002), while the other surveys the increase in genetic

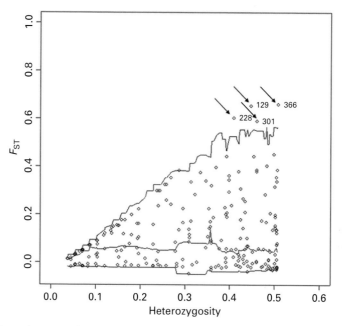

Figure 6.4. Example of outlier loci detection in the common frog (*Rana temporaria*) by comparison of empirical data with simulated neutral data. Here, for 392 AFLP markers, genetic differentiation (F_{ST}) between high-altitude and low-altitude populations is plotted against expected heterozygosity in order to detect loci potentially involved in adaptation to altitude. The lower, intermediate and higher lines represent the 5%, 50% and 95% confidence intervals of the neutral simulations, respectively. AFLP markers are represented by dots; most of them conform to the neutral expectations, except four loci that lie outside the 95% line and are pointed out by arrows (adapted from Bonin *et al.* 2006).

differentiation between populations at specific loci that could indicate the existence of distinct adaptive optima (Beaumont and Nichols 1996; Vitalis *et al.* 2001; Beaumont and Balding 2004) (see Fig. 6.4).

In a conservation biology context, where rapid decisions must often be made on the basis of incomplete information, genome surveys can potentially become a valuable tool for estimating the adaptive value of populations (Luikart *et al.* 2003; Bonin *et al.* 2007). They can now be conducted for almost any species with relatively limited effort and cost, at least with random markers such as AFLPs. For that matter, many taxa, including threatened species, should largely benefit in the near future from the achievement of the numerous ongoing genome projects (Kohn *et al.* 2006). Moreover, genome scans do not require additional information such as sequence, phenotypic or pedigree data that may be challenging to obtain rapidly for threatened species or populations (Storz 2005). As such, they also offer the possibility to reveal the genetic basis of 'cryptic adaptations' (Bonin *et al.* 2007); that is, traits whose involvement in adaptive mechanisms may not be particularly obvious at first sight (e.g. biochemical, metabolic or behavioural adaptations).

However, the analysis of genome scans is not without pitfalls. Even if a locus is identified as an outlier, this does not necessarily mean that it is involved in the expression of a selected trait. Other factors can mimic the variation patterns produced by selection, including genotyping errors, aberrant mutation or recombination rates (Luikart *et al.* 2003). In particular, unrealistic demographic or neutral hypotheses might lead to the characterization of false selection signatures (Akey *et al.* 2004; Nielsen 2005). In practice, the confounding effects of demography or neutral history can be ruled out by using independent experimental replicates (Wilding *et al.* 2001; Campbell and Bernatchez 2004; Bonin *et al.* 2006). In such replicates, selected loci are expected to display the same trend, in contrast to neutral loci whose behaviour is determined mainly by chance through genetic drift. Moreover, even if outlier loci are truly adaptive, they might not reflect the overall adaptive diversity of the screened genome (Luikart *et al.* 2003; Bonin *et al.* 2006; Bonin *et al.* 2007), since this will depend greatly on whether (1) the scan is dense enough (in terms of number of screened loci) to include all selected regions of the genome; (2) the scan samples each selected genomic region equally and does not over- or under-represent a portion of them; and (3) the detection methods used are powerful enough to pinpoint even weak selection signals. Theoretical and practical advances will soon ensure that such conditions will be met for any genome survey in the near future (Luikart *et al.* 2003; Beaumont 2005; Bonin *et al.* 2006). Until then, however, the outcomes of genome scans must be considered with caution unless they are confirmed by independent data (Storz 2005).

Incorporating population genomics into conservation: the Population Adaptive Index

While the contribution of population genomics to the understanding of adaptation is increasingly recognized (Luikart *et al.* 2003; Storz 2005), this approach has only recently begun to be adopted in conservation biology. Of special interest is the recent introduction by Bonin *et al.* (2007) of the population adaptive index (PAI), which is a parameter measuring the adaptive significance of a population in relative terms when compared to other populations. The PAI estimation requires preliminary identification of the most likely adaptive loci in a given set of populations, usually by means of a genome scan. For a population, the PAI is then calculated as the percentage of adaptive loci with allelic frequencies differing significantly from those in the other populations (Box 6.1). In other words, the PAI accounts for adaptive features, which appear to be unique to a given population among several others, and thereby justifies its relevance for conservation.

Bonin *et al.* (2007) illustrated the use of the PAI in the common frog (*Rana temporaria*), the most widespread amphibian in Europe. Six populations in the northern French Alps were sampled and a genome scan was carried out on about 400 ALFP markers, identifying 14 potentially adaptive loci. Based on these loci, the PAI was estimated in each population as detailed in Box 6.1. The two populations with the highest PAIs were found at a low- and a high-altitude site, respectively (Bonin *et al.* 2007). As adaptation to altitude is believed to be the primary cause of adaptive divergence between common frog populations in the Alps (Miaud and Merilä 2001), this example shows that the PAI was able to capture the adaptive differences existing between these populations, independently from any a priori quantitative genetics study.

The PAI appears to be a valuable index that will probably attract some attention in the conservation field very soon. However, its robustness still remains to be tested. In particular, the minimal number of markers that must be scanned in order to make the index reliable still needs to be estimated. Moreover, the PAI will never be as informative as quantitative genetic data when these are available.

TRANSCRIPTION PROFILING BASED ON A cDNA MICROARRAY

The major and ultimate challenge in the assessment of adaptive genetic diversity is to discover the actual genes involved, and to what extent mutations or patterns of standing genetic variation are responsible for generating adaptive genetic/phenotypic variance. As detailed above, the candidate gene approach has been successful in achieving this goal in several cases. However, phenotypes can rarely be explained by allelic variants at a single locus, and therefore, the task of elucidating the genetic basis of most adaptive phenotypic variance will be more daunting. Nonetheless, progress in the field of functional genomics is promising. For example, evolutionary biologists have now reached a consensus that many of the key genetic differences between organisms will not only be in the form of allelic variation

Box 6.1.   Calculation of the Population Adaptive Index (PAI) (after Bonin *et al.* 2007)

The calculation of the PAI can be divided into three major steps:

(1) *Detection of loci potentially under divergent selection*
Loci potentially under divergent selection can be detected by assessing the levels of genetic differentiation (e.g. F_{ST}) at each marker locus between each possible pair of sampled populations. For example, if six populations are considered, 15 pairwise single analyses must be performed. Loci diverging from neutral expectations are then of two kinds. If they are detected in a single pairwise comparison only, they might be false positives whose outlier behaviour is likely to be linked to the statistical confidence level chosen. If not, they can be more safely regarded as potentially adaptive (Campbell and Bernatchez 2004; Bonin *et al.* 2006). As a result, the PAI calculation accounts only for this last kind of outlier loci.

(2) *Identification of the population(s) systematically involved in the detection of each adaptive locus*
To calculate the PAI, it is important to determine which particular population(s) is (are) responsible for the outlier behaviour of each adaptive locus. This can be done by examining the pattern of outlier detection. If one (several) population(s) is (are) systematically involved in the pairwise analyses revealing the adaptive locus under consideration, this means that allelic frequencies in this (these) population(s) are statistically different from those in the remaining populations. In other words, this (these) population(s) probably displays (display) singular adaptive features deserving special attention in management plans.

(3) *PAI calculation in itself*
Once each adaptive locus can be assigned to one or a few particular populations, the PAI can be determined for any population by simply calculating the percentage of adaptive loci which are assigned to this population out of the total number of adaptive loci. The higher this percentage is, the more notable this population is in terms of adaptive uniqueness.

NB: The PAI is exclusively valid for a given set of populations and markers. If the data set is altered (e.g. new populations are included), the PAI needs to be recalculated.

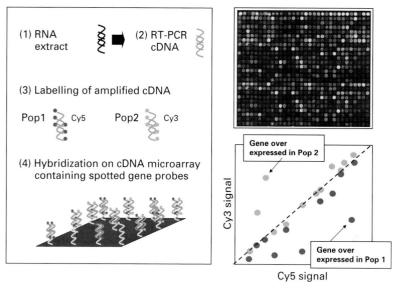

Figure 6.5. Basic principles of transcription profiling based on a cDNA microarray. The basic principles of cDNA microarray analysis consist in quantifying differential levels of gene expression by means of a competitive hybridization between RNA (retro-transcribed cDNA) from two experimental specimens. It involves four basic steps: (1) RNA is first extracted from each specimen to be analysed, (2) source RNA is retro-transcribed into complementary DNA (cDNA) and amplified using PCR, (3) cDNAs of both experimental specimens are marked with different fluorescent labels, and (4) labelled cDNAs of both specimens are simultaneously hybridized on a cDNA microarray usually containing probes for thousands of genes. Hybridized microarrays are then image-scanned and the differential intensity of both labels for each gene is translated into differential levels of expression between experimental specimens. (See also colour plate.)

at specific genes, but will also be manifested as changes in gene expression during development (Purugganan 1998; Streelman and Kocher 2000). Accordingly, technologies are being developed to characterize organismal transcriptomes that include the set of genes expressed in a particular tissue at a specific time. The method that is now most commonly used to define transcriptomes is termed 'DNA microarray' (Gibson 2002) (Fig. 6.5, and Plate 1, colour plate section). By comparing patterns of gene expression between populations, this method offers the possibility to identify the most significant shifts in gene expression involved in the adaptive divergence of populations. In fact, several recent studies have demonstrated the potential of transcription profiling to reveal differential gene expression between populations, and thus offer a tremendous opportunity for investigating the genomic basis of phenotypic divergence under various environmental conditions (Oleksiak *et al.* 2002; Bochdanovits *et al.* 2003; Oleksiak *et al.* 2005; Whitehead and Crawford 2006).

Even without prior information on which genes may be adaptive in a specific context, genome-wide expression employing natural populations permits the identification of specific genes potentially implicated in adaptive divergence and the direction of genetic divergence using a comparative approach. Evidence that the same subset of genes presents parallel, directional changes in expression among independently evolving populations of similar phenotype can also provide strong empirical support for the role of natural selection in shaping differential gene transcription profiling. Moreover, traits under strong selection are expected to display lower variance than traits under weaker selection. Since gene expression is a quantitative trait, testing for a significantly reduced variance in expression among individuals may further support the role of natural selection acting on those genes. One important constraint on the current use of this method in the context of conservation is that specific microarrays are currently available for only a handful of species, although the use of cross-specific arrays has proven feasible and useful (Renn *et al.* 2004). Also, it should be noted that microarrays only measure the steady state concentration of a gene's mRNA, and that mRNA translation, degradation, and protein turnover will also affect the active amount of a protein ultimately affecting a phenotypic change.

Rapid evolutionary changes of gene expression in farmed Atlantic salmon (*Salmo salar*): relevance for the conservation of wild populations

Selective breeding of Atlantic salmon (*Salmo salar*) was initiated in Norway some 35 years ago and is now intensively practised in Chile, the United Kingdom, the United States and Canada. At first, artificial selection was limited to the improvement of growth rate, but this practice now also targets traits such as age at sexual maturity, bacteria resistance, fat content and flesh colour. Moreover, phenotypic changes in traits that were not the focus of artificial selection have also been observed in Norwegian farmed salmon, including increased fat content in flesh (Rye and Gjerde 1996), poorer performance in natural conditions (Fleming *et al.* 2000; McGinnity *et al.* 2003) and morphological and behavioural changes (Fleming and Einum 1997), as well as a higher feeding rate and food conversion efficiency (Thodesen *et al.* 1999). The last decade has seen the world-wide production of farmed Atlantic salmon outstrip that of fisheries (FAO 2004); in addition, the number of farmed salmon escapees has reached alarming proportions, with about 2 million farmed salmon escaping annually from their sea cages (the natural populations in the North Atlantic are estimated to contain only 4 million individuals: McGinnity *et al.* 2003). Consequently knowledge of the heritable differences accumulated in salmon breeding strains is of considerable concern for the conservation of wild populations.

Roberge *et al.* (2006) have recently used a 3557-gene cDNA microarray (Rise *et al.* 2004) to compare levels of gene transcription in whole juveniles (fry stage) between farm populations and their population of origin from Norway and Canada. Their results showed that only five to seven generations of artificial selection led to significant heritable changes in gene expression, the average magnitude of the observed differences being approximately 20% for about 1.5% of the expressed genes at the juvenile stage. Up to 16% of all transcription profile changes between farmed and wild populations occurred in parallel for the same genes and in the same direction in both farmed populations from Norway and Canada, thus providing strong indirect evidence for the role of selection in shaping transcription profiles in those populations. Moreover, Roberge *et al.* (2006) also found that genetic drift and inadvertent selection may have caused undesirable evolutionary changes in farmed salmon. For example, their results revealed a 21% under-expression of the metallothionein (MT) A gene, which has been shown in mammals to be associated with obesity as a result of a higher food intake and abnormal energy balance (Beattie *et al.* 1998). MT is also a key factor in adaptation to heavy-metal environments (Posthuma and van Straalen 1993). Therefore, introgression of this reduced MT expression from farmed into wild populations could reduce resistance of the latter to environmental pollutants.

As mentioned before, the phenotypic importance of the transcription level changes observed in this study must be interpreted with caution since the link between transcription level and phenotypic expression is subject to other levels of regulation. Nonetheless, this approach suggests ways by which gene flow from farmed escapees may affect the genetic integrity and conservation of wild populations of salmon. Overall, the study of Roberge *et al.* (2006) provides a clear example of how new insights into the molecular basis of adaptive divergence can be acquired by comparing levels of gene transcription, and how these insights can be applied to conservation practice.

GENERAL DISCUSSION

This review focuses mainly on the adaptive diversity currently observed within species, and the different approaches to assessing this diversity. It deliberately omits a thorough discussion of neutral or nearly neutral poly-morphisms, although these are likely to include variants of adaptive impor-tance in the future (Moritz 1994; McKay and Latta 2002), and are also of major conservation importance.

We would also like to note two important challenges currently constrain-ing the search for adaptive polymorphisms in a conservation as well as in a more general framework. First, such a search is, in practice, often based on the strict assumption that selection, which directly acts on adaptive traits, will also have a discernible effect on the underlying QTLs or genes. In other words, most of the methods that aim at detecting adaptive loci assume that genetic differentiation is the immediate consequence of phenotypic differ-entiation. However, although this condition is respected a priori for traits with a simple genetic determinism, such as melanism in the rock pocket mouse, the situation is not so straightforward for polygenic traits. For such traits, it is indeed possible to observe a trait differentiation without any inter-population differentiation at the underlying QTLs (Latta 1998; Storz 2005). This is the case, for example, when the covariance of inter-population allelic frequencies at a QTLs is responsible for most of the trait variance (McKay and Latta 2002) (Fig. 6.6). Using simulated data, Le Corre and Kremer (2003) have even shown that among all the QTLs associated with a trait, only a small number of loci can be highly differentiated and thus contribute to most of the phenotypic variance between populations,

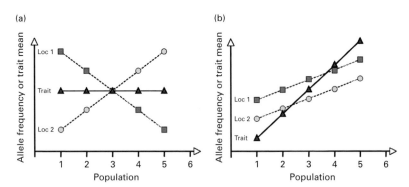

Figure 6.6. Expected effects of the among-population covariance of two QTLs (Loc 1 and Loc 2) on differentiation at the corresponding polygenic trait. If this covariance is negative (a), allelic frequencies at QTLs are strongly differentiated but the trait value remains similar over all populations, i.e. no trait differentiation is observed. If this covariance is positive (b), a strong trait differentiation is achieved with little allelic differentiation. This latter situation is likely to occur for polygenic traits under selection (adapted from McKay and Latta 2002).

independently from their individual effect on the trait value. In short, selection may fail to create a detectable signal on a subset, if not all, the loci involved in the expression of a particular adaptive phenotype (Rogers and Bernatchez 2005).

Secondly, sequence, genomic and/or quantitative data represent a basic and necessary starting point for many of the available approaches for assessing adaptive genetic diversity (Vasemägi and Primmer 2005) (Table 6.1). Therefore, the discovery of adaptive genetic variation remains a tedious task for non-model species, for which even basic prior knowledge is sparse and has to be collected at the outset (Storz 2005). Hopefully, data collection will require less effort in the near future, with the improvement of genotyping techniques and the increasing number of fully sequenced genomes. Nevertheless, very few studies have been able to reach a resolution power to the actual nucleotide level, even in the most extensively studied organisms (Flint and Mott 2001; Mackay 2001).

Consequently, identifying precisely which genes are under selection should be perceived for most species as a long-term prospect rather than as an immediately achievable goal. However, persevering in this direction is certainly not a futile exercise and may only require 'taking smaller steps', as expressed so elegantly by Vasemägi and Primmer (2005). Indeed, the methods addressed here can still offer valuable insights into the estimation of adaptive diversity and/or the understanding of its genetic basis, although none of them is likely to provide the key to every unanswered question. In this regard, experience has shown that additional information can be astutely collected by combining diverse research approaches (Emelianov et al. 2004; Rogers and Bernatchez 2005) or by investigating different functional units (i.e. DNA, mRNA, protein, phenotype: Schadt et al. 2003; Vasemägi et al. 2005).

In the search for adaptive genetic variation in wild populations, it is important to keep in mind that a specific discovery or pattern is relevant primarily to the biological background under consideration and cannot necessarily be extrapolated to other backgrounds (Hoekstra and Nachman 2003; Vasemägi and Primmer 2005). For example, adaptation is accompanied by a reduced genetic diversity within populations when directional selection promotes the spread of a particular allele (Storz 2005). In contrast, when balancing selection is the acting force, heterozygote genotypes are favoured, hence variability increases, as is the case for loci of the major histocompatibility complex (e.g. Aguilar et al. 2004). In short, information about adaptation is rarely interchangeable between case studies, and this represents a major constraint in the context of conservation. Time and/or practical limitations may indeed impose extrapolation between ecologically similar species.

We believe that the adaptive significance of populations will eventually become a parameter increasingly easier to access, with less effort, cost and time than it is presently the case (Kohn et al. 2006). Hopefully, these developments should help to resolve the still contentious issue of ESU definition. In particular, when prioritizing populations for protection purposes, the approaches outlined here could help resolve on a case-by-case basis when the neutral component of genetic diversity shaped by historical contingency ought to be favoured over its adaptive component influenced by contemporary determinism, and vice versa.

REFERENCES

Aguilar, A., Roemer, G., Debenham, S. et al. (2004). High MHC diversity maintained by balancing selection in an otherwise genetically monomorphic mammal. *Proceedings of the National Academy of Sciences USA*, **101**, 3490–3494.

Akey, J. M., Eberle, M. A., Rieder, M. J. et al. (2004). Population history and natural selection shape patterns of genetic variation in 132 genes. *PLoS Biology*, **2**, 1591–1599.

Beattie, J. H, Wood, A. M., Newman, A. M. et al. (1998). Obesity and hyperleptinemia in metallothionein (-I and -II) null mice. *Proceedings of the National Academy of Sciences USA*, **95**, 358–363.

Beaumont, M. A. (2005). Adaptation and speciation: what can F_{ST} tell us? *Trends in Ecology and Evolution*, **20**, 435–440.

Beaumont, M. A. and Balding, D. J. (2004). Identifying adaptive genetic divergence among populations from genome scans. *Molecular Ecology*, **13**, 969–980.

Beaumont, M. A. and Nichols, R. A. (1996). Evaluating loci for use in the genetic analysis of population structure. *Proceedings of the Royal Society Series B*, **263**, 1619–1626.

Bensch, S. and Akesson, M. (2005). Ten years of AFLP in ecology and evolution: why so few animals? *Molecular Ecology*, **14**, 2899–2914.

Bernatchez, L. (2004). Ecological theory of adaptive radiation: an empirical assessment from coregonine fishes (Salmoniformes). In *Evolution Illuminated: Salmon and their Relatives*, ed. A. P. Hendry and S. C. Stearns. Oxford: Oxford University Press, pp. 175–207.

Black, W. C., Baer, C. F., Antolin, M. F. and DuTeau, N. M. (2001). Population genomics: genome-wide sampling of insect populations. *Annual Review of Entomology*, **46**, 441–469.

Bochdanovits, Z., van der Klis, H. and de Jong, G. (2003). Covariation of larval gene expression and adult body size in natural populations of *Drosophila melanogaster*. *Molecular Biology and Evolution*, **20**, 1760–1766.

Bonin, A., Miaud, C., Taberlet, P. and Pompanon, F. (2006). Explorative genome scan to detect candidate loci for adaptation along a gradient of altitude in the common frog (*Rana temporaria*). *Molecular Biology and Evolution*, **23**, 773–783.

Bonin, A., Nicolè, F., Pompanon, F., Miaud, C. and Taberlet, P. (2007). The Population Adaptive Index: a new index to help measure intraspecific genetic diversity and prioritize populations for conservation. *Conservation Biology*, **21**, 697–708.

Bowen, B. W. (1999). Preserving genes, species, or ecosystems? Healing the fractured foundations of conservation policy. *Molecular Ecology*, **8**, S5–S10.

Campbell, D. and Bernatchez, L. (2004). Generic scan using AFLP markers as a means to assess the role of directional selection in the divergence of sympatric whitefish ecotypes. *Molecular Biology and Evolution*, **21**, 945–956.

Cheverud, J. M. (1988). A comparison of genetic and phenotypic correlations. *Evolution*, **42**, 958–968.

Crandall, K. A., Bininda-Emonds, O. R. P., Mace, G. M. and Wayne, R. K. (2000). Considering evolutionary processes in conservation biology. *Trends in Ecology and Evolution*, **15**, 290–295.

Emelianov, I., Marec, F. and Mallet, J. (2004). Genomic evidence for divergence with gene flow in host races of the larch budmoth. *Proceedings of the Royal Society Series B*, **271**, 97–105.

Falconer, D. S and Mackay, T. F. (1996). *Introduction to Quantitative Genetics*. Harlow, UK: Longman.

FAO (Food and Agriculture Organization of the United Nations) (2004). *FISHSTAT Plus*. http://www.fao.org/fi/statist/FISOFT/FISHPLUS.asp.

Fleming, I. A. and Einum, S. (1997). Experimental tests of genetic divergence of farmed from wild Atlantic salmon due to domestication. *ICES Journal of Marine Science*, **54**, 1051–1063.

Fleming, I. A., Hindar, K., Mjolnerod, I. B. *et al.* (2000). Lifetime success and interactions of farm salmon invading a native population. *Proceedings of the Royal Society Series B*, **267**, 1517–1523.

Flint, J. and Mott, R. (2001). Finding the molecular basis of quantitative traits: successes and pitfalls. *Nature Reviews Genetics*, **2**, 437–445.

Ford, M. J. (2002). Applications of selective neutrality tests to molecular ecology. *Molecular Ecology*, **11**, 1245–1262.

Frankham, R., Ballou, J. D. and Briscoe, D. A. (2002). *An Introduction to Conservation Genetics*. Cambridge: Cambridge University Press.

Fraser, D. J. and Bernatchez, L. (2001). Adaptive evolutionary conservation: towards a unified concept for defining conservation units. *Molecular Ecology*, **10**, 2741–2752.

Gibson, G. (2002). Microarrays in ecology and evolution: a preview. *Molecular Ecology*, **11**, 17–24.

Hendry, A. P. (2002). $Q_{ST} > = \neq < F_{ST}$? *Trends in Ecology and Evolution*, **17**, 502.

Hoekstra, H. E. and Nachman, M. W. (2003). Different genes underlie adaptive melanism in different populations of rock pocket mice. *Molecular Ecology*, **12**, 1185–1194.

Hoekstra, H. E., Drumm, K. E. and Nachman, M. W. (2004). Ecological genetics of coat color variation in pocket mice: geographic variation in selected and neutral genes. *Evolution*, **58**, 1329–1341.

Hoekstra, H. E., Krenz, J. G. and Nachman, M. W. (2005). Local adaptation in the rock pocket mouse (*Chaetodipus intermedius*): natural selection and phylogenetic history of populations. *Heredity*, **94**, 217–228.

Kawecki, T. J. and Ebert, D. (2004). Conceptual issues in local adaptation. *Ecology Letters*, **7**, 1225–1241.

Kohn, M. H., Pelz, H. J. and Wayne, R. K. (2000). Natural selection mapping of the warfarin-resistance gene. *Proceedings of the National Academy of Sciences USA*, 97, 7911–7915.

Kohn, M. H., Murphy, W. J., Ostrander, E. A. and Wayne, R. K. (2006). Genomics and conservation genetics. *Trends in Ecology and Evolution*, 21, 629–637.

Lande, R. (1992). Neutral theory of quantitative genetic variance in an island model with local extinction and colonization. *Evolution*, 46, 381–389.

Latta, R. G. (1998). Differentiation of allelic frequencies at quantitative trait loci affecting locally adaptive traits. *American Naturalist*, 151, 283–292.

Le Corre, V. and Kremer, A. (2003). Genetic variability at neutral markers, quantitative trait loci and trait in a subdivided population under selection. *Genetics*, 164, 1205–1219.

Luikart, G., England, P. R., Tallmon, D., Jordan, S. and Taberlet, P. (2003). The power and promise of population genomics: from genotyping to genome typing. *Nature Reviews Genetics*, 4, 981–994.

Lynch, M., Pfender, M., Spitze, K. *et al.* (1999). The quantitative and molecular genetic architecture of a subdivided species. *Evolution*, 53, 100–110.

Mackay, T. F. (2001). The genetic architecture of quantitative traits. *Annual Review of Genetics*, 35, 303–339.

McGinnity, P., Prodohl, P., Ferguson, A. *et al.* (2003). Fitness reduction and potential extinction of wild populations of Atlantic salmon, *Salmo salar*, as a result of interactions with escaped farm salmon. *Proceedings of the Royal Society Series B*, 270, 2443–2450.

McKay, J. K. and Latta, R. G. (2002). Adaptive population divergence: markers, QTL and traits. *Trends in Ecology and Evolution*, 17, 285–291.

Merilä, J. (1997). Quantitative trait and allozyme divergence in the greenfinch (*Carduelis chloris*, Aves : Fringillidae). *Biological Journal of the Linnean Society*, 61, 243–266.

Merilä, J. and Crnokrak, P. (2001). Comparison of genetic differentiation at marker loci and quantitative traits. *Journal of Evolutionary Biology*, 14, 892–903.

Miaud, C. and Merilä, J. (2001). Local adaptation or environmental induction? Causes of population differentiation in alpine amphibians. *Biota*, 2, 31–50.

Morin, P. A., Luikart, G., Wayne, R. K. and the Single Nucleotide Polymorphism Workshop Group (2004). SNPs in ecology, evolution and conservation. *Trends in Ecology and Evolution*, 19, 208–216.

Moritz, C. (1994). Defining 'Evolutionarily Significant Units' for conservation. *Trends in Ecology and Evolution*, 9, 373–375.

Moritz, C. (2002). Strategies to protect biological diversity and the evolutionary processes that sustain it. *Systematic Biology*, 51, 238–254.

Mousseau, T. A. and Fox, C. W. (1998). The adaptive significance of maternal effects. *Trends in Ecology and Evolution*, 13, 403–407.

Nachman, M. W., Hoekstra, H. E. and D'Agostino, S. L. (2003). The genetic basis of adaptive melanism in pocket mice. *Proceedings of the National Academy of Sciences USA*, 100, 5268–5273.

Nair, S., Williams, J. T., Brockman, A. *et al.* (2003). A selective sweep driven by pyrimethamine treatment in southeast Asian malaria parasites. *Molecular Biology and Evolution*, 20, 1526–1536.

Nielsen, R. (2005). Molecular signatures of natural selection. *Annual Review of Genetics*, 39, 197–218.

Oleksiak, M. F., Churchill. G. A. and Crawford, D. L. (2002). Variation in gene expression within and among natural populations. *Nature Genetics*, **32**, 261–266.

Oleksiak, M. F., Roach, J. L. and Crawford, D. L. (2005). Natural variation in cardiac metabolism and gene expression in *Fundulus heteroclitus*. *Nature Genetics*, **37**, 67–72.

O'Hara, R. B. and Merilä, J. (2005). Bias and precision in Q_{st} estimates: problems and some solutions. *Genetics*, **171**, 1331–1339.

Payseur, B. A., Cutter, A. D. and Nachman, M. W. (2002). Searching for evidence of positive selection in the human genome using patterns of microsatellite variability. *Molecular Biology and Evolution*, **19**, 1143–1153.

Phillips, P. C. (2005). Testing hypotheses regarding the genetics of adaptation. *Genetica*, **123**, 15–24.

Pigliucci, M. (2005). Evolution of phenotypic plasticity: where are we going now? *Trends in Ecology and Evolution*, **20**, 481–486.

Podolsky, R. H. and Holtsford, T. P. (1995). Population structure of morphological traits in *Clarkia dudleyana*. I. Comparison of F_{ST} between allozymes and morphological traits. *Genetics*, **140**, 733–744.

Posthuma, L. and van Straalen, N. M. (1993). Heavy-metal adaptation in terrestrial invertebrates: a review of occurrence, genetics, physiology and ecological consequences. *Comparative Biochemistry and Physiology Part C: Pharmacology, Toxicology and Endocrinology*, **106**, 11–38.

Purugganan, M. D. (1998). The molecular evolution of development. *BioEssays*, **20**, 700–711.

Reed, D. H. and Frankham, R. (2001). How closely correlated are molecular and quantitative measures of genetic variation? A meta-analysis. *Evolution*, **55**, 1095–1103.

Renn, S. C., Aubin-Horth, N. and Hofmann, H. A. (2004). Biologically meaningful expression profiling across species using heterologous hybridization to a cDNA microarray. *BMC Genomics*, **5**, 42.

Rise, M. L., von Schalburg, K. R., Brown, G. D. *et al.* (2004). Development and application of a salmonid EST database and cDNA microarray: data mining and interspecific hybridization characteristics. *Genome Research*, **14**, 478–490.

Roberge, C., Einum, S., Guderley, H. and Bernatchez, L. (2006). Rapid parallel evolutionary changes of gene transcription profiles in farmed Atlantic salmon. *Molecular Ecology*, **15**, 9–20.

Roff, D. A. (1995). The estimation of genetic correlations from phenotypic correlations: a test of Cheverud's conjecture. *Heredity*, **74**, 481–490.

Roff, D. A. (1996). The evolution of genetic correlations: an analysis of patterns. *Evolution*, **50**, 1392–1403.

Roff, D. A. and Mousseau, T. A. (1987). Quantitative genetics and fitness: lessons from *Drosophila*. *Heredity*, **58**, 103–118.

Rogers, M. R., Gagnon, V. and Bernatchez, L. (2002). Genetically based phenotype–environment association for swimming behavior in lake whitefish ecotypes (*Coregonus clupeaformis* Mitchill). *Evolution*, **56**, 2322–2329.

Rogers, S. M. and Bernatchez, L. (2005). Integrating QTL mapping and genome scans towards the characterization of candidate loci under parallel selection in the lake whitefish (*Coregonus clupeaformis*). *Molecular Ecology*, **14**, 351–361.

Rosenblum, E. B., Hoekstra, H. E. and Nachman, M. W. (2004). Adaptive reptile colour variation and the evolution of the *Mc1r* gene. *Evolution*, **58**, 1794–1808.

Ryder, O. A. (1986). Species conservation and systematics: the dilemma of subspecies. *Trends in Ecology and Evolution*, **1**, 9–10.

Rye, M. and Gjerde, B. (1996). Phenotypic and genetic parameters of body composition traits and flesh colour in Atlantic salmon, *Salmo salar* L. *Aquaculture Research*, **27**, 121–133.

Schadt, E. E., Monks, S. A. and Friend, S. H. (2003). A new paradigm for drug discovery: integrating clinical, genetic, genomic and molecular phenotype data to identify drug targets. *Biochemical Society Transactions*, **31**, 437–443.

Schlötterer, C. (2002). Towards a molecular characterization of adaptation in local populations. *Current Opinion in Genetics and Development*, **12**, 683–687.

Schlötterer, C. (2004). The evolution of molecular markers: just a matter of fashion? *Nature Reviews Genetics* **5**, 63–69.

Schluter, D. (1996). Adaptive radiation along genetic lines of least resistance. *Evolution*, **50**, 1766–1774.

Schluter, D. (2000). *The Ecology of Adaptive Radiation*. Oxford: Oxford University Press.

Slate, J. (2005). Quantitative trait locus mapping in natural populations: progress, caveats and future directions. *Molecular Ecology*, **14**, 363–379.

Spitze, K. (1993). Population structure in *Daphnia obtusa*: quantitative genetic and allozymic variation. *Genetics*, **135**, 367–374.

Storz, J. F. (2005). Using genome scans of DNA polymorphism to infer adaptive population divergence. *Molecular Ecology*, **14**, 671–688.

Streelman, J. T. and Kocher, T. D. (2000). From phenotype to genotype. *Evolution and Development*, **2**, 166–173.

Thodesen, J., Grisdale-Helland, B., Helland, S. J. and Gjerde, B. (1999). Feed intake, growth and feed utilization of offspring from wild and selected Atlantic salmon (*Salmo salar*). *Aquaculture*, **180**, 237–246.

van Tienderen, P. H., de Haan, A. A., van der Linden, C. G. and Vosman, B. (2002). Biodiversity assessment using markers for ecologically important traits. *Trends in Ecology and Evolution*, **17**, 577–582.

Vasemägi, A. and Primmer, C. R. (2005). Challenges for identifying functionally important genetic variation: the promise of combining complementary research strategies. *Molecular Ecology*, **14**, 3623–3642.

Vasemägi, A., Nilsson, J. and Primmer, C. R. (2005). Expressed sequence tag- linked microsatellites as a source of gene associated polymorphisms for detecting signatures of divergent selection in Atlantic salmon (*Salmo salar* L.). *Molecular Biology and Evolution*, **22**, 1067–1076.

Vigouroux, Y., McMullen, M., Hittinger, C. T. *et al.* (2002). Identifying genes of agronomic importance in maize by screening microsatellites for evidence of selection during domestication. *Proceedings of the National Academy of Sciences USA*, **99**, 9650–9655.

Vitalis, R., Dawson, K. and Boursot, P. (2001). Interpretation of variation across marker loci as evidence of selection. *Genetics*, **158**, 1811–1823.

Vos, P., Hogers, R, Bleeker, M. *et al.* (1995). AFLP: a new technique for DNA fingerprinting. *Nucleic Acids Research*, **23**, 4407–4414.

Waples, R. S. (1991). Pacific salmon, *Oncorhynchus* spp., and the definition of 'species' under the Endangered Species Act. *Marine Fisheries Reviews*, **53**, 11–22.

Whitehead, A. and Crawford, D. L. (2006). Neutral and adaptive variation in gene expression. *Proceedings of the National Academy of Sciences USA*, **103**, 5425–5430.

Wilding, C. S., Butlin, R. K. and Grahame, J. (2001). Differential gene exchange between parapatric morphs of *Littorina saxatilis* detected using AFLP markers. *Journal of Evolutionary Biology*, **14**, 611–619.

Wright, S. (1951). The genetical structure of populations. *Annals of Eugenics*, **15**, 323–354.

Wright, S. I. and Gaut, B. S. (2005). Molecular population genetics and the search for adaptive evolution in plants. *Molecular Biology and Evolution*, **22**, 506–519.

7

Monitoring and detecting translocations using genetic data

GIORGIO BERTORELLE, CHIARA PAPETTI,
HEIDI C. HAUFFE AND LUIGI BOITANI

Restocking is a common procedure for artificially increasing the population size of fish and game species in a particular geographical area. A similar intervention, which entails the (re)introduction of individuals from a source population (natural or captive) to a target area, is an important tool for ecosystem restoration, and is often essential to the recovery or rescue of endangered species or populations (Griffith *et al.* 1989; Frankham *et al.* 2002). In both cases, the principal aim of these so-called *translocations* is to establish stable and self-sustaining populations, taking care to preserve the original genetic structure and ecosystem dynamics of the particular species, while avoiding interference with natural evolutionary processes. But how can this goal be achieved in practice?

From an evolutionary and genetic perspective, these primary goals can be said to be attained when the introduced animals are successfully reproducing in the target environment, when negative selection pressures due to the effects of inbreeding or outbreeding depression are negligible, and when evolutionary potential is maintained (Moritz 1999; Frankham *et al.* 2002; Hufford and Mazer 2003; Tallmon *et al.* 2004). The challenge is to develop specific translocation plans which guarantee the achievement of all these objectives, and monitor the success of their implementation. For example, selecting animals or groups of animals appropriately adapted to a target environment is only possible by conducting a costly and long-term preliminary phase of fitness analysis. Similarly, the effects of inbreeding or outbreeding depression on the fitness of individuals in a translocated population can go undetected for extended periods of time. As this chapter will discuss, one solution to this dilemma is offered by the analysis of genetic markers; in fact, theoretical population and evolutionary genetics, together with empirical

Population Genetics for Animal Conservation, eds. G. Bertorelle, M. W. Bruford, H. C. Hauffe, A. Rizzoli and C. Vernesi. Published by Cambridge University Press. © Cambridge University Press 2009.

evidence suggest that levels and patterns of genetic variation within and between groups can be used, integrated with ecological studies, to plan and monitor translocations.

International conservation authorities have recognized that the pattern of genetic variation in introduced animals or plants should reflect, as far as possible, the patterns observed in the same or neighbouring geographical areas, or ecologically similar habitats (IUCN 1987, 1995). This approach is expected to mitigate the potential mismatch between the environmental conditions in the target area and the genetic background of the translocated animals (the consequences may be particularly serious if source animals are bred in captivity, e.g. Fleming and Gross 1993; Lynch and O'Hely. 2001). Moreover, if native individuals are still present in the target area, the genetic make-up of translocated individuals is also important because interbreeding between residents and translocated individuals could generate outbreeding depression effects or the loss of unique variants. Obviously, it is not possible to define an absolute threshold level of genetic divergence between the source individuals and residents of the target area (or populations neighbouring the target area) above which translocations are not recommended (for the debate about the best way to define groups to be managed independently, called evolutionarily significant units (ESUs) or management units (MUs), please see Fraser and Bernatchez 2001; Palsbøll *et al.* 2006). However, if statistically significant differences between source and the target groups are identified (for example by the analysis of molecular variance, or AMOVA: Excoffier *et al.* 1992, 2005), source and target populations cluster into different inferred parental groups (as indicated, for example, by STRUCTURE: Falush *et al.* 2003), and single individuals can be assigned with high confidence into genetically distinct units (e.g. using GENECLASS: Piry *et al.* 2004), then it seems reasonable to seriously reconsider a translocation plan. At the same time, the level of genetic variation within a group after the translocation(s) should be comparable to that observed in a stable and non-threatened population of the same or ecologically, demographically and/or phylogenetically closely related species in order to prevent negative inbreeding effects, and guarantee future adaptations to altered environmental conditions.

In practice, actual conservation decisions are often much more complicated than this, for at least four reasons. First, the two simple goals listed above concerning genetic variation within and between groups could be in conflict when a species or a small population with extremely low genetic variation is threatened, but a genetically similar source is not available. In this case, both inbreeding *and* outbreeding depression cannot be avoided.

Supplementing genetic variation using a genetically divergent source, thus focusing the intervention on the avoidance of inbreeding depression, seems the most appropriate choice in such extreme circumstances (Storfer 1999; Frankham *et al.* 2002; Tallmon *et al.* 2004). Second, in cases of a species-wide numerical decrease, it may be very difficult to define the 'natural' patterns and levels of genetic variation that could be restored through translocation. For example, it was only through the genetic analysis of several historical (museum) gray wolf (*Canis lupus*) samples in the USA that Leonard *et al.* (2005) were able to conclude that the current US Fish and Wildlife restoration targets of a few hundred individuals, selected according to the current subspecies classification, should be re-evaluated. In fact, the diversity of the eradicated western population was more than twice that of the extant population (corresponding to an estimated historical census size of 380 000 individuals), and gene flow was probably extensive across the subspecies range. Third, although we have been discussing goals in terms of genetic variation, in general, what we should be most concerned about conserving is adaptive genetic variation; that is, the variation at expressed fitness-related loci. However, in most conservation studies, genetic variation is almost exclusively quantified using neutral genetic markers, even though it has been claimed that their use as a proxy for selected loci is almost certainly an approximation (Fraser and Bernatchez 2001; Moran 2002). Fourth, although the statistical 'distinctiveness' of an individual or a group of individuals crucially depends on the number of markers and individuals analysed, the classical distinction between statistical and biological significance, as usual, cannot be neglected. Despite these problems and restrictions, until markers for a substantial number of fitness-related loci are more easily available, it will remain a useful and efficient approximation to define the primary aim of a translocation as defined by the IUCN (i.e. recovering populations at risk avoiding substantial modifications of the population structure) in terms of genetic variation at commonly analysed neutral genetic markers.

The link between genetic variation and the primary goal of a translocation plan implies that management plans can be monitored (and evaluated) by the statistical analysis of genetic variation (Schwartz *et al.* 2007). In principle, the success of a translocation can only be directly confirmed in the long term; that is, only an increase in population size (and possibly the attainment of equilibrium dynamics) and appropriate adaptation to environmental change (for example, climatic changes or variation in the pathogen community), provide indisputable evidence of the success of an intervention. However, in many cases, it is not possible to evaluate these criteria and the demographic parameters commonly used have proved to be insufficient substitutes (Fisher and Lindenmayer 2000; Goossens *et al.* 2002; Arrendal *et al.* 2004; Mock *et al.* 2004). For example, a rapid demographic increase just after a translocation event could be a transitory process with limited long-term significance, especially if an initial heterosis effect is followed by outbreeding depression (e.g. Marr *et al.* 2002), or if the initial relaxation of selective pressures (possibly caused by the human interventions inherent in the early phases of translocation) accidentally favours the dissemination of maladaptive variants or genes with a low level of polymorphism (Stockwell *et al.* 2003). In the worst scenario, a demographic increase could be the first step of a replacement process if, for example, the translocated individuals are different and much more prolific than the autoctonous individuals in the managed area. The study of genetic variation using living individuals and, if necessary, museum specimens as pre-translocation controls, represents an alternative, effective and relatively simple tool to indirectly evaluate the success of a translocation plan. Similarly, when past translocation events were poorly recorded in or absent from historical documents, the analyses of genetic variation can be used to detect them and evaluate their impact. In the next two sections, this chapter will focus on these two aspects: *monitoring* and *detecting translocations using genetic data*, considering examples from different species.

GENETIC ANALYSES AND MONITORING OF WELL-KNOWN TRANSLOCATION EVENTS

Table 7.1 summarizes the results of recent genetic studies of species for which translocations have been documented in the last 100 years. These articles were selected from public bibliographical databases using the keywords *translocation, restocking, introduction, genetic variation* and/or *genetic data*. Almost all studies report that these translocations result in a perturbation in the pattern of genetic variation creating new patterns that could negatively affect the fitness of the species concerned. Before describing and commenting on some of these studies, we will discuss an exception: the recent study by Dowling *et al.* (2005) of the razorback sucker (*Xyrauchen texanus*, a teleost fish), which represents, in our opinion, an appropriate integration of management practices and genetic analyses.

The last self-sustaining population of the razorback sucker is found in Lake Mohave, which is located in the states of Arizona and Nevada. In the last decade, the estimated census size of this species has declined by a factor

Table 7.1 *Some recent examples where the genetic variation effects of translocations plans have been evaluated.*

Reference	Species	The translocations considered started in:	Pre-translocation genetic data?	Markers[a]	Potentially negative effects of translocation on genetic variation
Hale et al. (2004)	Sciurus vulgaris (red squirrel)	Mid to late 1800s	Yes, from museum specimens	MtDNA CR (395 bp)	Introgression of continental European haplotypes, now dominant (and possibly selected) in northeastern Britain; loss of original phylogeographic pattern (present in Europe with high F_{ST} values)
DeYoung et al. (2003)	Odocoileus virginianus (white-tailed deer)	Early to mid-1900s	No	STRs (17 loci)	Some level of homogenization between differentiated groups, loss of isolation by distance pattern in some areas; high genetic variation but bottleneck effects still detectable
Lambert et al. (2005)	Philesturnus carunculatus rufusater (New Zealand saddleback)	1925, but especially in the 1960–1990 interval	No	Minisatellite fingerprinting + 1 isozyme + STRs (6 loci)	Slight reduction of variation in some translocated groups; slightly larger effects in second order translocations
Li et al. (2005)	Metasequoia glyptostroboides (dawn redwood)	1948	No	RAPD	Slight reduction of variation in translocated populations (only due to founder effects); translocation likely generated artificial population structure
Maudet et al. (2002)	Capra ibex (Alpine ibex)	1930	No	STRs (19 loci)	Bottleneck effects detected, but severe decrease of genetic variation only in groups with <10 founders; large differentiation in reintroduced groups

Reference	Species	Year	Ancient/museum DNA	Markers	Main findings
Vernesi et al. (2003)	Sus scrofa (wild boar)	1950	No	STRs (9 loci)	Some level of introgression (~15%) of allochthonous gene pools into local groups in some areas
Mock et al. (2004)	Meleagris gallopavo merriami (Merriam's turkey)	1951	No	STRs (9 loci)	Reduced variation in translocated groups
Drew et al. (2003)	Martes pennanti (Fisher)	1960	Yes, from museum specimens	MtDNA CR (300 bp)	Perturbation of the phylogeographic pattern in some populations
Susnik et al. (2004)	Thymallus thymallus (Adriatic grayling)	1960	No	STRs (15 loci)	High level of introgression (40–50%) of non-indigenous stocked animals; only few non-introgressed individuals can be identified
Bodkin et al. (1999)	Enhydra lutris (sea otter)	1960	No	MtDNA (using RFLP)	Reduced and increased variation in reintroduced groups from one or several source populations, respectively; artificial structure generated only in reintroduced groups with small number of founders
Lance et al. (2003)	Sciurus niger cinereus (Delmarva fox squirrel)	1968	No	MtDNA CR (330 bp) + STRs (3 loci)	Apparently none, but geographic pattern may be modified since in most cases, source squirrels were taken from multiple locations
Robichaux et al. (1997)	Argyroxiphium sandwicense (Mauna Kea silversword)	1970	No	RAPD	Large reduction (73%) of polymorphism in an outplanted population
Launey et al. (2006)	Esox lucius (pike)	1970	No	STRs (7 loci)	Mixing between local and introduced stocks in some rivers

Table 7.1 (*cont.*)

Reference	Species	The translocations considered started in:	Pre-translocation genetic data?	Markers[a]	Potentially negative effects of translocation on genetic variation
Van Houdt *et al.* (2005)	*Salmo trutta* (brown trout)	1970	No	MtDNA CR (using SSCP9 + RAPD fingerprinting)	Extensive contribution of hatcheries, with homogenization effects, in lower parts of the rivers
Hansen *et al.* (2000)	*Salmo trutta* (brown trout)	1980s	No	MtDNA ND gene (using RFLP) + STRs (7 loci)	Hybridization between resident and hatchery animals; major contribution of hatcheries in river trout (47%), minor in sea trout (7%)
Ruokonen *et al.* (2000)	*Anser erythropus* (lesser white-fronted goose)	1981	Yes, from museum specimens	MtDNA CR (221 bp)	Likely hybridization between different species (initially occurred in captivity before releases)
Latch *et al.* (2005)	*Meleagris gallopavo silvestris* (wild turkey)	1983	No	MtDNA CR (500 bp) + STRs (10 loci)	Artificial population structure generated by restocked groups
Arrendal *et al.* (2004)	*Lutra lutra* (Eurasian otter)	Late 1980s	No	MtDNA CR (1000 bp) with SSCP + STRs (6 loci)	Translocations were unsuccessful in one area and changed the genetic composition of the resident population in another
Dowling *et al.* (2005)	*Xyrauchen texanus* (razorback sucker)	1991	Yes	MtDNA cyt-b gene (using SSCP)	Apparently none: management was based on preliminary genetic analyses and genetic monitoring
Arnaud-Haond *et al.* (2004)	*Pinctada margaritifera cumingii* (pearl oyster)	1992	Yes	3 size polymorphism markers	Homogenization of local groups; F_{ST} reduced from 6% to 0%
Kraaijeveld-Smit *et al.* (2005)	*Alytes muletensis* (Mallorcan midwife toad)	1992	No	8 microsatellite loci	Apparently none

Source	Species	Year		Markers	Results
Hughes et al. (2003)	Paratya australiensis (freshwater shrimp)	1993	Yes	MtDNA COI gene (630 bp) + isozymes (3 loci)	Resident genotypes were completely extinct in one of the two translocation sites after 7 generations
Sigg (2006)	Onychogalea fraenata (bridled nailtail wallaby)	1993	No	STRs (7 loci)	Reduced variation and increased divergence in reintroduced groups
McGlaughlin et al. (2002)	Abronia umbellata (pink sand verbena)	1995	No	ISSR fingerprinting	Decreased diversity in reintroduced populations with size <1000
Krauss et al. (2002)	Grevillea scapigera (Corrigin grevillea)	1996	No	AFLP fingerprinting	Decreased variation (20% of heterozygosity reduction; 85% of all seeds were the product of 4 clones
Wilbur et al. 2005	Argopecten irradians (bay scallop)	1998	Yes	MtDNA ATP-synthase gene (1000 bp)	Apparently none, since likely divergent translocated individuals (bred in hatcheries) contributed only marginally to local genetic variation
Rampe et al. 2006	Lasthenia conjugens (Contra Costa goldfield)	1999	No	ISSR fingerprinting	Apparently none
Hare et al. 2006	Crassostrea virginica (Eastern oyster)	2002	Yes	mtDNA COI, COIII, NADH genes (using RFLP) + STRs (8 loci)	Apparently none, since likely divergent translocated individuals (artificially selected disease tolerant strains) contributed only marginally to local genetic variation

[a] COI, cytochrome oxidase I; CR, control region; ISSR, intersimple sequence repeats; NADH, reduced form of nicotinamide adenine dinucleotide.

of 60 to fewer than 1000 individuals. The cause of this acute decline has been identified as excessive predation on larvae and subsequent recruitment failure. In order to prevent the extinction of this population, a drastic and complex restocking programme was initiated in 1991, which included capturing larvae, rearing them to juveniles in hatchery facilities and to subadults in protected areas, and releasing these individuals back into the wild. The programme also involved genetic analysis of cytochrome *b* sequences. Almost 3000 samples were typed, from individuals collected before and after the reintroductions, larvae at different stages, and repatriated fish (marked with implanted transponder tags). Using genetic typing at different sites and in different groups of animals, and statistical tests based on non-parametric permutation approaches, Dowling *et al.* (2005) showed that levels and patterns of genetic variation were not significantly modified by this restocking programme. Therefore, contrary to initial fears, it appears that the large numbers of repatriates are not the progeny of only a small number of parents, and the original genetic structure has been preserved. Although Dowling and collaborators did not face several of the major problems normally associated with restocking, i.e. the choice of source population and the difficulties related to captive breeding, and although an independent support from nuclear markers would be desirable, they proved experimentally that genetic typing can be used to prevent negative genetic consequences of such programmes.

In contrast to the case of the razorback sucker, genetic analysis of many other species has been used only to unveil the impact of translocations that were planned without considering their potential genetic consequences (see Table 7.1). Unfortunately, Table 7.1 also shows that these translocations usually result in a decline of genetic variation and/or modification of population structure, either by producing large differentiations in small, reintroduced groups or by homogenizing naturally differentiated genetic pools. For example, the translocation of juvenile pearl oysters (*Pinctada margaritifera cumingii*) between the islands of French Polynesia, in an attempt to increase the hatchery production, reduced the level of genetic differentiation between populations from 6% to almost 0% over a few years (Arnaud-Haond *et al.* 2004). Similarly, a recovery plan for the near-extinct white-tailed deer (*Odocoileus virginianus*) in the state of Mississippi led to detectable effects of population, and possibly even sub-species, mixing in some areas, and the establishment of an artificial population structure in others (DeYoung *et al.* 2003). Figure 7.1 (also Plate 2, colour plate section) illustrates another clear example of admixture due to translocations, where the composite gene pool of a restocked population and the three 'pure'

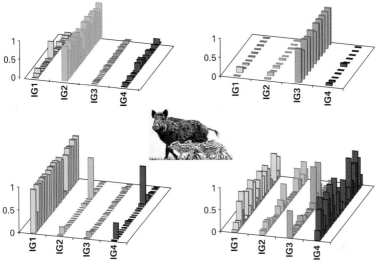

Figure 7.1. A simple situation in which wild boar (*Sus scrofa*) restocking in a central Italian region results in a complex pattern of genetic contributions in single individuals. The Bayesian method implemented in the software S TRUCTURE (Pritchard *et al.* 2000) inferred four groups (from IG1 to IG4), and the relative contributions (the bars) of each inferred group in several individuals (each individual is a row of bars in each histogram) is estimated in four different populations (the different histograms). Restocking is evident only in the lower right histogram, where most individuals show a hybrid genetic pool. Modified from Vernesi *et al.* (2003). (See also colour plate.)

indigenous or semi-indigenous populations of wild boar (*Sus scrofa*) can be easily distinguished (Vernesi *et al.* 2003).

The establishment of an artificial population structure is much more severe when only a few animals are used to repopulate empty or almost empty areas. For example, the release of some individuals of the roe deer (*Capreolus capreolus*) from eastern Europe into Italian alpine populations has modified the phylogeographic pattern in some regions, so that intro-duced mtDNA variants are surrounded by very different local sequences (Vernesi *et al.* 2002) (Fig. 7.2). Founder effects and genetic drift during and after the translocation event can also generate significant genetic diver-gence between the source and the reintroduced groups (e.g. Alpine ibex, *Capra ibex* (*ibex*): Maudet *et al.* 2002; New Zealand saddleback, *Philesturnus carunculatus rufusater*: Lambert *et al.* 2005), which tend to be magnified if individuals from translocated populations are themselves used as founders for subsequent management interventions (Clegg *et al.* 2002).

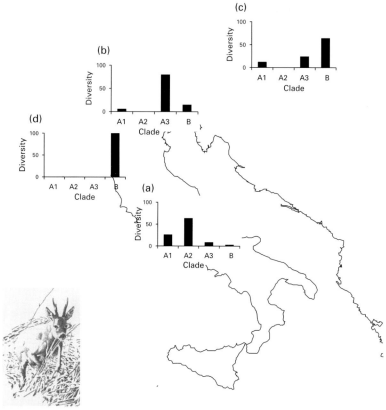

Figure 7.2. The frequency distribution of two major mitochondrial DNA clades, A and B (with A subdivided in to three subclades A1, A2, A3), in the roe deer (*Capreolus capreolus*) in central–southern Italy (histogram a), Eastern Alps (histogram b), central–eastern Europe (histogram c) and a sample of individuals from an area in the Western Alps where restocking with Eastern European animals was frequent in the past (histogram d). The genetic legacy of the restocked area with geographically distant populations is obvious (histogram d), and the phylogeographic pattern in Italy was evidently modified by translocations. The haplotype diversity is higher in the restocked group (0.82) compared to the averages computed from different populations in the other three areas (0.40, 0.67 and 0.76), suggesting that translocation was also accompanied by population mixing. Data from 244 individuals and 342 bp, from Vernesi *et al.* (2002) and unpublished material. Clades A and B correspond to the clades Central+West and East in Randi *et al.* (2004), respectively.

Experimental data confirm the theoretical prediction that the effects of a bottleneck during translocations include a reduction in the levels of genetic variation (e.g. Maudet *et al.* 2002; McGlaughlin *et al.* 2002; Mock *et al.* 2004). However, this expected pattern is not systematically observed in all populations and species, as a result, for example, of a lack of statistical power (e.g. too few loci or individuals in the sample), or of limiting the analysis to measures of genetic variation marginally affected by short bottlenecks (e.g. the expected heterozygosity). In addition, if comparable data for ecologically or phylogenetically similar species or other populations of the same species are not available, the genetic variation observed in a single sample cannot easily, or objectively, be considered 'low' or 'high'. Specific methods are available and should be applied in cases where such data are lacking to investigate if the results of the single sample support the hypothesis of a decline in variation (e.g. Beaumont 1999; Piry et al. 1999; Garza and Williamson 2001; Williamson-Natesan 2005). In any case, drastic reductions of variation are expected only when a few individuals are reintroduced (for an example, see Maudet et al. 2002 for Alpine ibex), or if post-translocation dynamics do not involve an increase in population size. In fact, a rapid demographic expansion after a reintroduction could prevent additional drift effects and mitigate the reduction of genetic variation. Therefore, past reintroductions followed by interventions favouring population expansion (e.g. strict legal protection, constant monitoring, feeding, etc.) will have unintentionally also safeguarded the genetic well-being of those particular populations.

In conclusion, past translocation programmes which did not specifically consider the maintenance of genetic variation and structure as one of their objectives usually modified levels and/or patterns of genetic variation, as measured by the most commonly used neutral markers, mtDNA and microsatellites (Table 7.1). This general conclusion, as already mentioned, cannot be directly translated into an overall evaluation of the management plan unless negative fitness effects or reduced potential for future evolutionary changes are also observed. This is a limitation of genetic studies based on neutral markers only. However, since demographic processes (such as random drift or admixture) affect the entire genome, we predict that translocations that modify patterns and levels of genetic variation at neutral markers will also have similar effects on adaptive genes. Some obvious translocation failures confirm this view. For example, released sea otter (Enhydra lutris) individuals, shown to be genetically divergent at neutral markers from the residents, left almost no descendents (Arrendal et al. 2004). With the exception of very extreme situations, where no alternative options are available to rescue a species or a population, theoretical arguments and experimental data suggest that a preliminary analysis of neutral genetic variation should be used to develop translocation plans. In addition, subsequent genetic monitoring is an indirect, but reliable, method to evaluate the management programme.

DETECTING A PAST TRANSLOCATION EVENT FROM GENETIC DATA

In the previous section we have shown that, in most cases, translocations without genetic planning have measurable consequences on the levels and patterns of genetic variation. This also means that genetic typing can be used to infer past translocation events when, as frequently occurs, historical data are fragmentary, imprecise or unavailable. Introduction and restocking of hunting/fishing species, in fact, were not appropriately recorded in the past, since the displacement of individuals was not considered detrimental. It is interesting to note that the more genetically incorrect the translocation, the easier its detection using genetic data. For example, detecting past introductions when the source individuals are strictly related among themselves, or are strongly differentiated from the target group, appears a relatively easy task. On the contrary, these investigations are much more complicated, but probably also less urgent, if the translocation produced limited genetic effects.

A substantial number of methods in population genetics, both at the population and at the individual level (see Excoffier and Heckel 2006 for a review), can be used to detect and statistically test for the typical signatures of a translocation; for example, Hardy–Weinberg and/or linkage equilibrium deviation, highly divergent and unexpected haplotypes or genotypes, highly reduced or increased levels of genetic variation, and incongruent patterns of genetic differentiation between neighbouring groups. Unfortunately, demographic processes not related to past translocation events can also produce similar patterns. However, the combined use of several methods, together with an appropriate choice of reference groups, will usually permit the identification, or the exclusion, of translocation events with significant genetic impact. Below we report three recent case studies in game species.

Frantz *et al.* (2006) revealed a case of illegal translocation of red deer (*Cervus elaphus*) into a hunting area in Luxembourg using genetic typing at 13 microsatellites by comparing the 'translocated' deer genotypes with a reference sample of about 400 individuals from the surrounding regions in Belgium, Germany, France, and Luxembourg itself. The geographic distribution of the reference samples was considered wide enough to include all possible sources of natural migrations. A preliminary Bayesian reconstruction of the most likely population structure in the reference sample suggested that at least three geographical groups with F_{ST} values between 5% and 10% can be inferred. Using both descriptive and inferential statistics, the authors concluded that the genotypes from the suspected animals could be excluded from the three local groups with high confidence. Therefore, in this case, an (illegal) translocation event was detected using the exclusion principle, although the origin and the level of ecological divergence of the introduced animals remained unknown.

Another interesting case of genetic data from several and geographically widespread populations allowing not only *translocation detection*, but also the identification of the source group, is that of the Alpine chamois (*Rupicapra rupicapra*) (Pecchioli *et al.* 2006). Some individuals sampled in two localities in the Eastern Italian Alps, not far from the region where the first PGAC meeting that motivated this book took place in 2003, had a puzzling mtDNA sequence. This sequence, in fact, was highly divergent from the vast majority of sequences observed in the same and in the surrounding regions. When 1500 bp including the control region and the cytochrome b gene were analysed, pairs of sequences commonly found in the Eastern Alps were in fact separated by no more than 55 substitutions, but the 'outlier sequence' showed, on average, 80 additional molecular changes. The individuals with the divergent haplotype had no similar level of differentiation at 11 nuclear microsatellites markers, but the two localities where these mtDNA haplotypes were found showed significant nuclear differentiation from the surrounding regions. Since translocation events from the Gran Paradiso National Park (Western Alps) in the two Eastern localities were documented in the 1970s, several individuals from this Park were sampled and analysed. Again, most Western chamois belonged to the typical Alpine genetic clade, but some individuals had the divergent sequence found in two Eastern sites. Finally, samples were also analysed from a related species, the Pyrenean chamois (*Rupicapra pyrenaica*), and it was concluded that divergent haplotype found in the Alps could be assigned to the Pyrenean species. The two species have a disjoint distribution separated by more than 400 kilometres, and recent natural migration events can be excluded. Therefore, genetic typing revealed that undocumented translocations of the Pyrenean chamois into the Western Alps occurred in the past, possibly associated with the exchange of gifts between the royal families of Spain and Italy. In fact, the Gran Paradiso National Park was a royal reserve until the 1920s. Successive (and documented) translocations in the Eastern Alps introduced not only Western Alpine chamois, but also previously translocated Pyrenean genes. Interestingly, as revealed by the nuclear microsatellite markers, hybridization between the two species has apparently taken place freely, although an analysis of hybrid fitness is necessary to confirm that such interbreeding has had no negative consequences.

Genetic typing can be very useful not only to detect translocation events, but also to exclude them. For example, wolves (*Canis lupus*) were eradicated from the Alps in the 1920s, and before legal protection in 1971, only about 100 individuals survived in central Italy. Since then, the central Italian population has expanded both demographically and geographically, reaching the southwestern Alps in the 1990s. This natural recolonization process was (and is) not entirely accepted by local communities, and the hypothesis that Alpine wolves were actually artificially reintroduced using eastern or northern European wolves, or dog–wolf hybrids, has sometimes been used to justify this negative attitude. When mtDNA sequences and microsatellite genotypes were analysed from wolf scats collected in the Western Italian Alps and compared by Lucchini *et al.* (2002) with the data from a reference sample of 270 wolves from Central Italy, Bulgaria, Croatia, Greece, Turkey, Israel, Latvia, Finland and Spain, and 100 dogs, assignment tests and a simple alignment of sequences confirmed definitively that the ongoing colonization of the Alps is a natural process and not the result of human-mediated translocations.

In conclusion, recent molecular and statistical methods offer a momentous opportunity, not only to understand the effects of documented translocations, but also to detect them when historical data are missing or individuals are displaced illegally. Future investigations are essential, not only for their forensic implications but, more importantly, to better reconstruct and understand the genetic effects of numerous uncontrolled and often arbitrary relocations of individuals that occurred in the last few centuries in many species. All of the above cases can be considered ongoing experiments of genetically inappropriate management plans, and their detection and study will be of great value for refining the development of future guidelines for translocations and reintroductions.

REFERENCES

Arnaud-Haond, S., Vonau, V., Bonhomme, F. *et al.* (2004). Spatio-temporal variation in the genetic composition of wild populations of pearl oyster (*Pinctada margaritifera cumingii*) in French Polynesia following 10 years of juvenile translocation. *Molecular Ecology*, **13**, 2001–2007.

Arrendal, J., Walker, C. W., Sundquist, A. K., Hellborg, L. and Vilà, C. (2004). Genetic evaluation of an otter translocation program. *Conservation Genetics*, **5**, 79–88.

Beaumont, M. A. (1999). Detecting population expansion and decline using microsatellites. *Genetics*, **153**, 2013–2029.

Bodkin, J. L., Ballachey, B. E., Cronin, M. A. and Scribner, K. T. (1999). Population demographics and genetic diversity in remnant and translocated populations of sea otters. *Conservation Biology*, **13**, 1378–1385.

Clegg, S. M., Degnan, S. M., Kikkawa, J. *et al.* (2002). Genetic consequences of sequential founder events by an island-colonizing bird. *Proceedings of the National Academy of Sciences USA*, **99**, 8127–8132.

DeYoung, R. W., Demarais, S., Honeycutt, R. L. *et al.* (2003). Genetic consequences of white-tailed deer (*Odocoileus virginianus*) restoration in Mississippi. *Molecular Ecology*, **12**, 3237–3252.

Dowling, T. E., Marsh, P. C., Kelsen, A. T. and Tibbets, C. A. (2005). Genetic monitoring of wild and repatriated populations of endangered razorback sucker (*Xyrauchen texanus*, Catostomidae, Teleostei) in Lake Mohave, Arizona–Nevada. *Molecular Ecology*, **14**, 123–135.

Drew, R. E., Hallett, J. G., Aubry, K. B. *et al.* (2003). Conservation genetics of the fisher (*Martes pennanti*) based on mitochondrial DNA sequencing. *Molecular Ecology*, **12**, 51–62.

Excoffier, L. and Heckel, G. (2006). Computer programs for population genetics data analysis: a survival guide. *Nature Review Genetics*, **7**, 745–758.

Excoffier, L., Smouse, P. E. and Quattro, J. M. (1992). Analysis of molecular variance inferred from metric distances among DNA haplotypes: application to human mitochondrial DNA restriction data. *Genetics*, **131**, 479–491.

Excoffier, L., Laval, G. and Schneider, S. (2005). ARLEQUIN v. 3.0: an integrated software package for population genetics data analysis. *Evolutionary Bioinformatics Online*, **1**, 4–50.

Falush, D., Stephens, M. and Pritchard, J. K. (2003). Inference of population structure using multilocus genotype data: linked loci and correlated allele frequencies. *Genetics*, **164**, 1567–1587.

Fischer, J. and Lindenmayer, D. B. (2000). An assessment of the published results of animal relocations. *Biological Conservation*, **96**, 1–11.

Fleming, I. A. and Gross, M. R. (1993). Breeding success of hatchery and wild coho salmon (*Oncorhyinchus kisutch*) in competition. *Ecological Applications*, **3**, 230–245.

Frankham, R., Ballou J. D. and Briscoe, D. A. (2002). *Introduction to Conservation Genetics*. Cambridge: Cambridge University Press.

Frantz, A. C., Pourtois, J. T., Heuertz, M. *et al.* (2006). Genetic structure and assignment tests demonstrate illegal translocation of red deer (*Cervus elaphus*) into a continuous population. *Molecular Ecology*, **15**, 3191–3203.

Fraser, D. J. and Bernatchez, L. (2001). Adaptive evolutionary conservation: towards a unified concept for defining conservation units. *Molecular Ecology*, **10**, 2741–2752.

Garza, J. C. and Williamson, E. G. (2001). Detection of reduction in population size using data from microsatellite loci. *Molecular Ecology*, **10**, 305–318.

Goossens, B., Funk, S. M., Vidal, C. *et al.* (2002). Measuring genetic diversity in translocation programmes: principles and application to a chimpanzee release project. *Animal Conservation*, **5**, 225–236.

Griffith, B., Scott, J. M., Carpenter, J. W. and Reed, C. (1989). Translocation as a species conservation tool: status and strategy. *Science* **245**, 477–480.

Hale, M. L., Lurz, P. W. W. and Wolff, K. (2004). Patterns of genetic diversity in the red squirrel (*Sciurus vulgaris* L.): footprints of biogeographic history and artificial introductions. *Conservation Genetics*, **5**, 167–179.

Hansen, M. M., Ruzzante, D. E., Nielsen, E. E. and Mensberg, K. L. D. (2000). Microsatellite and mitochondrial DNA polymorphism reveals life-history dependent interbreeding between hatchery and wild brown trout (*Salmo trutta* L.). *Molecular Ecology*, 9, 583–594.

Hare, M. P., Allen, S. K. Jr, Bloomer, P. *et al.* (2006). A genetic test for recruitment enhancement in Chesapeake Bay oysters, *Crassostrea virginica*, after population supplementation with a disease tolerant strain. *Conservation Genetics*, 7, 717–734.

Hufford, K. M. and Mazer, S. J. (2003). Plant ecotypes: genetic differentiation in the age of ecological restoration. *Trends in Ecology and Evolution*, 18, 147–155.

Hughes, J. M., Goudkamp, K., Hurwood, D. A., Hancock, M. and Bunn, S. E. (2003). Translocation causes extinction of a local population of the freshwater shrimp, *Paratya australiensis*. *Conservation Biology*, 17, 1007–1012.

IUCN–World Conservation Union (1987). *The IUCN Position Statement on Translocation of Living Organisms; Introductions, Reintroductions and Re-Stocking.* Gland, Switzerland: IUCN. (http://www.iucn.org/themes/ssc/publications/policy/transe.htm)

IUCN–World Conservation Union (1995). *Guidelines for Reintroduction.* Gland, Switzerland: IUCN. (http://www.iucn.org/themes/ssc/publications/policy/reinte.htm)

Kraaijeveld-Smit, F. J. L., Beebee, T. J. C., Griffiths, R. A., Moore, R. D. and Schley, L. (2005). Low gene flow but high genetic diversity in the threatened Mallorcan midwife toad *Alytes muletensis*. *Molecular Ecology*, 14, 3307–3315.

Krauss, S. L., Dixon, B. and Dixon, K. W. (2002). Rapid genetic decline in a translocated population of the endangered plant *Grevillea scapigera*. *Conservation Biology*, 16, 986–994.

Lambert, D. M., King, T., Shepherd, L. D. *et al.* (2005). Serial population bottlenecks and genetic variation: translocated populations of the New Zealand saddleback (*Philesturnus carunculatus rufusater*). *Conservation Genetics*, 6, 1–14.

Lance, S. L., Maldonado, J. E., Bocetti, C. I. *et al.* (2003). Genetic variation in natural and translocated populations of the endangered Delmarva fox squirrel (*Sciurus niger cinereus*). *Conservation Genetics*, 4, 707–718.

Latch, E. K. and Rhodes, O. E. (2005). The effects of gene flow and population isolation on the genetic structure of reintroduced wild turkey populations: are genetic signatures of source populations retained? *Conservation Genetics*, 6, 981–997.

Launey, S., Morin, J., Minery, S. and Laroche, J. (2006). Microsatellite genetic variation reveals extensive introgression between wild and introduced stocks, and a new evolutionary unit in French pike *Esox lucius* L. *Journal of Fish Biology*, 68, 193–216.

Leonard, J. A., Vilà, C. and Wayne, R. K. (2005). Legacy lost: genetic variability and population size of extirpated US gray wolves (*Canis lupus*). *Molecular Ecology*, 14, 9–17.

Li, Y. Y., Chen, X. Y., Zhang, X. *et al.* (2005). Genetic differences between wild and artificial populations of *Metasequoia glyptostroboides*: implications for species recovery. *Conservation Biology*, 19, 224–231.

Lucchini, V., Fabbri, E., Marucco, F. *et al.* (2002). Noninvasive molecular tracking of colonizing wolf (*Canis lupus*) packs in the western Italian Alps. *Molecular Ecology*, 11, 857–868.

Lynch, M. and O'Hely, M. (2001). Captive breeding and the genetic fitness of natural populations. *Conservation Genetics*, 2, 363–378.

Marr, A. B., Keller, L. F. and Arcese, P. (2002). Heterosis and outbreeding depression in descendants of natural immigrants to an inbred population of song sparrows (*Melospiza melodia*). *Evolution*, **56**, 131–142.

Maudet, C., Miller, C., Bassano, B. *et al.* (2002). Microsatellite DNA and recent statistical methods in wildlife conservation management: applications in Alpine ibex [*Capra ibex (ibex)*]. *Molecular Ecology*, **11**, 421–436.

McGlaughlin, M., Karoly, K. and Kaye, T. (2002). Genetic variation and its relationship to population size in reintroduced populations of pink sand verbena, *Abronia umbellata* subsp. *breviflora* (Nyctaginaceae). *Conservation Genetics*, **3**, 411–420.

Mock, K. E., Latch, E. K. and Rhodes, O. E. Jr (2004). Assessing losses of genetic diversity due to translocation: long-term case histories in Merriam's turkey (*Meleagris gallopavo merriami*). *Conservation Genetics*, **5**, 631–645.

Moran, P. (2002). Current conservation genetics: building an ecological approach to the synthesis of molecular and quantitative genetic methods. *Ecology of Freshwater Fish*, **11**, 30–55.

Moritz, C. (1999). Conservation units and translocations: strategies for conserving evolutionary processes. *Hereditas*, **130**, 217–228.

Palsbøll, P. J., Bérubé, M. and Allendorf, F. W. (2006). Identification of management units using population genetic data. *Trends in Ecology and Evolution*, **22**, 11–16.

Pecchioli, E., Vernesi, C., Crestanello, B. *et al.* (2006). *Progetto Faunagen. Conservazione e gestione della fauna: un approccio genetico*, Report No. 35. Trento, Italy: Centro di Ecologia Alpina.

Piry, S., Luikart, G. and Cornuet, J. M. (1999). BOTTLENECK: a computer program for detecting recent reductions in the effective population size using allele frequency data. *Journal of Heredity*, **90**, 502–503.

Piry, S., Alapetite, A., Cornuet, J. M. *et al.* (2004). GENECLASS2: a software for genetic assignment and first-generation migrant detection. *Journal of Heredity*, **95**, 536–539.

Pritchard, J. K., Stephens, M. and Donnelly, P. (2000). Inference of population structure using multilocus genotype data. *Genetics*, **155**, 945–959.

Ramp, J. M., Collinge, S. K. and Ranker, T. A. (2006). Restoration genetics of the vernal pool endemic *Lasthenia conjugens* (Asteraceae). *Conservation Genetics*, **7**, 631–649.

Randi, E., Alves, P. C., Carranza, J. *et al.* (2004). Phylogeography of roe deer (*Capreolus capreolus*) populations: the effects of historical genetic subdivisions and recent nonequilibrium dynamics. *Molecular Ecology*, **13**, 3071–3083.

Robichaux, R. H., Friar, E. A. and Mount, D. W. (1997). Molecular genetic consequences of a population bottleneck associated with reintroduction of the Mauna Kea silversword (*Argyroxiphium sandwicense* ssp. *sandwicense* [Asteraceae]). *Conservation Biology*, **11**, 1140–1146.

Ruokonen, M., Kvist, L., Tegelström, H. and Lumme, L. (2000). Goose hybrids, captive breeding and restocking of the Fennoscandian populations of the Lesser White-fronted Goose (*Anser erythropus*). *Conservation Genetics*, **1**, 277–283.

Schwartz, M. K., Luikart, G. and Waples, R. S. (2007). Genetic monitoring as a promising tool for conservation and management. *Trends in Ecology and Evolution*, **22**, 25–33.

Sigg, D. P. (2006). Reduced genetic diversity and significant genetic differentiation after translocation: comparison of the remnant and translocated populations of bridled nailtail wallabies (*Onychogalea fraenata*). *Conservation Genetics*, **7**, 577–589.

Stockwell, C. A., Hendry, A. P. and Kinnison, M. T. (2003). Contemporary evolution meets conservation biology. *Trends in Ecology and Evolution*, **18**, 94–101.

Storfer, A. (1999). Gene flow and endangered species translocations: a topic revisited. *Biological Conservation*, **87**, 173–180.

Susnik, S., Berrebi, P., Dovc, P., Hansen, M. M. and Snoj, A. (2004). Genetic introgression between wild and stocked salmonids and the prospects for using molecular markers in population rehabilitation: the case of the Adriatic grayling (*Thymallus thymallus* L. 1785). *Heredity*, **93**, 273–282.

Tallmon, D. A., Luikart, G. and Waples, R. S. (2004). The alluring simplicity and complex reality of genetic rescue. *Trends in Ecology and Evolution*, **19**, 489–496.

Van Houdt, J. K. J., Pinceel, J., Flamand, M. C. *et al.* (2005). Migration barriers protect indigenous brown trout (*Salmo trutta*) populations from introgression with stocked hatchery fish. *Conservation Genetics*, **6**, 175–191.

Vernesi, C., Pecchioli, E., Caramelli, D. *et al.* (2002). The genetic structure of natural and reintroduced roe deer (*Capreolus capreolus*) populations in the Alps and Central Italy, with reference to the mitochondrial DNA phylogeography of Europe. *Molecular Ecology*, **11**, 1285–1297.

Vernesi, C., Crestanello, B., Pecchioli, E. *et al.* (2003). The genetic impact of population decline and reintroduction in the wild boar (*Sus scrofa*): a microsatellite analysis. *Molecular Ecology*, **12**, 585–595.

Wilbur, A. E., Seyoum, S., Bert, T. M. and Arnold, W. S. (2005). A genetic assessment of bay scallop (*Argopecten irradians*) restoration efforts in Florida's Gulf of Mexico coastal waters (USA). *Conservation Genetics*, **6**, 111–122.

Williamson-Natesan, E. (2005). Comparison of methods for detecting bottlenecks from microsatellite loci. *Conservation Genetics*, **6**, 551–562.

8

Non-invasive genetic analysis in conservation

BENOÎT GOOSSENS AND MICHAEL W. BRUFORD

INTRODUCTION

A key component of the emergence of conservation genetics as a recognis-
able subdiscipline of conservation biology over the last ten years has been the
development of methods to genetically assess and monitor populations of
endangered species non-invasively. The rapid development of methodologies for ensuring the accurate capture of molecular data from elusive, easily
stressed or potentially dangerous (!) organisms and concerns over the accuracy of the data produced have prompted a number of excellent reviews on the
subject in recent times (e.g. Taberlet *et al.* 1999; Taberlet and Luikart 1999;
Piggott and Taylor 2003; Woodruff 2003; Wayne and Morin 2004). Here, we
will review the issues and wide-ranging applications of non-invasive genetic
analysis without focusing on the molecular technicalities in great detail.

Why non-invasive genetics?
Before the advent of non-invasive genetics and the use of animal by-products
such as faeces, shed hair, feathers, bones, fish scales, teeth, etc., obtaining
genetic material from wild populations was often ethically (in particular for
species listed as endangered and critically endangered under CITES regulations) and logistically extremely difficult. Now such analysis is increasingly
possible and the sampling of large populations without visual/physical
contact is particularly beneficial for endangered species or if the species
studied can transmit or are susceptible to disease (e.g. great ape species
whose pathogens are often extremely similar to those of the researchers
studying them). In a century where linking behaviour, social structure,
dispersal and population genetic structure has become a new challenge for
conservation geneticists, the development of non-invasive sampling and
genotyping has provided the opportunity to explore these links and has
dramatically opened new areas for research.

Population Genetics for Animal Conservation, eds. G. Bertorelle, M. W. Bruford, H. C. Hauffe, A. Rizzoli
and C. Vernesi. Published by Cambridge University Press. © Cambridge University Press 2009.

The characterisation of non-invasive material using molecular markers allows biologists to identify and count individuals in wild populations, identify the sex of those individuals and determine their movement patterns, infer parentage, kinship and relatedness, and assess pathogens and diet (see Kohn and Wayne 1997 for a review). The possibility of this approach first came to attention when Higuchi and colleagues demonstrated that single human hair roots could provide enough DNA for molecular genetic analysis (Higuchi *et al.* 1988), thanks to the polymerase chain reaction (PCR). Following this discovery, Morin and colleagues were the first to apply this method to hair collected in nature, studying paternity in a wild population of chimpanzees (*Pan troglodytes*) in Tanzania (Morin and Woodruff 1992; Morin *et al.* 1994c) and gene flow between chimpanzee populations in Africa (Morin *et al.* 1994b). Then, very quickly, during the 1990s, alternative DNA sources began to be exploited in a diversity of organisms: for example faeces in terrestrial (Gerloff *et al.* 1995) and marine mammals (e.g. Reed *et al.* 1997; Valsecchi *et al.* 1998), birds (Pearce *et al.* 1997) and reptiles (Bricker *et al.* 1996).

Subsequently, several studies started to combine and compare the data from different types of non-invasive samples: e.g. shed/plucked hair and faeces (Taberlet *et al.* 1997; Bayes *et al.* 2000; Smith *et al.* 2000; Constable *et al.* 2001; Vigilant *et al.* 2001). Surprisingly, however, to date relatively few studies have been carried out using exhaustive population sampling with only non-invasive material, possibly due to the financial and logistical constraints involved in collecting and processing very large numbers of samples in this way (but see, for example, Buchan *et al.* 2003; Goossens *et al.* 2005; Zhan *et al.* 2006). Nonetheless, such approaches remain perfectly feasible and will become common in the literature in the near future.

What are the applications?

At and below the population level, the use of genetic markers such as the major histocompatibility complex (MHC) (Knapp 2005), amplified fragment length polymorphism (AFLP), microsatellites, minisatellites and mitochondrial DNA with non-invasive samples can be applied in a myriad of contexts (see Table 8.1 for a more exhaustive list of examples), such as individual identification (reviews in: Taberlet and Luikart 1999; Waits *et al.* 2001), species identification (Symondson 2002 for a review; Teletchea *et al.* 2005), exclusion and assignment of parentage (Jones and Ardren 2003 for a review), relatedness and kinship patterns (Ross 2001 for a review), dispersal patterns and individual movements (genotyping in space and time; Gagneux *et al.* 2001), inferring population structure

Table 8.1 *Applications of non-invasive genetics in conservation with examples taken from the literature*

Applications	Examples and references
Individual identification	Hairy-nosed wombat (Sloane *et al.* 2000)
Species identification	Mustelid species (Hansen and Jacobsen 1999; Lopez-Giraldez *et al.* 2005); canid species (Paxinos *et al.* 1997; Davison *et al.* 2002; Dalén *et al.* 2004; Reed *et al.* 2004); seal species (Reed *et al.* 1997); deer species (Galan *et al.* 2005); macropods (Alacs *et al.* 2003); Chinese tiger (Wan *et al.* 2003)
Parentage	Sumatran orang-utan (Utami *et al.* 2002); chimpanzee (Morin *et al.* 1994b, c; Gagneux *et al.* 1997a, 1999; Constable *et al.* 2001; Vigilant *et al.* 2001); black rhinoceros (Garnier *et al.* 2001); Asian elephant (Fernando *et al.* 2000)
Relatedness and kinship	Eastern imperial eagle (Rudnick *et al.* 2005); bonobos (Gerloff *et al.* 1999); chimpanzees (Mitani *et al.* 2000; Vigilant *et al.* 2001)
Dispersal system	Common wombat (Banks *et al.* 2002); bonobos (Gerloff *et al.* 1999); chimpanzees (Morin *et al.* 1994b; Gagneux *et al.* 2001)
Individual movements	Brown bear (Taberlet *et al.* 1997); Hanuman langur (Launhardt *et al.* 2001)
Population structure	*Antechinus agilis* (Kraaijeveld-Smit *et al.* 2002)
Population assignment	Black-footed albatross (Walsh and Edwards 2005); wolves (Randi and Lucchini 2002); Alpine ibex (Maudet *et al.* 2002); African elephant (Wasser *et al.* 2004)
Phylogeography	African elephant (Eggert *et al.* 2002); Asian elephant (Fernando *et al.* 2000); brown bear (Taberlet and Bouvet 1994)
Effective population size	Brown bear (Bellemain *et al.* 2005); Louisiana black bear (Triant *et al.* 2004); gray wolf (Creel *et al.* 2003); chinook salmon (Shrimpton and Heath 2003)
Censusing and capture/recapture	Savannah baboon (Storz *et al.* 2002); badger (Wilson *et al.* 2003); coyote (Kohn *et al.* 1999; Prugh *et al.* 2005); Northern hairy-nosed wombat (Banks *et al.* 2003); black rhinoceros (Cunningham *et al.* 2001)
Hybridization effects and hybridization monitoring	Red wolf and coyote (Adams *et al.* 2003); Canada lynx and bobcat (Schwartz *et al.* 2004)
Identification of ESUs	Bornean elephant (Fernando *et al.* 2003); Larch mountain salamander (Wagner *et al.* 2005)
Reconstruction of phylogenetic relationships	Brown bear (Taberlet and Bouvet 1994)

Table 8.1 (*cont.*)

Applications	Examples and references
Impact of habitat fragmentation and reduced gene flow	Desert bighorn sheep (Epps *et al.* 2005)
Sex determination	Eurasian otter (Dallas *et al.* 2000); red deer (Huber *et al.* 2002); canid species (Ortega *et al.* 2004; Seddon 2005); felid species (Pilgrim *et al.* 2005); wolverine (Hedmark *et al.* 2004); Asian elephant (Vidya *et al.* 2003); birds (Miyaki *et al.* 1998)
Molecular tracking or genetic tagging, genetic monitoring	Brown bear (Taberlet *et al.* 1997; Lorenzini *et al.* 2004; Tallmon *et al.* 2004); mountain lion (Ernest *et al.* 2000); gray wolf (Lucchini *et al.* 2002); chimpanzees (Goossens *et al.* 2002); humpback whales (Palsbøll *et al.* 1997); black and brown bears (Woods *et al.* 1999), wolverine (Flagstad *et al.* 2004)
Disease status	Chimpanzees (Santiago *et al.* 2003); chimpanzees and gorillas (Makuwa *et al.* 2003); carnivores (Steinel *et al.* 2000)
Forensics (DNA barcodes) and legal actions	Birstein *et al.* 1998; Palumbi and Cipriano 1998; Pank *et al.* 2001; Carr *et al.* 2002; Fang and Wan 2002; Manel *et al.* 2002; Shivji *et al.* 2002; Chapman *et al.* 2003; Hebert *et al.* 2003, 2004; Moritz and Cicero 2004; Will and Rubinoff 2004; Barrett and Hebert 2005; Prendini 2005; Schander and Willassen 2005

(Pritchard *et al.* 2000), population assignment (Blanchong *et al.* 2002 for a review) and phylogeography (Avise 2000), determination of effective population size (Kohn *et al.* 1999), population census (capture/recapture) and population size estimation (Schwartz *et al.* 1999; Mills *et al.* 2000 for a review; McKelvey and Schwartz 2004a, b; Paetkau 2004; Lukacs and Burham 2005a, b and Miller *et al.* 2005 for reviews), detection of hybridization events and monitoring of hybridization (Schwartz *et al.* 2004; Willis *et al.* 2004), phylogenetic species designation and the identification of evolutionarily significant units (Moritz 1994; Li *et al.* 2004), evaluation of the impact of habitat fragmentation and reduced gene flow among populations (e.g. Stow *et al.* 2001), molecular tracking (e.g. Taberlet *et al.* 1997), sex determination (e.g. Shaw *et al.* 2003), disease status and evolutionary study of viral genomes from faecal samples (e.g. Whittier *et al.* 2004), forensics and legal actions (Birstein *et al.* 1998; Palumbi and Cipriano 1998; Dalebout *et al.* 2002; Manel *et al.* 2002) and dietary analysis (Farrell *et al.* 2000; Fedriani and Kohn 2001; Deagle *et al.* 2005; Parsons *et al.* 2005).

METHODS AND SAMPLE SOURCES

Sample sources

DNA sample sources that have been used to study wild populations include shed hairs (collected in night nests) from great apes (Morin *et al.* 1994b, c; Constable *et al.* 2001; Goossens *et al.* 2005) and plucked hairs from wombats (Sloane *et al.* 2000; Banks *et al.* 2002b, 2003), Alpine marmots (*Marmota marmota*) (Goossens *et al.* 1998a), capuchin monkeys (*Cebus olivaceus*) (Valderrama *et al.* 1999) and bears (Taberlet *et al.* 1997; Woods *et al.* 1999; Triant *et al.* 2004). Another valuable source of DNA is epithelial material from the digestive tract, which is found in and around the surface of faecal material. Using PCR, DNA was first successfully amplified from a bear faecal sample by Höss *et al.* (1992). Since then, such DNA has been analysed from a variety of mammals including bats (Vege and McCracken 2001); common wombats (*Vombatus ursinus*) (Banks *et al.* 2002a); marine mammals (Tikel *et al.* 1996) including dolphins (Parsons *et al.* 1999; Parsons 2001) and seals (Reed *et al.* 1997); ungulates (Flagstad *et al.* 1999; Huber *et al.* 2002); African (*Loxodonta africana*) (Eggert *et al.* 2002) and Asian elephants (*Elephas maximus*) (Fernando *et al.* 2000); black rhinos (*Diceros bicornis*) (Garnier *et al.* 2001); pine martens (*Martes martes*) (Davison *et al.* 2002), Eurasian badgers (*Meles meles*) (Frantz *et al.* 2003; Wilson *et al.* 2003) and Eurasian otters (*Lutra lutra*) (Dallas *et al.* 2003; Hung *et al.* 2004); kit foxes (Paxinos *et al.* 1997), coyotes (*Canis latrans*) (Kohn *et al.* 1999; Prugh *et al.* 2005), wolves (*Canis lupus*) (Lucchini *et al.* 2002; Creel *et al.* 2003) and wolverines (*Gulo gulo*) (Flagstad *et al.* 2004); mountain lions (*Puma concolor*) (Ernest *et al.* 2000) and Iberian lynx (*Lynx pardinus*) (Palomares *et al.* 2002; Pires and Fernandes 2003); and primates (Constable *et al.* 1995; Launhardt *et al.* 1998; Gerloff *et al.* 1999; Launhardt *et al.* 2001; Oka and Takenaka 2001; Vigilant *et al.* 2001; Utami *et al.* 2002; Lukas *et al.* 2004; Goossens *et al.* 2005). Faeces have been also used in bird species (Segelbacher and Steinbrück 2001), mainly in the great bustard (*Otis tarda*) (Broderick *et al.* 2003; Ydaghdour *et al.* 2003). The most common non-invasive sample used in birds is feathers (Pearce *et al.* 1997; Segelbacher 2002). Taberlet (1991) first showed that a single plucked feather contained enough DNA for genetic analysis, then Morin *et al.* (1994a) amplified DNA from hornbill feathers. In a more recent study, Rudnick *et al.* (2005) used naturally shed feathers to identify Eastern imperial eagle (*Aquila heliaca*) individuals, generate parentage data and monitor a wild population in Kazakhstan. Other sources of DNA for birds are eggshells (Pearce *et al.* 1997), egg membranes (Nuechterlein and Buitron 2000) and urine (Nota and Takenaka 1999). Urine has also been used for genetic analyses in gray wolf (*Canis lupus*) (Valière and Taberlet 2000) and wolverine (*Gulo gulo*) (Hedmark *et al.* 2004). Recently, Yasuda *et al.* (2003) described a simple method of DNA extraction and micro-satellite typing from urine samples using a DNA/RNA extraction kit that should open avenues for new studies using urine. In fish, old scale samples can be useful as a source of DNA (Nielsen *et al.* 1999). Collections of fish scales can be found in many fisheries in the world and comprehensive genetic studies are consequently being carried out with a temporal perspec-tive on many fish species.

More unusual sources of DNA are chimpanzee wadges (chewed food remnants) containing buccal cells (Sugiyama *et al.* 1993; Takenaka *et al.* 1993), sloughed skin in whales (Valsecchi *et al.* 1998) and snakes (Bricker *et al.* 1996), skin swabbing in dolphins (Harlin *et al.* 1999), eggs in sea turtles (Moore *et al.* 2003), skin mucus in fish (Livia *et al.* 2006) and buccal swabs in amphibians (Pidancier *et al.* 2003). Other biological materials such as teeth and scrimshaw from sperm whales (*Physeter macrocephalus*) (Pichler *et al.* 2001), old teeth in red fox (*Vulpes vulpes*) (Wandeler *et al.* 2003), ivory in elephants (Comstock *et al.* 2003; Wasser *et al.* 2004), meat from whales (Baker *et al.* 1996; Palumbi and Cipriano 1998; Baker *et al.* 2000), dolphins (Baker *et al.* 1996), Chinese alligators (*Alligator sinensis*) (Yan *et al.* 2005), ostriches (Abdulmawjood and Buelte 2002) and sea turtles (Moore *et al.* 2003), body parts and remains in sharks (Hoelzel 2001; Pank *et al.* 2001; Shivji *et al.* 2002; Chapman *et al.* 2003) and whales (Carr *et al.* 2002; Dalebout *et al.* 2002), sturgeon caviar (Wolf *et al.* 1999), carcasses in deer species (Fang and Wan 2002) can all give reliable results for DNA analysis and are very useful in trade monitoring and wildlife poaching detection of endangered species.

Storage of samples
Hair samples
There are two kinds of hairs that can be used as non-invasive DNA sources: plucked and shed. Plucked hairs are by far the best source of hair DNA for both mitochondrial and nuclear DNA analysis while shed hairs can often be problematic for nuclear DNA analysis. Single plucked hairs with root material should provide enough DNA for genetic analysis, providing adequate storing conditions are used (see below). However, we recommend

collecting more than 10 hairs per individual (see Goossens *et al.* 1998b). Valderrama *et al.* (1999) described four methods of collecting fresh hair samples from wild and captive mammals: (1) shooting a rolled strip of duct tape, pressed onto the flat tip of a plastic syringe, from an air-powered dart pistol; (2) making a corral by enclosing a small area with duct tape; (3) wrapping a bait (i.e. to a tree) with duct tape; (4) wrapping inverted tape around the tip of a stick and touching the animal (only for captive animals). Hair traps based on barbed wire around trees (for bears) and sticky tape in rodent tubes can also be useful. Plucked hairs have been used for free-ranging primates (Valderrama *et al.* 1999), wombats (Sloane *et al.* 2000; Banks *et al.* 2003), brown bears (*Ursus arctos*) (Taberlet *et al.* 1994, 1997; Woods *et al.* 1999) and Alpine marmots (Goossens *et al.* 1998a).

For shed hairs, the roots have usually undergone apoptosis before shedding (telogen phase), and much of the nuclear DNA is degraded (Jeffery *et al.* 2007). However, epithelial tissue may be attached to the root of freshly shed hairs and provides a source of undegraded nuclear DNA (Linch *et al.* 1998). Shed hairs are commonly used for great ape studies (see Morin *et al.* 1994b, c; Gagneux *et al.* 1997a; Goossens *et al.* 2005) but can show unreliable results (Gagneux *et al.* 1997b).

Roon *et al.* (2005a) evaluated the optimal storage methods and DNA degradation rates for hair samples. Hair samples from brown bears were preserved using silica desiccation and –20 °C freezing over a 1-year period. Amplification success rates decreased significantly after a 6-month time point, regardless of storage method. It is therefore important to minimize delays between hair collection and extraction if we want to maximize the amplification success rate. However, hair samples are usually stored in clean paper envelopes (Goossens *et al.* 1998a; Woods *et al.* 1999; Sloane *et al.* 2000), since plastic bags produce static that make hair manipulation difficult and increase contamination.

Faecal samples

For the last 10 years, different storage methods have been tested for faecal samples of different species. It is vital that DNA degradation by nucleases is minimized as much as possible. This can be done by dehydrating the sample by air-drying (Flagstad *et al.* 1999 for sheep and reindeer (*Rangifer tarandus*); Farrell *et al.* 2000 for mountain lions and jaguars (*Panthera onca*)), by silica gel beads drying (Bradley and Vigilant 2002 for gorillas), by –20 °C-freezing (Ernest *et al.* 2000 for mountain lions), by alcohol (ethanol) treatment (Gerloff *et al.* 1999 for bonobos (*Pan paniscus*); Fernando *et al.* 2000 for Asian elephant; Constable *et al.* 2001 for chimpanzees; Goossens *et al.* 2005 for orang-utans (*Pongo pygmaeus*)), or by saturating the sample in a buffer (DETs: see below) containing high concentrations of salts or other chemicals interfering with enzymes (Frantzen *et al.* 1998). Frantzen *et al.* (1998) evaluated the effectiveness of these four methods for preserving fresh baboon faeces: drying, −20 °C freezing, 70% ethanol and DMSO/EDTA/Tris/salt solution (DETs). The latter was the most effective for preserving nuclear DNA and the three other methods performed equally well for mitochondrial DNA analysis and for short microsatellite fragments (less than 200 bp) showing that amplification success is dependent on storage method, PCR product size and molecular marker used. In another study, Piggott and Taylor (2003a) evaluated the same preservation methods (together with different extraction methods) but for faecal samples from two Australian marsupial herbivores (*Dasyurus maculatus* and *D. viverrinus*) and one intro-duced carnivore (*V. vulpes*). Their results showed that the highest amplifica-tion and lowest genotyping error rates were obtained with dried faecal samples extracted via a surface wash followed by Qiagen spin column purification. More recently, Roeder *et al.* (2004) compared faecal sample storage in ethanol and silica with a two-step method: soaking of the samples in ethanol followed by desiccation with silica. While the samples stored in silica showed the lowest DNA concentration, the two-step method yielded significantly more DNA in high quality samples. The ethanol and the two-step methods performed equally for lower quality samples. Nsubuga *et al.* (2004) obtained significantly higher amounts of DNA from wild mountain gorillas (*Gorilla beringei beringei*) and chimpanzees faecal samples using the protocol deve-loped by Roeder and colleagues. Moreover, they showed a small negative correlation between temperature at time of collection and the amount of DNA amplified. RNA*later* (see next paragraph) preservation solution did not produce better results than silica gel beads storage.

In 2002, Murphy *et al.* tested five preservation methods on brown bear faeces (90% ethanol, DETs buffer, silica-dried, oven-dried stored at room temperature, and oven-dried stored at −20 °C) at different time points (1 week, 1 month, 3 months and 6 months) and for two different genetic markers (mtDNA and nDNA). The ethanol-preserved samples had the highest success rates for both mtDNA and nDNA. The authors recommen-ded preservation of faecal samples in 90% ethanol when feasible and the drying method when collecting in remote field conditions. In a previous study, Murphy *et al.* (2000) evaluated four drying methods for brown bear faeces, with the freeze-drying and oven-drying producing the best amplifi-cation success rates. A recent tissue storage reagent, called RNA*later*® (Ambion, Inc.), has been successfully used to store faecal samples in our laboratory and those of others. It is an aqueous, non-toxic reagent that rapidly permeates most tissues to stabilize and protect RNA in fresh specimens. DNA (and obviously RNA) can be isolated from RNA*later*-stored samples with very good reliability in genotyping results. The only problem is that it is an expensive reagent. Faeces from birds can be stored at −20 °C in 90% ethanol (Broderick *et al.* 2003; Idaghdour *et al.* 2003).

Other samples

Urine Urine samples can be used for carnivores and can be easily collected on snow (see Valière and Taberlet 2000; Hedmark *et al.* 2004). It can also be collected in plastic sheets placed under sleeping nests of great apes. Individual chimpanzees or orang-utans often urinate from the side of the nest on awakening and urine can be collected and transferred to storage vials using disposable plastic pipettes. Unfortunately, fermentation and DNA degradation in cells may occur rapidly after urination (Hayakawa and Takenaka 1999), it is therefore recommended to collect as large a volume as possible and transfer it into two volumes of 95% ethanol. Urine can also be used for birds (Nota and Takenaka 1999), and fixed with 70–90% ethanol and stored at −20 °C.

Feathers In general, feathers are stored in envelopes or plastic bags and stored dry until analysis (see Segelbacher 2002). In a study on the Eastern imperial eagle, Rudnick *et al.* (2005) used naturally moulted adult feathers collected from nesting sites. He also plucked developing blood feathers from chicks. Adult feathers were stored dry at room temperature. Developing chick feathers were stored in a lysis buffer (see Rudnick *et al.* 2005 for details).

Wadges Buccal cells from food items (wadges) can be successfully extracted and mtDNA and nDNA can be amplified from the DNA extracted (Hashimoto *et al.* 1996 in chimpanzees). Wadge samples must be transferred to a sterile 50-ml polypropylene tube filled with 90% ethanol and 1 mM Na$_3$EDTA, to avoid bacterial and enzymatic degradation of the DNA.

Extraction kits and methods

Hair The most popular method for extracting DNA from hairs is the Chelex-100® and proteinase K method developed by Walsh *et al.* (1991). However, Vigilant (1999) obtained better results using Taq polymerase PCR buffer as the extraction buffer (see Allen *et al.* 1998). In our laboratory, we have used the latter and have found that using PCR buffer, water and proteinase K in a small extraction volume works very well for shed hairs (see Goossens *et al.* 2005; Jeffery 2007).

Faeces Cells containing DNA are not uniformly spread throughout faeces, and two or three extracts should be made per sample (see Goossens *et al.* 2000). It is also important to use a method that involves fewer steps and sample transfers, although the removal of substances that may inhibit PCR usually requires repeated purification steps involving several centrifugation steps. We recommend using the QIAamp Stool mini kit (QIAGEN) which has given reliable results in primates (gorillas: Bradley and Vigilant 2002; orang-utans: Utami *et al.* 2002; Goossens *et al.* 2005; baboons: Bayes *et al.* 2000) and other mammals (black rhinoceros: Garnier *et al.* 2001; brown bear: Bonin *et al.* 2004; wolverine: Hedmark *et al.* 2004). Other methods have been described in the literature and include: silica-based method (Boom *et al.* 1990), magnetic beads (Flagstad *et al.* 1999), diatomaceous earth method (Gerloff *et al.* 1995), Chelex-100® (Walsh *et al.* 1991), and surface wash followed by spin column purification (Piggott and Taylor 2003a). A pilot study is recommended as one extraction technique may work for some species but may not work for others. Extraction (and storing) methods will depend on the field conditions, location, season, size and age of the samples (see Taberlet *et al.* 1999; Piggott 2004). For bird faeces, Broderick *et al.* (2003) used a modification of Milligan's (1998) silica and guanidine isothiocyanate-based protocol (for a detailed protocol, see Idaghdour *et al.* 2003).

Urine Protocols using the QIAamp DNA stool mini kit (GmbH, Hilden, Germany) to extract DNA from urine collected in snow are well described in Valière and Taberlet (2000) for canids and in Hedmark *et al.* (2004) for wolverine. For birds, a detailed protocol is described in Nota and Takenaka (1999).

Feathers Eguchi and Eguchi (2000) developed a simple method to extract DNA from bird feathers, and also from snake cast-off skin using collagenase. Jensen *et al.* (2003) describe a technique to extract DNA from feathers using Chelex-100® (also used by Morin *et al.* (1994a) for hornbills).

Wadges Different methods can be used to extract DNA from wadges and can be found in Takenaka *et al.* (1993) and in Hashimoto *et al.* (1996).

RECENT INNOVATIONS

Multiplex PCR (Henegariu *et al.* 1997) systems for comparative genotyping are well developed in human forensics (Wallin *et al.* 2002; Shewale *et al.* 2004) and are now developed for other animal species such as cervids (Eld deer and swamp deer, Rusa and Vietnamese sika deers: Bonnet *et al.* 2002;

roe deer: Galan *et al.* 2003); primates (orang-utans: Immel *et al.* 1999; Roeder *et al.* 2006), fish (great white shark (*Carcharodon carcharias*): Chapman *et al.* 2003). Piggott *et al.* (2004) developed a multiplex pre-amplification method to improve microsatellite amplification and error rates when using faecal DNA. Qiagen have developed a multiplex kit, which is commonly used for genotyping of DNA extracted from non-invasive samples such as faeces and hair in our laboratory with reliable results.

In addition, the recent establishment of whole-genome amplification such as multiple displacement amplification (MDA) (Dean *et al.* 2002) promises to revolutionize non-invasive genetic analysis since in principle it allows the production of large quantities of whole-genomic DNA from minute sources, such as are routinely produced from non-invasive studies. MDA allows the generation of thousands of copies of whole genomes of up to 10 kilobase pairs (kb) in length (Dean *et al.* 2002). The isothermal MDA reaction utilizes the highly processive bacteriophage phi29 DNA polymerase and its DNA strand-displacing activity. In the MDA reaction, random hexamer primers annealed to denatured genomic DNA are extended by the phi29 DNA polymerase to form products up to 100 kb. As the DNA polymerase encounters another newly synthesized DNA strand downstream, it displaces it and thus creates a new single-stranded DNA template for priming. Strand displacement leads to hyperbranched primer extension reactions that may yield milligram amounts of DNA product from just a few nanograms of genomic DNA. Owing to its $3'-5'$ proofreading activity, the fidelity of the phi29 DNA polymerase is very high with an error rate of $<10^{-6}$ (Esteban *et al.* 1993), which in turn requires exonuclease-protected primers to achieve a high yield. As the reaction involves no thermal cycling and high molecular weight copies of genomic DNA are produced, the genomic coverage of MDA products is higher than that of the PCR-based whole-genome amplification methods degenerate oligonucleotide-primed PCR (DOP-PCR) and primer extension preamplification (PEP) (Dean *et al.* 2002). Rönn *et al.* (2006) recently tested this approach to assess its efficacy on a variety of primate DNA, including non-invasively collected samples, and found broadly that MDA template DNA produced equivalent genotype accuracy to unamplified DNA.

Molecular markers

The choice of a molecular marker will depend on the question of interest. Each marker has its own appropriate use and the costs and difficulty of genetic typing must be taken in consideration. The two most commonly used markers used in non-invasive genetics today are mitochondrial DNA and nuclear microsatellites. The specific attributes of these markers will not be discussed here, but their behaviour and likely information content in a non-invasive genetic analysis context will be alluded to. However, it is likely in the future that single nucleotide polymorphisms (SNPs) will become the genetic marker of choice to study the ecology and conservation of wild populations because they allow access to variability across the whole genome. Although examples remain scarce to date, in one study Seddon *et al.* (2005) addressed ecological and conservation issues in recolonized Scandinavian and Finnish wolf populations using 24 SNP loci. These loci were able to differentiate individual wolves and differentiate populations using assignment tests. Compared to microsatellites, SNPs allow the amplification of extremely small fragments, which makes them very useful for population and conservation genetics using non-invasive samples, and are much easier to automate, for example on microarrays (see chapter by Vernesi and Bruford, this volume). SNPs have the advantage that a range of different sequence types can be chosen, to give information on both neutral markers and those under selection (for example the major histocompatibility complex; Smulders *et al.* 2003).

Sex chromosomes in mammals (Fernando and Melnick 2001; Bryja and Konecny 2003; Hellborg and Ellegren 2003; Erler *et al.* 2004; Hedmark *et al.* 2004) and other vertebrates (birds: Griffiths *et al.* 1998; and fish: Matsuda 2003) can provide DNA sequence useful for the identification of an animal's gender. Using both Y-chromosomal DNA and an autosomal or X-linked marker is useful in providing sex information (Griffiths and Tiwari 1993; Sloane *et al.* 2000; Huber *et al.* 2002). The amelogenin locus can also be used to identify gender, e.g. in great apes (Bradley *et al.* 2001; Matsubura *et al.* 2005), bears (Yamamoto *et al.* 2002) and felids (Pilgrim *et al.* 2005). In birds, for example, Sacchi *et al.* (2004) used the CHD (Chromo-Helicase-DNA-Binding) sex-linked gene and feathers to differentiate males and females of the endangered short-toed eagle (*Circaetus gallicus*). Russello and Amato (2001) described a PCR-based test, using feather DNA, to identify the sex in an endangered parrot species, *Amazona guildingii*.

Amplified fragment length polymorphisms (AFLPs) are dominant markers that can be used in parentage, population assignment (Campbell *et al.* 2003), gene flow and migration, although they are less adequate for reconstructing past events and historic patterns of variation (Wayne and Morin 2004; Bensch and Åkesson 2005). However, their use in non-invasive analysis is likely to be limited due to the requirement for quite large amounts of template DNA and large fragment sizes.

TECHNICAL CHALLENGES

Non-invasive genetic analysis, despite its obvious advantages for studying wild populations of elusive and endangered species, has its own limitations and pitfalls that must be seriously taken into consideration. Samples collected non-invasively are far less reliable than invasive samples such as blood and tissue biopsies. DNA can be highly fragmented and sometimes PCR may be inhibited by co-extracted compounds present in the material. Contamination from humans (particularly for primate species) and cross-contamination between samples are common and must be avoided. Therefore, precautions need to be taken such as using a laboratory room dedicated to non-invasive sample storage and handling. DNA extracted from non-invasive samples can be of low quantity and quality and therefore analyses need to be rigorously done and checked (Taberlet and Luikart 1999; Taberlet *et al.* 1999; Bonin *et al.* 2004; McKelvey and Schwartz 2004a). DNA extraction has to be highly efficient (quick and avoiding unnecessary steps) and several new methods and ever-sophisticated and high-yielding extraction kits are now available to expedite rapid extraction and minimal liquid handling. DNA extraction has also to be able to remove inhibitory material during purification (Eggert *et al.* 2005) and whereas this used to be a laborious process, required reagents are now included in many of the commercially available kits.

Low template DNA copy number and PCR inhibition have led to several phenomena being observed in non-invasive genotyping. First, PCR products may be extremely difficult to generate and the resultant fragments may not be sufficient for analysis. We advise the use of more PCR cycles (up to 40–50) or a second round of PCR, using the fragments generated in the first to 'seed' the reaction. Decreasing the annealing temperature may also help. Increasing the number of cycles may, however, have a negative impact if a copying error is introduced, producing false polymorphisms. Therefore, replicate PCRs are imperative to confirm the results (see Taberlet *et al.* 1996; Goossens *et al.* 2000).

False data may occur in DNA sequences (artificial point mutations) or in microsatellite fragments (false allele lengths due to DNA polymerase slippage during PCR). DNA polymerase slippage is a general phenomenon in microsatellite PCRs, but can usually be compensated for by recording only the one (for homozygotes) or two (for heterozygotes) most intensely amplified fragments. False alleles may confuse this process, although such artifact fragments are usually weakly amplified, and should in any case be replicated (Bradley and Vigilant 2002). Furthermore, the stochastic non-amplification of one of the two potential alleles at a microsatellite locus can occur ('allelic dropout') because of low template copy number or DNA degradation. The latter may be a special problem for loci exhibiting a wide range of allele lengths, because longer alleles may not be amplifiable if their length exceeds the maximum fragment size present in the degraded template DNA. Repeated amplifications using several independent DNA extractions (see Navidi *et al.* 1992; Taberlet *et al.* 1996; Goossens *et al.* 1998b; Taberlet *et al.* 1999 for a review; Goossens *et al.* 2000) are a minimum requirement in such studies. Software such as GIMLET (Genetic Identification with MultiLocus Tags, http://pbil.univ-lyon1.fr/software/ Gimlet/gimlet.htm) can assist in the identification of false homozygotes and false alleles by comparing the repeated genotypes and the associated consensus genotype for each sample (Valière 2002).

THE NEED FOR PILOT STUDIES

We strongly advocate carrying out preliminary experimental protocols and critical pilot data evaluation before starting a full-scale study on a new species or population. If working with faecal samples, an environmental decay experiment can be extremely useful to establish the likely success of DNA extraction from faeces found in field conditions. Fresh samples always produce better DNA, but sometimes these may be impossible to find. Piggott (2004) investigated the effect of sample age (and seasonality) on the amplification and genotyping reliability of microsatellite loci from faecal DNA of a marsupial herbivore (the brush-tailed rock wallaby, *Petrogale penicillata*) and a carnivore (the red fox). The author compared DNA profiles from 1 day to 6 months for both species and found that as the age of the samples increased there was progressively poorer quality DNA present on the surface of the faeces. This resulted in significantly lower amplification rates and higher genotyping error. This problem is likely to be most severe in tropical environments, where rainfall is very frequent and the risk of washing the DNA off the outer layer (mucus) of the faeces is high. Therefore, there is a need to know how DNA yield correlates with the age of your sample (decay rate experiment) and the weather conditions, such as rain. It is also important to consider the diet of the species studied (Murphy *et al.* 2003). Problems are often met with leaf-eater species, probably due to vegetal inhibitory material. We also suggest liaising with biochemists for identifying, depending on some specific biological features of the study species, which compounds are expected to be co-extracted with DNA from sources such as faeces and hair. This knowledge could aid the choice of more efficient procedures for eliminating these molecules thus improving the quality of the extracted DNA.

Quantitation of DNA in non-invasive genetic samples has often proved problematic by conventional means due to the degraded nature of the DNA present, contamination with exogenous DNA and the presence of RNA. One reliable quantitation approach, described by Morin *et al.* (2001), uses a quantitative PCR assay with appropriate standards. This method has proven reliable in samples such as DNA extracted from previously autoclaved fox teeth (Wandeler *et al.* 2003) and provides a major positive development in the field. Once the pilot data have been produced, the software GEMINI (Genotyping Errors and Multitube Approach for Individual Identification, http://pbil.univ-lyon1.fr/software/gemini.html; Valière *et al.* 2002) allows the user to evaluate and quantify the effects of genotyping errors on the genetic identification of individuals. It also allows simulation of the effectiveness of a specific multitubes approach to correct these errors.

REQUIREMENTS

To summarize, before starting any non-invasive genetic study on a specific species, and especially when using faecal samples, it is necessary to:

(1) identify the genetic markers that you will need (i.e. if you use microsatellites, check available published markers)
(2) carry out a pilot study on the effect of age and season on the reliability of microsatellite genotyping (particularly for tropical species and if you work with faeces)
(3) select the most appropriate sample preservation method (check the literature or test it if necessary)
(4) select the most appropriate DNA extraction method (check the literature or test it if necessary)
(5) test the effects of genotyping errors and multitubes approach using software such as GEMINI (Valière *et al.* 2002)
(6) during collection, try to sample the same faeces at least twice, and always sample the outer layer of the faeces (mucus).

ANALYSIS

Reliability

Different methods have recently been published to check the integrity of the genotypes produced during a study and to ensure that the multi-locus genotypes are correct. The method most commonly used so far is the multitubes approach (originally developed by Navidi *et al.* 1992) formalized by Taberlet *et al.* in 1996. Since then, most studies using non-invasive DNA carry out three to seven replicate PCRs per sample for each locus. However, such a number of replicates considerably increases the cost and time of such studies. Therefore, the pre-screening methods described above (Morin *et al.* 2001) or computer packages (Valière 2002; Valière *et al.* 2002; van Oosterhout *et al.* 2004; Roon *et al.* 2005b) are highly recommended before starting any non-invasive work. MICRO-CHECKER (van Oosterhout *et al.* 2004, http://www.microchecker.hull.ac.uk/) is software that tests for genotyping errors due to null or false alleles and for allelic dropout. It can also be used to discriminate between Hardy–Weinberg deviations caused by null alleles and those caused by demographic factors such as consanguinity. More recently, Roon *et al.* (2005b) tested the effectiveness of several methods for error-checking non-invasive genetic data and cautioned against using non-comprehensive data filters in non-invasive genetic studies and suggested the combination of data filters with careful technique and thoughtful non-invasive study design. Wilberg *et al.* (2004) have produced software (GENECAP) to facilitate the analysis of multilocus genotype data in non-invasive DNA sampling and genetic capture–recapture studies. It uses multilocus genetic data to match samples with identical genotypes, calculate frequency of alleles, identify sample genotypes that differ by one and two alleles, calculate probabilities of identity, and match probabilities for matching samples.

Demographic information

Alongside previously mentioned software, such as GIMLET (Valière *et al.* 2002), increasingly sophisticated approaches, such as the likelihood-based methods implemented in API-CALC 1.0 (Ayres and Overall 2004) allow the user to calculate probabilities of identity (individualize from non-invasive genetic data) allowing for complications such as genetic substructure, inbreeding and the presence of close relatives.

A large number of software packages have been designed in the last 10 years to assign parentage. The strengths and weaknesses of these methods have been reviewed by Jones and Ardren (2003). We strongly recommend assessing their merits before selecting any recent parentage software. There are four approaches for calculating parentage: (1) exclusion (based on the Mendelian rules of inheritance), which uses incompatibilities between offspring and parents to reject a particular parent/offspring pair and assumes all potential parents are sampled and no genotype errors; (2) categorical allocation, which uses likelihood-based (LOD score) approaches to select the most likely parent from a pool of non-excluded parents and allows the user to include a genotyping error rate and incomplete sampling of potential parents (Marshall *et al.* 1998; Slate *et al.* 2000); (3) fractional allocation, which assigns some fraction (between 0 and 1) of each offspring to all non-excluded candidate parents (see Devlin *et al.* 1988); (4) parental reconstruction, which uses the multilocus genotypes of parents and offspring to reconstruct the genotypes of unknown parents contributing gametes to a progeny array for which one parent is known a priori (Jones 2001). Table 8.2 provides a list of the most recent parentage software used in the literature, with case studies and, for most of the examples, implications for conservation. The most common software used for parentage analysis is CERVUS (Marshall *et al.* 1998).

A number of software packages can be used to estimate relatedness in wild populations. The most commonly used are RELATEDNESS (Queller and Goodnight 1989) and KINSHIP (Goodnight and Queller 1999). RELATEDNESS estimates pairwise relatedness between individuals or average pairwise relatedness between groups using regression, while KINSHIP tests pedigree relationships using likelihood methods. Another package, DELRIOUS (Stone and Björklund 2001), analyses molecular marker data and calculates delta and relatedness estimates with confidence limits. Finally, IDENTIX (Belkhir *et al.* 2002) tests relatedness in a population using permutation methods.

There are several packages available that allow the identification of a source population for a specific dispersing individual. Eldridge *et al.* (2001) used assignment tests in the programs STRUCTURE (Pritchard *et al.* 2000) and GENECLASS (Cornuet *et al.* 1999) to identify the source population of rock wallaby (*Petrogale lateralis*) individuals. Berry *et al.* (2004) examined the accuracy of assignment tests to measure dispersal in the grand skink (*Oligosoma grande*) and suggested the use of Bayesian assignment methods. Hansson *et al.* (2003) used GENECLASS software to assign immigrants of specific cohorts of great reed warblers (*Acrocephalus arundinaceus*) and revealed female-biased dispersal in that species. Möller and Beheregaray (2004) used GENECLASS and RELATEDNESS to identify male-biased dispersal patterns in bottlenose dolphins (*Tursiops aduncus*). Isolation by distance calculated with Mantel tests (Liedloff 1999) can also be used to estimate dispersal in wild populations. Many examples now exist where these approaches have been used on invasive samples, but only a few studies have used non-invasive DNA sampling to assign gene flow or dispersal patterns in animal species (Broderick *et al.* 2003 in great bustard; Launhardt *et al.* 2001 in langurs (*Semnopithecus entellus*); Gerloff *et al.* 1999 in bonobos (*Pan paniscus*); Morin *et al.* 1994b and Gagneux *et al.* 2001 in chimpanzees).

Table 8.2 *List of the most recent parentage software used in the literature*

Software	Authors	Web site	Examples
CERVUS	Marshall et al. (1998)	helios.bto.ed.ac.uk/evolgen/cervus/cervus.html	Baker et al. (2004) (red foxes); Garnier et al. (2001) (black rhinoceros); Utami et al. (2002) (orang-utans)
FAMOZ	Gerber et al. (2003)	www.pierroton.inra.fr/genetics/labo/Software/Famoz	
GERUD 1.0	Jones (2001)	www.bio.tamu.edu/USERS/ajones/JonesLab.htm	Chapman et al. (2004) (hammerhead shark, *Sphyrna tiburo*)
GERUD 2.0	Jones (2005)	www.bio.tamu.edu/USERS/ajones/JonesLab.htm	
KINSHIP	Goodnight and Queller (1999)	www.gsoftnet.us/GSoft.html	Clinchy et al. (2004) (brushtail possums, *Trichosurus vulpecula*);
NEWPAT	Worthington Wilmer et al. (1999)	www.zoo.cam.ac.uk/zoostaff/amos/newpat.html	Hoffman and Amos (2005) (Antarctic fur seals, *Arctocephalus gazella*)
PAPA	Duchesne et al. (2002)	www.bio.ulaval.ca/louisbernatchez/	
PARENTAGE 1.0	Emery et al. (2001)	maths.abdn.ac.uk/~ijw/downloads/download.htm	
PARENTE	Cercueil et al. (2002)	www2.ujf-grenoble.fr/leca/membres/manel.html	
PASOS	Duchesne et al. (2005)	www.bio.ulaval.ca/louisbernatchez/	
PATRI	Signorovitch and Nielsen (2002)	www.biom.cornell.edu/Homepages/Rasmus_Nielsen/files.html	
PROBMAX	Danzmann (1997)	www.uoguelph.ca/~rdanzman/software/probmax/	
RELATEDNESS	Queller and Goodnight (1989)	www.gsoftnet.us/GSoft.html	

Finally, and perhaps most excitingly, in the last 5–10 years, the use of non-invasive genetic sampling for capture–recapture population census studies on several animal species has begun; for example, in painted turtles (*Chrysemys picta*) (Pearse *et al.* 2001); whales (Palsbøll *et al.* 1997); bears (Woods *et al.* 1999; Dobey *et al.* 2005); and African elephants (Eggert *et al.* 2003); coyotes (Kohn *et al.* 1999; Prugh *et al.* 2005), with parallel development of new methods to estimate the size of populations using molecular mark–recapture data (Mills *et al.* 2000; Waits and Leberg 2000; Paetkau 2003; McKelvey and Schwartz 2004a, b; Paetkau 2004; Waits 2004; McKelvey and Schwartz 2005; Miller *et al.* 2005). There are a number of capture–recapture methods now available for use with non-invasive DNA-based capture–recapture data. These methods are highlighted in a recent review by Lukacs and Burnham (2005b). The most recent software (CAPWIRE: Miller *et al.* 2005) allows the application of a number of models of population aggregation and faecal deposition rates, and this has recently been applied with success, for example, to giant pandas (*Ailuropoda melano-leuca*) (Zhan *et al.* 2006).

PERSPECTIVE

Non-invasive analysis is becoming the only acceptable method for retrieving genetic data from many endangered species. Earlier problems with reliability are being rapidly resolved and technical innovations such as multiplex PCR kits and whole-genome amplification may soon make this type of analysis the norm. However, care is still necessary and experimental designs must allow full verification of the data, both by the researchers themselves and others wishing to replicate their work.

ACKNOWLEDGEMENTS

We would like to thank the Darwin Initiative for the Survival of Species (Defra, UK) for funding and Universiti Malaysia Sabah for facilities during the writing of this article.

REFERENCES

Abdulmawjood, A. and Buelte, M. (2002). Identification of ostrich meat by restriction fragment length polymorphism (RFLP) analysis of cytochrome *b* gene. *Journal of Food Science*, **5**, 1688–1691.

Adams, J. R., Kelly, B. T. and Waits, L. P. (2003). Using faecal DNA sampling and GIS to monitor hybridization between red wolves (*Canis rufus*) and coyotes (*Canis latrans*). *Molecular Ecology*, **12**, 2175–2186.

Alacs, E., Alpers, D., de Tores, P. J., Dillon, M. and Spencer, P. B. S. (2003). Identifying the presence of quokkas (*Setonix brachyurus*) and other macropods using cytochrome *b* analyses from faeces. *Wildlife Research*, **30**, 41–47.

Allen, M., Engström, A. S., Meyers, S. *et al.* (1998). Mitochondrial DNA sequencing of shed hairs and saliva on robbery caps: sensitivity and matching probabilities. *Journal of Forensic Sciences*, **43**, 453–464.

Avise, J. C. (2000). *Phylogeography: The History and Formation of Species*. Cambridge, Mass.: Harvard University Press.

Ayres, K. L. and Overall, A. D. J. (2004). API-CALC 1.0: a computer program for calculating the average probability of identity for substructure, inbreeding and the presence of close relatives. *Molecular Ecology Notes*, **4**, 315–318.

Baker, C. S., Cipriano, F. and Palumbi, S. R. (1996). Molecular genetic identification of whale and dolphin products from commercial markets in Korea and Japan. *Molecular Ecology*, **5**, 671–685.

Baker, C. S., Lento, G. M., Cipriano, F. and Palumbi, S. R. (2000). Predicted decline of protected whales based on molecular genetic monitoring of Japanese and Korean markets. *Proceedings of the Royal Society Series B*, **267**, 1191–1199.

Baker, P. J., Funk, S. M., Bruford, M. W. and Harris, S. (2004) Polygynandry in a red fox population: implications for the evolution of group living in canids? *Behavioral Ecology*, **15**, 766–778.

Banks, M. A. and Eichert, W. (2000). WHICHRUN (v. 3.2): a computer program for population assignment of individuals based on multilocus genotype data. *Journal of Heredity* **91**, 87–89.

Banks, S. C., Piggott, M. P., Hansen, B. D., Robinson, N. A. and Taylor, A. C. (2002a). Wombat coprogenetics: enumerating a common wombat population by microsatellite analysis of faecal DNA. *Australian Journal of Zoology*, **50**, 193–204.

Banks, S. C., Skerratt, L. F. and Taylor, A. C. (2002b). Female dispersal and relatedness structure in common wombats (*Vombatus ursinus*). *Journal of Zoology*, **256**, 389–399.

Banks, S. C., Hoyle, S. D., Horsup, A., Sunnucks, P. and Taylor, A. C. (2003). Demographic monitoring of an entire species (the northern hairy-nosed wombat, *Lasiorhinus krefftii*) by genetic analysis of non-invasively collected material. *Animal Conservation*, **6**, 101–107.

Barrett, R. D. H. and Hebert, P. D. N. (2005). Identifying spiders through DNA barcodes. *Canadian Journal of Zoology*, **83**, 481–491.

Bayes, M. K., Smith, K. L., Alberts, S. C., Altmann, J. and Bruford, M. W. (2000). Testing the reliability of microsatellite typing from faecal DNA in the savannah baboon. *Conservation Genetics*, **1**, 173–176.

Belkhir, K., Castric, V. and Bonhomme, F. (2002). IDENTIX, a software to test for relatedness in a population using permutation methods. *Molecular Ecology Notes*, **2**, 611–614.

Bellemain, E., Swenson, J. E., Tallmon, D., Brunberg, S. and Taberlet, P. (2005). Estimating population size of elusive animals with DNA from hunter-collected feces: comparing four methods for brown bears. *Conservation Biology*, **19**, 150–161.

Bensch, S. and Åkesson, M. (2005). Ten years of AFLP in ecology and evolution: why so few animals? *Molecular Ecology*, **14**, 2899–2914.

Berry, O., Tocher, M. D. and Sarre, S. D. (2004). Can assignment tests measure dispersal? *Molecular Ecology*, **13**, 551–561.

Birstein, V. J., Doukakis, P., Sorkin, B. and DeSalle, R. (1998). Population aggregation analysis of three caviar-producing species of sturgeons and implications for the species identification of black caviar. *Conservation Biology*, **12**, 766–775.

Blanchong, J. A., Scribner, K. T. and Winterstein, S. R. (2002). Assignment of individuals to populations: Bayesian methods and multi-locus genotypes. *Journal of Wildlife Management*, **66**, 321–329.

Bonin, A., Bellemain, E., Eidesen, P. B. *et al.* (2004). How to track and assess genotyping errors in population genetics studies. *Molecular Ecology*, **13**, 3261–3273.

Bonnet, A., Thevenon, S., Maudet, F. and Maillard, J. C. (2002). Efficiency of semi-automated fluorescent multiplex PCRs with 11 microsatellite markers for genetic studies of deer populations. *Animal Genetics*, **33**, 343–350.

Boom, R., Sol, C. J. A., Salimans, M. M. M. *et al.* (1990). Rapid and simple method for purification of nucleic acids. *Journal of Clinical Microbiology*, **28**, 495–603.

Bradley, B. J. and Vigilant, L. (2002). False alleles derived from microbial DNA pose a potential source of error in microsatellite genotyping of DNA from faeces. *Molecular Ecology Notes*, **2**, 602–605.

Bradley, B. J., Chambers, K. E. and Vigilant, L. (2001). Accurate DNA-based sex identification of apes using non-invasive samples. *Conservation Genetics*, **2**, 179–181.

Bricker, J., Bushar, L. M., Reinert, H. K. and Gelbert, L. (1996). Purification of high quality DNA from shed skins. *Herpetological Review*, **27**, 133–134.

Broderick, D., Idaghdour, Y., Korrida, A. and Hellmich, J. (2003). Gene flow in great bustard populations across the Straits of Gibraltar as elucidated from excremental PCR and mtDNA sequencing. *Conservation Genetics*, **4**, 793–800.

Bryja, J. and Konecny, A. (2003). Fast sex identification in wild mammals using PCR amplification of the *Sry* gene. *Folia Zoologica*, **52**, 269–274.

Buchan, J. C., Alberts, S. C., Silk, J. B. and Altmann, J. (2003). True paternal care in a multi-male primate society. *Nature*, **425**, 179–181.

Campbell, D., Duchesne, P. and Bernatchez, L. (2003). AFLP utility for population assignment studies: analytical investigation and empirical comparison with microsatellites. *Molecular Ecology*, **12**, 1979–1991.

Carr, S. M., Marshall, H. D., Johnstone, K. A., Pynn, L. M. and Stenson, G. B. (2002). How to tell a sea monster: molecular discrimination of large marine mammals of the North Atlantic. *Biological Bulletin*, **202**, 1–5.

Cercueil, A., Bellemain, E. and Manel, S. (2002). Parente: computer program for parentage analysis. *Journal of Heredity*, **93**, 458–459.

Chapman, D. D., Abercrombie, D. L., Douady, C. J. *et al.* (2003). A streamlined, bi-organelle, multiplex PCR approach to species identification, application to global conservation and trade monitoring of the great white shark, *Carcharodon carcharias. Conservation Genetics*, **4**, 415–425.

Chapman, D. D., Prodöhl, P. A., Gelsleichter, J., Manire, C. A. and Shivji, M. S. (2004). Predominance of genetic monogamy by females in a hammerhead shark, *Sphyrna tiburo*: implications for shark conservation. *Molecular Ecology*, **13**, 1965–1974.

Clinchy, M., Taylor, A. C., Zanette, L. Y., Krebs, C. J. and Jarman, P. J. (2004). Body size, age and paternity in common brushtail possums (*Trichosurus vulpecula*). *Molecular Ecology*, **13**, 195–202.

Comstock, K. E., Ostrander, E. A. and Wasser, S. K. (2003). Amplifying nuclear and mitochondrial DNA from African elephant ivory: a tool for monitoring the ivory trade. *Conservation Biology*, **17**, 1–4.

Constable, J. L., Packer, C., Collins, D. A. and Pusey, A. E. (1995). Nuclear DNA from primate dung. *Nature*, **373**, 393.

Constable, J. L., Ashley, M. V., Goodall J. and Pusey, A. E. (2001). Noninvasive paternity assignment in Gombe chimpanzees. *Molecular Ecology*, **10**, 1279–1300.

Cornuet, J. M., Piry, S., Luikart, G., Estoup, A. and Solignac, M. (1999). New methods employing multilocus genotypes to select or exclude populations as origins of individuals. *Genetics*, **153**, 1989–2000.

Creel, S., Spong, G., Sands, J. L. *et al.* (2003). Population size estimation in Yellowstone wolves with error-prone noninvasive microsatellite genotypes. *Molecular Ecology*, **12**, 2003–2009.

Cunningham, J., Morgan-Davies, A. M. and O'Ryan, C. (2001). Counting rhinos from dung: estimating the number of animals in a reserve using microsatellite DNA. *South African Journal of Science*, **97**, 293–294.

Currat, M., Ray, N. and Excoffier, L. (2004). Splatche: a program to simulate genetic diversity taking into account environmental heterogeneity. *Molecular Ecology Notes* **4**, 139–142.

Dalebout, M. L., Lento, G. M., Cipriano, F., Funahashi, N. and Baker, C. S. (2002). How many protected minke whales are sold in Japan and Korea? A census by microsatellite profiling. *Animal Conservation*, **5**, 143–152.

Dalén L., Götherström, A. and Angerbjörn, A. (2004). Identifying species from pieces of faeces. *Conservation Genetics*, **5**, 109–111.

Dallas, J. F., Carss, D. N., Marshall, F. *et al.* (2000). Sex identification of the Eurasian otter *Lutra lutra* by PCR typing of spraints. *Conservation Genetics*, **1**, 181–183.

Dallas, J. F., Coxon, K. E., Sykes, T. *et al.* (2003). Similar estimates of population genetic composition and sex ratio derived from carcasses and faeces of Eurasian otter *Lutra lutra*. *Molecular Ecology*, **12**, 275–282.

Danzmann, R. G. (1997). Probmax: a computer program for assigning unknown parentage in pedigree analysis from known genotypic pools of parents and progeny. *Journal of Heredity*, **88**, 333.

Davison, A., Birks, J. D. S., Brookes, R. C., Braithwaite, T. C. and Messenger, J. E (2002). On the origin of faeces: morphological versus molecular methods for surveying rare carnivores from their scats. *Journal of Zoology* **257**, 141–143.

Deagle, B., Tollit, D. J., Jarman, S. N. *et al.* (2005). Molecular scatology as a tool to study diet: analysis of prey in scats from captive Steller sea lions. *Molecular Ecology*, **14**, 1831–1842.

Dean, F. B., Hosono, S., Fang, L. *et al.* (2002). Comprehensive human genome amplification using multiple displacement amplification. *Proceedings of the National Academy of Sciences USA*, **99**, 5261–5266.

Devlin, B., Roeder, K. and Ellstrand, N. C. (1988). Fractional paternity assignment: theoretical development and comparison to other methods. *Theoretical and Applied Genetics*, **76**, 369–380.

Dobey, S., Masters, D. V., Scheick, B. K. *et al.* (2005). Ecology of Florida black bears in the Okefenokee–Osceola ecosystem. *Wildlife Monographs*, **158**, 1–41.

Duchesne, P., Godbout, M.-H. and Bernatchez, L. (2002). PAPA (package for the analysis of parental allocation): a computer program for simulated and real parental allocation. *Molecular Ecology Notes*, 2, 191–193.

Duchesne, P., Castric, T. and Bernatchez, L. (2005). PASOS (parental allocation of singles in open systems): a computer program for individual parental allocation with missing parents. *Molecular Ecology Notes*, 5, 701–704.

Eggert, L. S., Rasner, C. A. and Woodruff, D. S. (2002). The evolution and phylogeography of the African elephant (*Loxodonta africana*) inferred from mitochondrial DNA sequence and nuclear microsatellite markers. *Proceedings of the Royal Society Series B*, 269, 1993–2006.

Eggert, L. S., Eggert, J. A. and Woodruff, D. S. (2003). Estimating population sizes for elusive animals: the forest elephants of Kakum National Park, Ghana. *Molecular Ecology*, 12, 1389–1402.

Eggert, L. S., Maldonado, J. E. and Fleischer, R. C. (2005). Nucleic acid isolation from ecological samples: animal scat and other associated materials. *Molecular Evolution: Producing the Biochemical Data*, Part B, *Methods in Enzymology*, 395, 73–87.

Eguchi, T. and Eguchi, Y. (2000). High yield DNA extraction from the snake cast-off skin or bird feathers using collagenase. *Biotechnology Letters*, 22, 1097–1100.

Eldridge, M. D. B., Kinnear, J. E. and Onus, M. L. (2001). Source population of dispersing rock wallabies (*Petrogale lateralis*) identified by assignment tests on multilocus genotypic data. *Molecular Ecology*, 10, 2867–2876.

Emery, A. M., Wilson, I. J., Craig, S., Boyle, P. R. and Noble, L. R. (2001). Assignment of paternity groups without access to parental genotypes: multiple mating and developmental plasticity in squid. *Molecular Ecology*, 10, 1265–1278.

Epps, C. W., Pasbøll, P. J., Wehausen, J. D. *et al.* (2005). Highways block gene flow and cause a rapid decline in genetic diversity of desert bighorn sheep. *Ecology Letters*, 8, 1029–1038.

Erler, A., Stoneking, M. and Kayser, M. (2004). Development of Y-chromosomal microsatellite markers for nonhuman primates. *Molecular Ecology*, 13, 2921–2930.

Ernest, H. B., Penedo, M. C. T., May, B. P., Syvanen, M. and Boyce, W. M (2000). Molecular tracking of mountain lions in the Yosemite Valley region in California: genetic analysis using microsatellites and faecal DNA. *Molecular Ecology*, 9, 433–441.

Esteban, J. A., Salas, M. and Blanco, L. (1993). Fidelity of phi29 DNA polymerase: comparison between protein-primed initiation and DNA polymerization. *Journal of Biological Chemistry*, 268, 2719–2726.

Fang, S. G. and Wan, Q. H. (2002). A genetic fingerprinting test for identifying carcasses of protected deer species in China. *Biological Conservation*, 103, 371–373.

Farrell, L. E., Roman, J. and Sunquist, M. E. (2000). Dietary separation of sympatric carnivores identified by molecular analysis of scats. *Molecular Ecology*, 9, 1583–1590.

Fedriani, J. M. and Kohn, M. H. (2001). Genotyping faeces links individuals to their diet. *Ecology Letters*, 4, 477–483.

Fernando, P. and Lande, R. (2000). Molecular genetic and behavioral analysis of social organization in the Asian elephant (*Elephas maximus*). *Behavioral Ecology and Sociobiology*, 48, 84–91.

Fernando, P. and Melnick, D. J. (2001). Molecular sexing eutherian mammals. *Molecular Ecology Notes*, 1, 350–353.

Fernando, P., Pfrender, M. E., Encalada, S. E. and Lande, R. (2000). Mitochondrial DNA variation, phylogeography and population structure of the Asian elephant. *Heredity*, **84**, 362–372.

Fernando, P., Vidya, T. N. C., Rajapakse, C., Dangolla, A. and Melnick, D. J. (2003). Reliable noninvasive genotyping: fantasy or reality? *Journal of Heredity*, **94**, 115–123.

Flagstad, Ø., Roed, K., Stacy, J. E. and Jakobsen, K. S. (1999). Reliable noninvasive genotyping based on excremental PCR of nuclear DNA purified with a magnetic bead protocol. *Molecular Ecology*, **8**, 879–883.

Flagstad, Ø., Hedmark, E. and Landa, A. *et al.* (2004). Colonization history and noninvasive monitoring of a reestablished wolverine population. *Conservation Biology*, **18**, 676–688.

Frantz, A. C., Pope, L. C., Carpenter, P. J. *et al.* (2003). Reliable microsatellite genotyping of the Eurasian badger (*Meles meles*). using faecal DNA. *Molecular Ecology*, **12**, 1649–1661.

Frantzen, M. A. J., Silk, J. B., Ferguson, J. W. H., Wayne, R. K. and Kohn, M. H. (1998). Empirical evaluation of preservation methods for faecal DNA. *Molecular Ecology*, **7**, 1423–1428.

Gagneux, P., Boesch, C. and Woodruff, D. S. (1997a). Furtive mating in female chimpanzees. *Nature*, **387**, 358–359.

Gagneux, P., Boesch, C. and Woodruff, D. S. (1997b). Microsatellite scoring errors associated with noninvasive genotyping based on nuclear DNA amplified from shed hair. *Molecular Ecology*, **6**, 861–868.

Gagneux, P., Boesch, C. and Woodruff, D. S. (1999). Female reproductive strategies, paternity and community structure in wild West African chimpanzees. *Animal Behaviour*, **57**, 19–32.

Gagneux, P., Gonder, M. K., Goldberg, T. L. and Morin, P. A. (2001). Gene flow in wild chimpanzee populations: what genetic data tell us about chimpanzee movement over space and time. *Philosophical Transactions of the Royal Society Series B*, **356**, 889–897.

Galan, M., Cosson, J. F., Aulagnier, S. *et al.* (2003). Cross-amplification tests of ungulate primers in roe deer (*Capreolus capreolus*) to develop a multiplex panel of 12 microsatellite loci. *Molecular Ecology Notes*, **3**, 142–146.

Galan, M., Baltzinger, C., Hewison, A. J. M. and Cosson, J. F. (2005). Distinguishing red and roe deer using DNA extracted from hair samples and the polymerase chain reaction (PCR) method. *Wildlife Society Bulletin*, **33**, 204–211.

Garnier, J. N., Bruford, M. W. and Goossens, B. (2001). Mating system and reproductive skew in the black rhinoceros. *Molecular Ecology*, **10**, 2031–2042.

Gerber, S., Chabrier, P. and Kremer, A. (2003). FAMOZ: a software for parentage analysis using dominant, codominant and uniparentally inherited markers. *Molecular Ecology Notes*, **3**, 479–481.

Gerloff, U., Schlötterer, C., Rassmann, K. *et al.* (1995). Amplification of hypervariable simple sequence repeats (microsatellites) from excremental DNA of wild living bonobos (*Pan paniscus*). *Molecular Ecology*, **4**, 515–518.

Gerloff, U., Hartung, B., Fruth, B., Hohmann, G. and Tautz, D. (1999). Intracommunity relationships, dispersal pattern and paternity success in a wild living community of bonobos (*Pan paniscus*) determined from DNA analysis of faecal samples. *Proceedings of the Royal Society Series B*, **266**, 1189–1195.

Goodnight, K. F. and Queller, D. C. (1999). Computer software for performing likelihood tests of pedigree relationship using genetic markers. *Molecular Ecology*, **8**, 1231–1234.

Goossens, B., Graziani, L., Waits, L. P. *et al.* (1998a). Extra-pair paternity in the monogamous Alpine marmot revealed by nuclear DNA microsatellite analysis. *Behavioral Ecology and Sociobiology*, **43**, 281–288.

Goossens, B., Waits, L. and Taberlet, P. (1998b). Plucked hair samples as a source of DNA: reliability of dinucleotide microsatellite genotyping. *Molecular Ecology*, **7**, 1237–1241.

Goossens, B., Chikhi, L., Utami, S. S., de Ruiter, J. and Bruford, M. W. (2000). A multi-sample, multi-extracts approach for microsatellite analysis of faecal samples in an arboreal ape. *Conservation Genetics*, **1**, 157–162.

Goossens, B., Funk, S. M., Vidal, C. *et al.* (2002). Measuring genetic diversity in translocation programmes: principles and application to a chimpanzee release project. *Animal Conservation*, **5**, 225–236.

Goossens, B., Chikhi, L., Jalil, M. F. *et al.* (2005). Patterns of genetic diversity and migration in increasingly fragmented and declining orang-utan (*Pongo pygmaeus*) populations from Sabah, Malaysia. *Molecular Ecology*, **14**, 441–456.

Griffiths, R. and Tiwari, B. (1993). Primers for the differential amplification of the sex-determining region Y gene in a range of mammal species. *Molecular Ecology*, **2**, 405–406.

Griffiths, R., Double, M. C., Orr, K. and Dawson, R. J. G. (1998). A DNA test to sex most birds. *Molecular Ecology*, **7**, 1071–1075.

Hansen, M. M. and Jacobsen, L. (1999). Identification of mustelid species: otter (*Lutra lutra*), American mink (*Mustela vison*) and polecat (*Mustela putorius*), by analysis of DNA from faecal samples. *Journal of the Zoological Society of London*, **247**, 177–181.

Hansson, B., Bensch, S. and Hasselquist, D. (2003). A new approach to study dispersal: immigration of novel alleles reveals female-biased dispersal in great reed warblers. *Molecular Ecology*, **12**, 631–637.

Harlin, A. D., Wursig, B., Baker, C. S. and Markowitz, T. M. (1999). Skin swabbing for genetic analysis: application to dusky dolphins (*Lagenorhynchus obscurus*). *Marine Mammal Science*, **15**, 409–425.

Hashimoto, C., Furuichi, T. and Takenaka, O. (1996). Matrilineal kin relationship and social behaviour of wild bonobos (*Pan paniscus*): sequencing the D-loop region of mitochondrial DNA. *Primates*, **37**, 305–318.

Hayakawa, S. and Takenaka, O. (1999). Urine as another potential source for template DNA in polymerase chain reaction (PCR). *American Journal of Primatology*, **48**, 299–304.

Hebert, P. D. N., Cywinska, A., Ball, S. L. and DeWaard, J. R. (2003). Biological identifications through DNA barcodes. *Proceedings of the Royal Society Series B*, **270**, 313–321.

Hebert, P. D. N., Stoeckle, M. Y., Zemlak, T. S. and Francis, C. M. (2004). Identification of birds through DNA barcodes. *PLoS Biology*, **2**, 1657–1663.

Hedmark, E., Flagstad, O., Segerstrom, P. *et al.* (2004). DNA-based individual and sex identification from wolverines (*Gulo gulo*) faeces and urine. *Conservation Genetics*, **5**, 405–410.

Hellborg, L. and Ellegren, H. (2003). Y chromosome conserved anchored tagged sequences (YCATS) for the analysis of mammalian male-specific DNA. *Molecular Ecology*, **12**, 283–291.

Henegariu, O., Heerema, N. A., Dlouhy, S. R., Vance, G. H. and Vogt, P. H. (1997). Multiplex PCR, critical parameters and step-by-step protocol. *Bio Techniques*, **23**, 504–511.

Higuchi, R., Von Beroldingen, C. H., Sensabaugh, G. F. and Erlich, H. A. (1988). DNA typing from single hairs. *Nature*, **332**, 543–546.

Hoelzel, A. R. (2001). Shark fishing in fin soup. *Conservation Genetics*, **2**, 69–72.

Hoffman, J. I. and Amos, W. (2005). Microsatellite genotyping errors: detection approaches, common sources and consequences for paternal exclusion. *Molecular Ecology*, **14**, 599–612.

Höss, M., Kohn, M., Pääbo, S., Knauer, F. and Schroder, W. (1992). Excrement analysis by PCR. *Nature*, **359**, 199.

Huber, S., Bruns, U. and Arnold, W. (2002). Sex determination of red deer using polymerase chain reaction of DNA from feces. *Wildlife Society Bulletin*, **30**, 208–212.

Hung, C. M., Li, S. H. and Lee, L. L. (2004). Faecal DNA typing to determine the abundance and spatial organization of otters (*Lutra lutra*) along two stream systems in Kinmen. *Animal Conservation*, **7**, 301–311.

Idaghdour, Y., Broderick, D. and Korrida, A. (2003). Faeces as a source of DNA for molecular studies in a threatened population of great bustards. *Conservation Genetics*, **4**, 789–792.

Immel, U. D., Hummel, S., Herrmann, B. (1999). DNA profiling of orangutan (*Pongo pygmaeus*) feces to prove descent and identity in wildlife animals. *Electrophoresis*, **20**, 1768–1770.

Jeffery, K. J., Abernethy, K. A., Tutin, C. E. G. and Bruford, M. W. (2007). Biological and environmental degradation of gorilla hair and microsatellite amplification success in hair extracted DNA. *Biological Journal of the Linnean Society*, **91**, 281–294.

Jensen, T., Pernasetti, F. M. and Durrant, B. (2003). Conditions for rapid sex determination in 47 avian species by PCR of genomic DNA from blood, shell-membrane blood vessels, and feathers. *Zoo Biology*, **22**, 561–571.

Jones, A. G. (2001). GERUD 1.0: a computer program for the reconstruction of parental genotypes from progeny arrays using multilocus DNA data. *Molecular Ecology Notes*, **1**, 215–218.

Jones, A. G. (2005). GERUD 2.0: a computer program for the reconstruction of parental genotypes from half-sib progeny arrays with known or unknown parents. *Molecular Ecology Notes*, **5**, 708–711.

Jones, A. G. and Ardren, W. R. (2003). Methods of parentage analysis in natural populations. *Molecular Ecology*, **12**, 2511–2523.

Knapp, L. A. (2005). Facts, faeces and setting standards for the study of MHC genes using noninvasive samples. *Molecular Ecology*, **14**, 1597–1599.

Kohn, M. K. and Wayne, R. K. (1997). Facts from feces revisited. *Trends in Ecology and Evolution*, **12**, 223–227.

Kohn, M., York, E. C., Kamradt, D. A., Haught, G., Sauvajot, R. M. and Wayne, R. K. (1999). Estimating population size by genotyping feces. *Proceedings of the Royal Society Series B*, **266**, 657–663.

Kraaijeveld-Smit, F. J. L., Lindenmayer, D. B. and Taylor, A. C. (2002). Dispersal patterns and population structure in a small marsupial, *Antechinus agilis*, from two forests analysed using microsatellite markers. *Australian Journal of Zoology*, **50**, 325–338.

Launhardt, K., Epplen, C., Epplen, J. T. and Winkler, P. (1998). Amplification of microsatellites adapted from human systems in faecal DNA of wild Hanuman langurs (*Presbytis entellus*). *Electrophoresis*, **19**, 1356–1361.

Launhardt, K., Borries, C., Hardt, C., Epplen, J. T. and Winkler, P. (2001). Paternity analysis of alternative male reproductive routes among the langurs (*Semnopithecus entellus*) of Ramnagar. *Animal Behaviour*, **61**, 53–64.

Li, M., Wei, F. W. and Goossens, B. *et al.* (2004). Mitochondrial phylogeography and subspecific variation in the red panda (*Ailurus fulgens*): implications for conservation. *Molecular Phylogenetics and Evolution*, **36**, 78–89.

Liedloff, A. (1999). Mantel, v. 2: Mantel nonparametric test calculator. http://www.sci.qut.edu.au/nrs/mantel.htm.

Linch, C. A., Smith, S. L. and Prahlow, J. A. (1998). Evaluation of the human hair root for DNA typing subsequent to microscopic comparison. *Journal of Forensic Sciences*, **43**, 305–314.

Livia, L., Antonella, P., Hovirag, L., Mauro, N. and Panara, F. A. (2006). Nondestructive, rapid, reliable and inexpensive method to sample, store and extract high-quality DNA from fish body mucus and buccal cells. *Molecular Ecology Notes*, **6**, 257–260.

Lopez-Giraldez, F., Gomez-Moliner, B. J., Marmi, J. and Domingo-Roura, X. (2005). Genetic distinction of American and European mink (*Mustela vison* and *M. lutreola*) and European polecat (*M. putorius*) hair samples by detection of a species-specific SINE and a RFLP assay. *Journal of Zoology*, **265**, 405–410.

Lorenzini, R., Posillico, M., Lovari, S. and Petrella, A. (2004). Non-invasive genotyping of the endangered Apennine brown bear: a case study not to let one's hair down. *Animal Conservation*, **7**, 199–209.

Lucchini, V., Fabbri, E., Marucco, F. *et al.* (2002). Noninvasive molecular tracking of colonizing wolf (*Canis lupus*) packs in the western Italian Alps. *Molecular Ecology*, **11**, 857–868.

Lukacs, P. M. and Burham, K. P. (2005a). Estimating population size from DNA-based closed capture–recapture data incorporating genotyping error. *Journal of Wildlife Management*, **69**, 396–403.

Lukacs, P. M. and Burhnam, K. P (2005b). Review of capture–recapture methods applicable to noninvasive genetic sampling. *Molecular Ecology*, **14**, 3909–3920.

Lukas, D., Bradley, C. J., Nsubuga, A. M. *et al.* (2004). Major histocompatibility complex and microsatellite variation in two populations of wild gorillas. *Molecular Ecology*, **13**, 3389–3402.

Makuwa, M., Souquière, S. and Telfer, P. *et al.* (2003). Occurrence of hepatitis viruses in wild-born non-human primates: a 3-year (1998–2001) epidemiological survey in Gabon. *Journal of Medical Primatology*, **32**, 307–314.

Manel, S., Berthier, P. and Luikart, G. (2002). Detecting wildlife poaching: identifying the origin of individuals with Bayesian assignment tests and multilocus genotypes. *Conservation Biology*, **16**, 650–659.

Marshall, T. C., Slate, J., Kruuk, L. E. B. and Pemberton, J. M. (1998). Statistical confidence for likelihood-based paternity inference in natural populations. *Molecular Ecology*, **7**, 639–655.

Matsubura, M., Basabose, A. K., Omari, I. *et al.* (2005). Species and sex identification of western lowland gorillas (*Gorilla gorilla gorilla*), eastern lowland gorillas (*Gorilla beringei graueri*) and humans. *Primates*, **46**, 199–202.

Matsuda, M. (2003). Sex determination in fish: lessons from the sex-determining gene of the teleost medaka, *Oryzias latipes*. *Development Growth and Differentiation*, **45**, 397–403.

Maudet, C., Miller, C., Bassano, B. *et al.* (2002). Microsatellite DNA and recent statistical methods in wildlife conservation management: applications in Alpine ibex [*Capra ibex (ibex)*]. *Molecular Ecology*, **11**, 421–436.

McKelvey, K. S. and Schwartz, M. K (2004a). Genetic errors associated with population estimation using non-invasive molecular tagging: problems and new solutions. *Journal of Wildlife Management*, **68**, 439–448.

McKelvey, K. S. and Schwartz, M. K. (2004b). Providing reliable and accurate capture–mark–recapture estimates in a cost-effective way. *Journal of Wildlife Management*, **68**, 453–456.

McKelvey, K. S. and Schwartz, M. K. (2005). Dropout: a program to identify problem loci and samples for non-invasive genetic samples in a capture–mark–recapture framework. *Molecular Ecology Notes*, **5**, 716–718.

Miller, S. D., Joyce, P. and Waits, L. P (2005). A new method for estimating the size of small populations from genetic mark–recapture data. *Molecular Ecology*, **14**, 1991–2005.

Milligan, B. G. (1998). Total DNA isolation. In *Molecular Genetic Analysis of Populations: A Practical Approach*, ed. A. R. Hoelzel. Oxford: IRL Press, pp. 50–52.

Mills, L. S., Citta, J. J., Lair, K., Schwartz, M. and Tallmon, D. (2000). Estimating animal abundance using non-invasive DNA sampling: promise and pitfalls. *Ecological Applications*, **10**, 283–294.

Mitani J. C., Merriwether D. A. and Zhang, C. B. (2000). Male affiliation, cooperation and kinship in wild chimpanzees. *Animal Behaviour*, **59**, 885–893.

Miyaki, C. Y., Griffiths, R., Orr, K. *et al.* (1998). Sex identification of parrots, toucans, and curassows by PCR: perspectives for wild and captive population studies. *Zoo Biology*, **17**, 415–423.

Möller, L. M. and Beheregaray, L. B. (2004). Genetic evidence for sex-biased dispersal in resident bottlenose dolphins (*Tursiops aduncus*). *Molecular Ecology*, **13**, 1607–1612.

Moore, M. K., Nemiss, J. A., Rice, S. M., Quattro, J. M. and Woodley, C. M. (2003). Use of restriction fragment length polymorphisms to identify sea turtle eggs and cooked meats to species. *Conservation Genetics*, **4**, 95–103.

Morin, P. A. and Woodruff, D. S. (1992). Paternity exclusion using multiple hypervariable microsatellite loci amplified from nuclear DNA of hair cells. In *Paternity in Primates: Genetic Tests and Theories*, ed. R. D. Martin, A. F. Dixson and E. J. Wickings. Basel: Karger, pp. 63–81.

Morin, P. A., Messier, J. and Woodruff, D. S. (1994a). DNA extraction, amplification, and direct sequencing from hornbill feathers. *Journal of the Science Society of Thailand*, **30**, 31–41.

Morin, P. A., Moore, J. J., Chakraborty, R. *et al.* (1994b). Kin selection, social structure, gene flow, and the evolution of chimpanzees. *Science*, **265**, 1193–1201.

Morin, P. A., Wallis, J., Moore, J. J. and Woodruff, D. S. (1994c). Paternity exclusion in a community of wild chimpanzees using hypervariable simple sequence repeats. *Molecular Ecology*, **3**, 469–478.

Morin, P. A., Chambers, K. E., Boesch, C. and Vigilant, L. (2001). Quantitative PCR analysis of DNA from noninvasive samples for accurate microsatellite genotyping of wild chimpanzes (*Pan troglodytes verus*). *Molecular Ecology*, **10**, 1835–1844.

Moritz, C. (1994). Applications of mitochondrial DNA analysis in conservation: a critical review. *Molecular Ecology*, **3**, 401–411.

Moritz, C. and Cicero, C. (2004). DNA barcoding: promise and pitfalls. *PLoS Biology*, **2**, 1529–1531.

Mowat, G. and Strobeck, C. (2000). Estimating population size of grizzly bears using hair capture, DNA profiling, and mark–recapture analysis. *Journal of Wildlife Management*, **64**, 184–193.

Murphy, M. A., Waits, L. P. and Kendall, K. C. (2000). Quantitative evaluation of fecal drying methods for brown bear DNA analysis. *Wildlife Society Bulletin*, **28**, 951–957.

Murphy, M. A., Waits, L. P., Kendall, K. C. *et al.* (2002). An evaluation of long-term preservation methods for brown bear (*Ursus arctos*) faecal DNA samples. *Conservation Genetics*, **3**, 435–440.

Murphy, M. A., Waits, L. P. and Kendall, K. C. (2003). The influence of diet on faecal DNA amplification and sex identification in brown bears (*Ursus arctos*). *Molecular Ecology*, **12**, 221–226.

Navidi, W., Arnheim, N. and Waterman, M. S. (1992). A multiple-tubes approach for accurate genotyping of very small DNA samples by using PCR: statistical considerations. *American Journal of Human Development*, **50**, 347–359.

Nielsen, E. E., Hansen, M. M. and Loeschcke, V. (1999). Analysis of DNA from old scale samples: technical aspects, applications and perspectives for conservation. *Hereditas*, **130**, 265–276.

Nota, Y. and Takenaka, O. (1999). DNA extraction from urine and sex identification in birds. *Molecular Ecology*, **8**, 1235–1238.

Nsubuga, A. M., Robbins, M. M., Roeder, A. D. *et al.* (2004). Factors affecting the amount of genomic DNA extracted from ape faeces and the identification of an improved storage method. *Molecular Ecology*, **13**, 2089–2094.

Nuechterlein, G. L. and Buitron, D. (2000). A field technique for extracting blood from live bird eggs for DNA analysis. *Waterbirds*, **23**, 121–124.

Oka, T. and Takenaka, O. (2001). Wild gibbons' parentage tested by non-invasive DNA sampling and PCR-amplified polymorphic microsatellites. *Primates*, **42**, 67–73.

Ortega, J., Franco, M. D., Adams, B. A., Ralls, K. and Maldonado, J. E. (2004). A reliable, non-invasive method for sex determination in the endangered San Joaquin kit fox (*Vulpes macrotis mutica*) and other canids. *Conservation Genetics*, **5**, 715–718.

Paetkau, D. (2003). An empirical exploration of data quality in DNA-based population inventories. *Molecular Ecology*, **12**, 1375–1387.

Paetkau, D. (2004). The optimal number of markers in genetic capture–mark–recapture studies. *Journal of Wildlife Management*, **68**, 449–452.

Palomares, F., Godoy, J. A., Piriz, A., O'Brien, S. J. and Johnson, W. E. (2002). Faecal genetic analysis to determine the presence and distribution of elusive carnivores: design and feasibility for the Iberian lynx. *Molecular Ecology*, **11**, 2171–2182.

Palsbøll, P., Allen, J., Brube, M. *et al.* (1997). Genetic tagging of humpback whales. *Nature*, **388**, 676–679.

Palumbi, S. R. and Cipriano, F. (1998). Species identification using genetic tools: the value of nuclear and mitochondrial gene sequences in whale conservation. *Journal of Heredity*, **89**, 459–464.

Pank, M., Stanhope, M., Natanson, L., Kohler, N. and Shivji, M. (2001). Rapid and simultaneous identification of body parts from the morphologically similar sharks *Carcharhinus obscurus* and *Carcharhinus plumbeus* (Carcharhinidae) using multiplex PCR. *Marine Biotechnology*, **3**, 231–240.

Parsons, K. M. (2001). Reliable microsatellite genotyping of dolphin DNA from faeces. *Molecular Ecology Notes*, **1**, 341–344.

Parsons, K. M., Dallas, J. F. and Claridge, D. E. *et al.* (1999). Amplifying dolphin mitochondrial DNA from faecal plumes. *Molecular Ecology*, **8**, 1753–1768.

Parsons, K. M., Piertney, S. B., Middlemas, S. J., Hammond, P. S. and Armstrong, J. D. (2005). DNA-based identification of salmonid prey species in seal faeces. *Journal of Zoology*, **266**, 275–281.

Paxinos, E., McIntosh, C., Ralls, K. and Fleischer, R. (1997). A noninvasive method for distinguishing among canid species: amplification and enzyme restriction of DNA from dung. *Molecular Ecology*, **6**, 483–486.

Pearce, J. M., Fields, R. L. and Scribner, K. T. (1997). Nest materials as a source of genetic data for avian ecological studies. *Journal of Field Ornithology*, **68**, 471–481.

Pearse, D. E., Eckerman, C. M., Janzen, F. J. and Avise, J. C. (2001). A genetic analogue of 'mark–recapture' methods for estimating population size: an approach based on molecular parentage assessments. *Molecular Ecology*, **10**, 2711–2718.

Pichler, F. B., Dalebout, M. L. and Baker, C. S. (2001). Nondestructive DNA extraction from sperm whale teeth and scrimshaw. *Molecular Ecology Notes*, **1**, 106–109.

Pidancier, N., Miquel, C. and Miaud, C. (2003). Buccal swabs as a non-destructive tissue sampling method for DNA analysis in amphibians. *Herpetological Journal*, **13**, 175–178.

Piggott, M. P. (2004). Effect of sample age and season of collection on the reliability of microsatellite genotyping of faecal DNA. *Wildlife Research*, **31**, 485–493.

Piggott, M. P. and Taylor, A. C. (2003a). Remote collection of animal DNA and its applications in conservation management and understanding the population biology of rare and cryptic species. *Wildlife Research*, **30**, 1–13.

Piggott, M. P. and Taylor, A. C. (2003b). Extensive evaluation of faecal preservation and DNA extraction methods in Australian native and introduced species. *Australian Journal of Zoology*, **51**, 341–355.

Piggott, M. P., Bellemain, E., Taberlet, P. and Taylor, A. (2004). A multiplex pre-amplification method that significantly improves microsatellite amplification and error rates for faecal DNA in limiting conditions. *Conservation Genetics*, **5**, 417–420.

Pilgrim, K. L., McKelvey, K. S., Riddle, A. E. and Schwartz, M. K. (2005). Felid sex identification based on noninvasive genetic samples. *Molecular Ecology Notes*, **5**, 60–61.

Pires, A. E. and Fernandes, M. L. (2003). Last lynxes in Portugal? Molecular approaches in a pre-extinction scenario. *Conservation Genetics*, **4**, 525–532.

Prendini, L. (2005). Identifying spiders through DNA barcodes. *Canadian Journal of Zoology*, **83**, 498–504.

Pritchard, J. K., Stephens, M. and Donnelly, P. (2000). Inference of population structure using multilocus genotype data. *Genetics*, **155**, 945–959.

Prugh, L. R., Ritland, C. E., Arthur, S. M. and Krebs, C. J. (2005). Monitoring coyote population dynamics by genotyping faeces. *Molecular Ecology*, **14**, 1585–1596.

Queller, D. C. and Goodnight, K. F. (1989). Estimating relatedness using genetic markers. *Evolution*, **43**, 258–275.

Randi, E. and Lucchini, V. (2002). Detecting rare introgression of domestic dog genes into wild wolf (*Canis lupus*) populations by Bayesian admixture analyses of microsatellite variation. *Conservation Genetics*, **3**, 31–45.

Reed, J. E., Baker, R. J., Ballard, W. B. and Kelly, B. T. (2004). Differentiating Mexican gray wolf and coyote scats using DNA analysis. *Wildlife Society Bulletin*, **32**, 685–692.

Reed, J. Z., Tollit, D. J., Thompson, P. M. and Amos, W. (1997). Molecular scatology: the use of molecular genetic analysis to assign species, sex and individual identity to seal faeces. *Molecular Ecology*, **6**, 225–234.

Roeder, A. D., Archer, F. I., Poinar, H. N. and Morin, P. A. (2004). A novel method for collection and preservation of faeces for genetic studies. *Molecular Ecology Notes*, **4**, 761–764.

Roeder, A. D., Jeffery, K. and Bruford, M. W. (2006). A universal microsatellite multiplex kit for genetic analysis of great apes. *Folia Primatologica*, **77**, 240–245.

Rönn, A.-C., Andrès, O. and Bruford, M. W. *et al.* (2006). Multiple displacement amplification for generating an unlimited source of DNA for genotyping in nonhuman primate species. *International Journal of Primatology*, **27**, 1145–1169.

Roon, D. A., Thomas, M. E., Kendall, K. C. and Waits, L. P. (2005a). Evaluating mixed samples as a source of error in non-invasive genetic studies using microsatellites. *Molecular Ecology*, **14**, 195–201.

Roon, D. A., Waits, L. P. and Kendall, K. C. (2005b). A simulation test of the effectiveness of several methods for error-checking non-invasive genetic data. *Animal Conservation*, **8**, 203–215.

Ross, K. G. (2001). Molecular ecology of social behaviour: analyses of breeding systems and genetic structure. *Molecular Ecology*, **10**, 265–284.

Rudnick, J. A., Katzner, T. E., Bragin, E. A., Rhodes, O. E. and Dewoody, J. A. (2005). Using naturally shed feathers for individual identification, genetic parentage analyses, and population monitoring in an endangered Eastern imperial eagle (*Aquila heliaca*) population from Kazakhstan. *Molecular Ecology*, **14**, 2959–2967.

Russello, M. A. and Amato, G. (2001). Application of a noninvasive, PCR-based test for sex identification in an endangered parrot, *Amazona guildingii*. *Zoo Biology*, **20**, 41–45.

Sacchi, P., Soglia, D., Maione, S. *et al.* (2004). A non-invasive test for sex identification in short-toed eagle (*Circaetus gallicus*). *Molecular and Cellular Probes*, **18**, 193–196.

Santiago, M. L., Bibollet-Ruche, F., Bailes, E. *et al.* (2003). Amplification of a complete simian immunodeficiency virus genome from fecal RNA of a wild chimpanzee. *Journal of Virology*, **77**, 2233–2242.

Schander, C. and Willassen, E. (2005). What can biological barcoding do for marine biology? *Marine Biology Research*, **1**, 79–83.

Schwartz, M. C., Tallmon, D. and Luikart, G. (1999). Using genetics to estimate the size of wild populations: many methods, much potential, uncertain utility. *Animal Conservation*, 2, 321–323.

Schwartz, M. K., Pilgrim, K. L., McKelvey, K. S. *et al.* (2004). Hybridization between Canada lynx and bobcats: genetic results and management implications. *Conservation Genetics*, 5, 349–355.

Seddon, J. M. (2005). Canid-specific primers for molecular sexing using tissue or non-invasive samples. *Conservation Genetics*, 6, 147–149.

Seddon, J. M., Parker, H. G., Ostrander, E. A. and Ellegren, H. (2005). SNPs in ecological and conservation studies: a test in the Scandinavian wolf population. *Molecular Ecology*, 14, 503–511.

Segelbacher, G. (2002). Noninvasive genetic analysis in birds: testing reliability of feather samples. *Molecular Ecology Notes*, 2, 367–369.

Segelbacher, G. and Steinbrück, G. (2001). Birds faeces for sex identification and microsatellite analysis. *Vogelwarte*, 41, 139–142.

Shaw, C. N., Wilson, P. J. and White, B. N. (2003). A reliable molecular method of gender determination for mammals. *Journal of Mammalogy*, 84, 123–128.

Shewale, J. G., Nasir, H., Schneida, E. *et al.* (2004). Y-chromosome STR system, Y-PLEX ™ 12, for forensic casework: development and validation. *Journal of Forensic Sciences*, 49, 1278–1290.

Shivji, M., Clarke, S., Pank, M. *et al.* (2002). Genetic identification of pelagic shark body parts for conservation and trade monitoring. *Conservation Biology*, 16, 1037–1047.

Shrimpton, J. M. and Heath, D. D. (2003). Census vs. effective population size in Chinook salmon: large- and small-scale environmental perturbation effects. *Molecular Ecology*, 12, 2571–2583.

Signorovitch, J. and Nielsen, R. (2002). PATRI: paternity inference using genetic data. *Bioinformatics*, 18, 341–342.

Slate, J., Marshall, T. and Pemberton, J. (2000). A retrospective assessment of the accuracy of the paternity inference program CERVUS. *Molecular Ecology*, 9, 801–808.

Sloane, M. A., Sunnucks, P., Alpers, D., Beheregaray, B. and Taylor, A. C. (2000). Highly reliable genetic identification of individual hairy-nosed wombats from single remotely collected hairs: a feasible censusing method. *Molecular Ecology*, 9, 1233–1240.

Smith, K. L., Alberts, S. C., Bayes, M. K. *et al.* (2000). Cross-species amplification, non-invasive genotyping, and non-Mendelian inheritance of human STRPs in savannah baboons. *American Journal of Primatology*, 51, 219–227.

Smulders, M. J. M., Snoek, L. B., Booy, G. and Vosman, B. (2003). Complete loss of MHC genetic diversity in the Common Hamster (*Cricetus cricetus*) population in The Netherlands: consequences for conservation strategies. *Conservation Genetics*, 4, 441–451.

Steinel, A., Munson, L., van Vuuren, M. and Truyen, U. (2000). Genetic characterization of feline parvovirus sequences from various carnivores. *Journal of General Virology*, 81, 345–350.

Stone, J. and Björklund, M. (2001). DELRIOUS: a computer program designed to analyse molecular marker data and calculate delta and relatedness estimates with confidence. *Molecular Ecology Notes*, 1, 209–212.

Storz, J. F., Ramakrishnan, U. and Alberts, S. C. (2002). Genetic effective size of a wild primate population: influence of current and historical demography. *Evolution*, **56**, 817–829.

Stow, A. J., Sunnucks, P., Briscoe, D. A. and Gardner, M. G. (2001). The impact of habitat fragmentation on dispersal of Cunningham's skink (*Egernia cunninghami*): evidence from allelic and genotypic analyses of microsatellites. *Molecular Ecology*, **10**, 867–878.

Sugiyama, Y., Kawamoto, S., Takenaka, O., Kumizaki, K. and Norikatsu, W. (1993). Paternity discrimination and inter-group relationships of chimpanzees at Bossou. *Primates*, **34**, 545–552.

Symondson, W. O. C. (2002). Molecular identification of prey in predator diets. *Molecular Ecology*, **11**, 627–641.

Taberlet, P. (1991). A single plucked feather as a source of DNA for bird genetics. *Auk*, **108**, 959–960.

Taberlet, P. and Bouvet, J. (1994). Mitochondrial DNA polymorphism, phylogeography, and conservation genetics of the brown bear (*Ursus arctos*) in Europe. *Proceedings of the Royal Society Series B*, **255**, 195–200.

Taberlet, P. and Luikart, G. (1999). Non-invasive genetic sampling and individual identification. *Biological Journal of the Linnean Society*, **68**, 41–55.

Taberlet, P., Griffin, S., Goossens, B. *et al.* (1996). Reliable genotyping of samples with very low DNA quantities using PCR. *Nucleic Acids Research*, **24**, 3189–3194.

Taberlet, P., Camarra, J.-J., Griffin, S. *et al.* (1997). Nonivasive genetic tracking of the endangered Pyrenean brown bear population. *Molecular Ecology*, **6**, 869–876.

Taberlet, P., Waits, L. P. and Luikart, G. (1999). Noninvasive genetic sampling: look before you leap. *Trends in Ecology and Evolution*, **14**, 323–327.

Takenaka, O., Takashi, H., Kawamoto, S., Arakawa, M. and Takenaka, A. (1993). Polymorphic microsatellite DNA amplification customised for chimpanzee paternity testing. *Primates*, **34**, 27–35.

Tallmon, D. A., Bellemain, E., Swenson, J. E. and Taberlet, P. (2004). Genetic monitoring of Scandinavian brown bear effective population size and immigration. *Journal of Wildlife Management*, **68**, 960–965.

Teletchea, F., Maudet, C. and Hänni, C. (2005). Food and forensic molecular identification: update and challenges. *Trends in Biotechnology*, **23**, 359–366.

Tikel, D., Blair, D. and Marsh, D. (1996). Marine mammal faeces as a source of DNA. *Molecular Ecology*, **5**, 456–457.

Triant, D. A., Pace, R. M. and Stine, M. (2004). Abundance, genetic diversity and conservation of Louisiana black bears (*Ursus americanus luteolus*) as detected through noninvasive sampling. *Conservation Genetics*, **5**, 647–659.

Utami, S. S., Goossens, B., Bruford, M. W., de Ruiter, J. and van Hooff, J. A. R. A. M. (2002). Male bimaturism and reproductive success in Sumatran orangutans. *Behavioral Ecology*, **13**, 643–652.

Valderrama, X., Karesh, W. B., Wildman, D. E. and Melnick, D.J. (1999). Noninvasive methods for collecting fresh hair tissue. *Molecular Ecology*, **8**, 1749–1752.

Valière, N. (2002). Gimlet: a computer program for analyzing genetic individual identification data. *Molecular Ecology Notes*, **2**, 377–379.

Valière, N. and Taberlet, P (2000). Urine collected in the field as a source of DNA for species and individual identification. *Molecular Ecology*, **9**, 2150–2152.

Valière, N., Berthier, P., Mouchiroud, D. and Pontier, D. (2002). GEMINI: software for testing the effects of genotyping errors and multitubes approach for individual identification. *Molecular Ecology Notes*, **2**, 83–86.

Valsecchi, E., Glockned-Ferrari, D., Ferrari, M. and Amos, W. (1998). Molecular analysis of the efficiency of sloughed skin sampling in whale population genetics. *Molecular Ecology*, **7**, 1419–1422.

Van Oosterhout, C., Hutchinson, W. F., Wills, D. P. M. and Shipley, P. (2004). MICRO-CHECKER: software for identifying and correcting genotyping errors in microsatellite data. *Molecular Ecology Notes*, **4**, 535–538.

Vege, S. and McCracken, G. F. (2001). Microsatellite genotypes of big brown bats (*Eptesicus fuscus*: Vespertilionidae, Chiroptera) obtained from their feces. *Acta Chiroptera*, **3**, 237–244.

Vidya, T. N. C., Kumar, V. R., Arivazhagan, C. and Sukumar, R. (2003). Application of molecular sexing to free-ranging Asian elephant (*Elephas maximus*) populations in southern India. *Current Science*, **85**, 1074–1077.

Vigilant, L. (1999). An evaluation of techniques for the extraction and amplification of DNA from naturally shed hairs. *Biological Chemistry*, **380**, 1329–1331.

Vigilant, L., Hofreiter, M., Siedel, H. and Boesch, C. (2001). Paternity and relatedness in wild chimpanzee communities. *Proceedings of the National Academy of Sciences USA*, **98**, 12890–12895.

Wagner, R. S., Miller, M. P., Crisafulli, C. M. and Haig, S. M. (2005). Geographic variation, genetic structure, and conservation unit designation in the Larch Mountain salamander (*Plethodon larselli*). *Canadian Journal of Zoology*, **83**, 396–406.

Waits, L. P. (2004). Using noninvasive genetic sampling to detect and estimate abundance of rare wildlife species. In *Sampling Rare or Elusive Species: Concepts, Designs, and Techniques for Estimating Population Parameters*, ed. W. L. Thompson. Washington, DC: Island Press, pp. 211–228.

Waits, L. P. and Leberg, P. L. (2000). Biases associated with population estimation using molecular tagging. *Animal Conservation*, **3**, 191–199.

Waits, L. P., Luikart, G. and Taberlet, P. (2001). Estimating the probability of identity among genotypes in natural populations: cautions and guidelines. *Molecular Ecology*, **10**, 249–256.

Wallin, J. M., Holt, C. L., Lazaruk, K. D., Nguyen, T. H. and Walsh, P. S. (2002). Constructing universal multiplex PCR systems for comparative genotyping. *Journal of Forensic Sciences*, **47**, 52–65.

Walsh, H. E. and Edwards, S. V. (2005). Conservation genetics and Pacific fisheries bycatch: mitochondrial differentiation and population assignment in black-footed albatrosses (*Phoebastria nigripes*). *Conservation Genetics*, **6**, 289–295.

Walsh, P. S., Metzger, D. A. and Higuchi, R. (1991). Chelex-100 as a medium for simple extraction of DNA for PCR-based typing from forensic material. *Biotechniques*, **10**, 506–513.

Wan, Q. H. and Fang, S. G. (2003). Application of species-specific polymerase chain reaction in the forensic identification of tiger species. *Forensic Science International*, **131**, 75–78.

Wan, Q. H., Fang, S. G., Chen, G. F. *et al.* (2003). Use of oligonucleotide fingerprinting and faecal DNA in identifying the distribution of the Chinese tiger (*Panthera tigris amyyensis* Hilzheimer). *Biodiversity and Conservation*, **12**, 1641–1648.

Wandeler, P., Smith, S., Morin, P. A., Pettifor, R. A. and Funk, S. M. (2003). Patterns of nuclear DNA degeneration over time: a case study in historic teeth samples. *Molecular Ecology*, **12**, 1087–1093.

Wasser, S. K., Shedlock, A. M., Comstock, K. *et al.* (2004). Assigning African elephant DNA to geographic region of origin: applications to the ivory trade. *Proceedings of the National Academy of Sciences USA*, **101**, 14847–14852.

Wayne, R. K. and Morin, P. A. (2004). Conservation genetics in the new molecular age. *Frontiers in Ecology and Environment*, **2**, 89–97.

Whittier, C. A., Horne, W., Slenning, B., Loomis, M. and Stoskopf, M. K. (2004). Comparison of storage methods for reverse-transcriptase PCR amplification of rotavirus RNA from gorilla (*Gorilla g. gorilla*) fecal samples. *Journal of Virological Methods*, **116**, 11–17.

Wilberg, M. J. and Dreher, B. P. (2004). GENECAP: a program for analysis of multilocus genotype data for non-invasive sampling and capture–recapture population estimation. *Molecular Ecology Notes*, **4**, 783–785.

Will, K. W. and Rubinoff, D. (2004). Myth of the molecule: DNA barcodes for species cannot replace morphology for identification and classification. *Cladistics*, **20**, 47–55.

Willis, P. M., Cresp, B. J., Dill, L. M., Baird, R. W. and Hanson, M. B. (2004). Natural hybridization between Dall's porpoises (*Phocoenoides dalli*) and harbour porpoises (*Phocoena phocoena*). *Canadian Journal of Zoology*, **82**, 828–834.

Wilson, G. J., Frantz, A. C., Pope, L. C. *et al.* (2003). Estimation of badger abundance using faecal DNA typing. *Journal of Applied Ecology*, **40**, 658–666.

Wolf, C., Hubner, P. and Luthy, J. (1999). Differentiation of sturgeon species by PCR-RFLP. *Food Research International*, **32**, 699–705.

Woodruff, D. S. (2003). Non-invasive genotyping and field studies of free-ranging non-human primates. In *Kinship and Behavior in Primates*, ed. B. Chapais and C. Berman. Oxford: Oxford University Press, pp. 46–68.

Woods, J. G., Paetkau, D., Lewis, D. *et al.* (1999). Genetic tagging of free-ranging black and brown bears. *Wildlife Society Bulletin*, **27**, 616–627.

Worthington Wilmer, J., Allen, P. J., Pomeroy, P. P., Twiss, S. D. and Amos, W. (1999). Where have all the fathers gone? An extensive microsatellite analysis of paternity in the grey seal (*Halichoerus grypus*). *Molecular Ecology*, **8**, 1417–1429.

Yamamoto, K., Tsubota, T., Komatsu, T. *et al.* (2002). Sex identification of Japanese black bear, *Ursus thibetanus japonicus*, by PCR based on amelogenin gene. *Journal of Veterinary Medical Science*, **64**, 505–508.

Yan, P., Wu, X. B., Shi, Y. *et al.* (2005). Identification of Chinese alligators (*Alligator sinensis*) meat by diagnostic PCR of the mitochondrial cytochrome *b* gene. *Biological Conservation*, **121**, 45–51.

Yasuda, T., Iida, R., Takeshita, H. *et al.* (2003). A simple method of DNA extraction and STR typing from urine samples using a commercially available DNA/RNA extraction kit. *Journal of Forensic Sciences*, **48**, 108–110.

Ydaghdour, Y., Broderick, D. and Korrida, A. (2003). Faeces as a source of DNA for molecular studies in a threatened population of great bustards. *Conservation Genetics*, **4**, 789–792.

Zhan, X., Li, M., Zhang, Z. *et al.* (2006). Molecular censusing increases key giant panda population estimate by more than 100%. *Current Biology*, **16**, R451-R452.

9

The role of ancient DNA in conservation biology

JON BEADELL, YVONNE CHAN AND ROBERT FLEISCHER

INTRODUCTION

A central goal of conservation is the maintenance of ecosystems, species or populations at their current state, or the restoration of biological systems to some former state. In cases of recent ecological collapse, such as the decline of a population due to the introduction of disease or an invasive competitor, or due to over-hunting or habitat destruction, we may have monitored the process from start to finish and the former state may be sufficiently well-described to give us a target for restoration. In most cases, though, serious monitoring only begins after a decline is identified, and only anecdotal evidence is available to guide our reconstruction of the past. In addition, the mechanisms that have driven changes in ecosystems are typically unknown. Are these changes the result of natural processes acting over many millennia, or has human activity drastically altered the natural trajectory? The recent application of genetics to conservation has allowed us to describe more fully the current status of populations by quantifying such properties as levels of inbreeding, effective population sizes, levels of genetic variation, and gene flow (Fleischer 1998; DeSalle and Amato 2004). Through the application of coalescent models, population genetics has also given us insight into the historical status of populations, whether such properties as size and growth of a population have changed and on what time scale these changes have occurred. Unfortunately, the stochastic nature of the coalescent process and the effects of selection often impair our ability to confidently reconstruct historical states. With the relatively recent development of ancient DNA (aDNA) techniques, however, we can now step directly backwards in time to characterize historical genetic diversity and to better understand the processes that have generated current levels of genetic diversity and population structure. Our ability to travel back in time using aDNA has allowed us to view conservation issues with a broader

Population Genetics for Animal Conservation, eds. G. Bertorelle, M. W. Bruford, H. C. Hauffe, A. Rizzoli and C. Vernesi. Published by Cambridge University Press. © Cambridge University Press 2009.

temporal perspective and has provided a better framework for understanding the impact of humans in shaping contemporary animal populations. The possibility of extracting information from aDNA was first recognized in 1984 when DNA was extracted from a 140-year old museum specimen of the quagga (*Equus quagga*), an extinct member of the horse family (Higuchi *et al.* 1984). Since then, interest has increased exponentially and the development of aDNA research has been reviewed by many authors (Wayne *et al.* 1999; Hofreiter *et al.* 2001b; Shapiro and Cooper 2003; Pääbo *et al.* 2004; van Tuinen *et al.* 2004; Willerslev and Cooper 2005). Few of these reviews focus on aDNA in conservation (but see Wayne and Morin 2004) and therefore, the purpose of this chapter will be to examine applications of aDNA that have immediate significance to conservation biology. Broadly, these applications fall into the following categories: (1) identifying evolutionarily significant units, (2) establishing former ranges, (3) systematics and forensics, (4) interpreting modern genetic variation, and (5) assessing effects of environmental change. As with any technique, unique laboratory and analytical challenges await those intending to use aDNA and therefore, we will briefly discuss these issues before delving into conservation applications.

For the purposes of this chapter, we will broadly define aDNA as DNA extracted from any non-living source. These sources include, but are not limited to, toe pads and teeth from museum specimens, permafrost bones, owl pellets, old blood smears, soil cores and coprolites (fossilized faeces). Although non-invasive sampling of relatively modern non-living sources of DNA such as excrement samples or hair has proven extremely useful in conservation research and requires many of the same rigorous laboratory controls, we will exclude this topic from the present discussion as it is described in more depth in Goossens and Bruford (this volume).

METHODOLOGICAL CHALLENGES

Laboratory

Unless DNA is preserved under conditions of rapid desiccation, freezing or high salt concentration (Lindahl 1993), it is subject to numerous and varied biochemical transformations. Therefore, extreme caution and special techniques are required when accessing the information it contains. Immediately upon cell death, the complex DNA-repair machinery that typically protects DNA begins to degrade, exposing DNA to immediate damage by endogenous nucleases and microbes as well as other longer-acting chemical processes such as oxidation, spontaneous hydrolysis and cross-linking (reviewed in Pääbo *et al.* 2004). As a result of these processes, aDNA studies are probably limited to samples that are no more than about 1 million years old (Willerslev and Cooper 2005), and the oldest authenticated amplifications are from material dating to 300–400 000 years before present (ybp) (Willerslev and Cooper 2005). Even in younger specimens, only small quantities of accessible DNA remain and this DNA tends to be short in length. To overcome this limitation, the vast majority of aDNA studies to date have targeted small fragments of the mitochondrial genome, which occurs in higher copy number relative to nuclear DNA. The future of aDNA, however, is hinted at by studies that have begun to target entire mitochondrial genomes (Krause *et al.* 2006) as well as neutral and phenotypically important nuclear loci (Bunce *et al.* 2003; Huynen *et al.* 2003 moa; Jaenicke-Després *et al.* 2003 maize) also on a genome-wide scale (Noonan *et al.* 2005; Noonan *et al.* 2006; Poinar *et al.* 2006).

The polymerase chain reaction (PCR) has proven an invaluable tool in aDNA work because of its ability to create millions of DNA copies from just a single template molecule. Because of this power, however, PCR creates further complications in the analysis of aDNA. First, DNA damage is not limited to fragmentation but can also include nucleotide modification, such as the hydrolytic deamination of bases (Hofreiter *et al.* 2001a). Subsequent amplification of these damaged templates by DNA polymerase during PCR can result in the insertion of incorrect bases (typically the substitution of A for G and T for C: Stiller *et al.* 2006). Furthermore, due to the low copy number of template molecules and variable damage present in a single aDNA sample, template switching during PCR can result in the production of chimeric sequences (Olson and Hassanin 2003), which could lead to artificial variation across a population. Finally, and perhaps most frustrating, PCR is extremely sensitive to the presence of contaminating DNA. Modern sources of DNA, including even extraneous DNA co-purified with commercial preparations of dNTPs (Leonard *et al.* 2006), can easily overwhelm ancient template in a PCR reaction even when primers are designed to be taxon-specific.

In order to overcome these problems, rigorous protocols have been developed to ensure that aDNA results are real (Cooper and Poinar 2000; Pääbo *et al.* 2004; Willerslev and Cooper 2005). Suggested rules include the use of dedicated DNA extraction laboratories that are physically isolated from PCR labs, cloning and sequencing of the products from multiple PCR amplifications, and replication by an independent lab. Only with these standards for authentication can we begin to interpret the results of aDNA. Maintaining these standards will be particularly important in the sphere of conservation where management decisions can depend on the outcome of aDNA studies.

Data analysis

As will be described below, simply providing a genetic characterization of an ancient specimen is often sufficient to inform conservation decisions, for example in expanding the known range of rare species or in controlling the trade of protected species. But questions such as 'Was genetic variation greater in the past than today?' and 'Did the population go through a bottleneck?' are common in conservation, and although aDNA can provide insight into this type of question, more complex analyses are required. Analysis of the data obtained from aDNA is not as straightforward as applying population genetic and phylogenetic analyses that were developed for a single time point to temporal data. Lumping samples from different time periods can artificially increase variation and introduce unknown biases. Furthermore, explicitly considering the temporal component of the sampling may increase the statistical power of the analyses (Drummond et al. 2003). This is particularly important for aDNA studies that are often limited by small sample sizes.

Fortunately, there have been recent advances in analysis techniques, such as the serial coalescent (Rodrigo and Felsenstein 1999), combined with Bayesian or likelihood methods, that provide tools for conservation biologists interested in reconstructing demographic history and testing hypotheses with aDNA. The coalescent is a stochastic process that can be used to model demographic history and provides a statistical framework for data analysis (Rosenberg and Nordborg 2002). The serial coalescent, developed by Rodrigo and Felsenstein (1999), expanded the coalescent to include samples taken from multiple time points. Methods to reconstruct historical demography include Bayesian Markov chain Monte Carlo (MCMC) methods that allow parameters such as historical population size and mutation rates to be inferred from constant population size or exponential growth models using the serial coalescent (e.g. BEAST: Drummond and Rambaut 2003). Parameter estimation from more complex models, such as identifying the timing and severity of a population bottleneck, can be addressed with approximate Bayesian computation (see Fig. 9.1) (Chan et al. 2006). If an a priori population model is unknown, Bayesian skyline plots can reconstruct complex demographic histories (Drummond et al. 2005). Hypotheses of drift or gene flow in response to climatic change can be tested with simulations using the computer package Serial SIMCOAL (Anderson et al. 2005; Ramakrishnan and Hadly unpublished data).

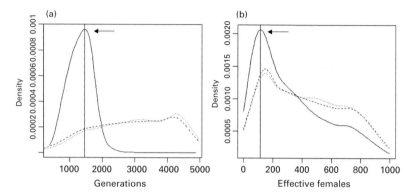

Figure 9.1. Posterior density curves indicating the timing (a) and severity (b) of a bottleneck in the South American rodent *Ctenomys sociabilis*, based on models incorporating data from 16 time points distributed over the last 10 000 years (solid line), modern data only (dashed line) and data from just two time points, modern and 10 000 ybp (dotted line). (Adapted from Chan *et al.* 2006.)

These techniques allow researchers to take full advantage of the unique information found in aDNA.

CONSERVATION APPLICATIONS

Evolutionarily significant units
The field of conservation genetics is frequently confronted with the apparently contradictory tasks of managing species to minimize inbreeding and at the same time, preserving populations that represent historically distinct gene pools (Haig 1998; Hedrick 2001; Hofreiter *et al.* 2004). Ancient DNA can alleviate the tension inherent to these tasks by evaluating the time scale on which populations have been distinct and thereby producing a valid benchmark for deciding whether to manage populations separately or as a single unit. In many cases, aDNA has revealed that modern population structure, which may define potential evolutionarily significant units (ESUs: Moritz 2002), has arisen only recently (Hofreiter *et al.* 2004). In a landmark study, Leonard *et al.* (2000), using DNA extracted from permafrost-preserved bones, showed that geographically disjunct populations of North American brown bear (*Ursus arctos*), which currently exhibit distinct mitochondrial haplotypes, probably derived from a single population as recently as 36 000 years ago. Modern population structure, therefore, probably reflects only founder events and the force of drift acting relatively recently on small populations that were isolated during the last glacial maximum.

A similar case is that of the threatened northeastern beach tiger beetle (*Cicindela dorsalis dorsalis*). This species, which was formerly widely distributed across the eastern seaboard of the United States, is now only known from a few populations. Under strict interpretation of the phylogenetic species concept (Nixon and Wheeler 1991), these populations could be classified as distinct species diagnosable by a single nucleotide change within the cytochrome oxidase III gene. However, aDNA analyses using museum specimens collected throughout the former range indicated that substantial and overlapping polymorphism existed in both populations prior to recent declines (Goldstein and DeSalle 2003). Thus, while current genetic polymorphism might have suggested the establishment of independent management units, aDNA revealed that this polymorphism is probably only an artifact of recent habitat fragmentation and human-mediated extinction of intermediate populations.

Delineation of management units may be complicated by recent habitat fragmentation and also by human activities that create zones of contact between previously allopatric groups (Perry *et al.* 2002; Ray *et al.* 2004). Evidence of hybridization can call into question whether previously separate populations or even species are truly distinct units. Here too, aDNA can shed light on historical levels of gene flow between populations prior to human disturbance. An excellent example of this is work performed by Wayne and colleagues (Wayne and Jenks 1991; Roy *et al.* 1996; reviewed in Roy *et al.* 1994). Coincident with a substantial decline in their numbers and changes to their habitat, evidence arose that red wolves (*Canis rufus*) were hybridizing with coyotes (*Canis latrans*), and by the mid-1900s only a single putatively pure red wolf population remained. Debate regarding the taxonomic status of the red wolf led to analysis of mtDNA and microsatellites from red wolf skins collected prior to the population decline and prior to potential human-mediated alteration of breeding behaviour. When compared to gray wolf and coyote genotypes, the historic red wolves exhibited no unique mtDNA haplotypes and genetic distances were smaller than would be expected if red wolves were a distinct canid species. This information, though somewhat controversial, will be invaluable in informing conservation decisions regarding the reintroduction of red wolf populations and the need to preserve extant 'pure' gene pools.

Establishing prior range

Information concerning the former range of an extirpated species, population or ESU is often important to justify reintroduction programmes, especially those carried out with species protected under the US Endangered Species

Act of 1973. Under this legislation, known prior range is usually sufficient justification for reintroduction, whereas introduction into areas not known to be prior range, even for rapid rescue or recovery of the species, requires additional permissions from the Department of the Interior and extensive environmental impact documentation under environmental quality regulations of the National Environmental Policy Act (NEPA Act of 1969). While collections of subfossil bones or museum skins may help to establish prior range, analysis of aDNA from these specimens can often resolve native range limits with more confidence and at a finer taxonomic scale by establishing genetic identity or similarity to the existing endangered population. In some cases, the taxon found in the prior range may prove to be substantially different, and prior range status can be muddled.

The first published use of aDNA to document prior range involved the highly endangered Laysan duck (*Anas laysanensis*), an endemic species currently restricted to tiny Laysan Island in the northwest Hawaiian Islands (Cooper *et al.* 1996; Rhymer 2001). The species nearly went extinct following devegetation of most of the island in the early 1900s, but recovered to a small, but mostly stable population numbering in the hundreds (Rhymer 2001). A related duck, the koloa (*Anas wyvilliana*), is known from all of the main Hawaiian Islands and cannot be clearly differentiated from the Laysan duck on the basis of skeletal morphology. To determine whether late Holocene subfossil bones recovered from lava tubes on the main island of Hawaii derived from koloa or Laysan duck, mtDNA control region sequences from the bones were compared to sequences from modern representatives. The sequences matched, nearly identically, those of the Laysan duck. These results, and the discovery of similar fossils found on other main and northwest Hawaiian Islands, indicate that the Laysan duck exhibited a substantially wider distribution prior to human impacts. Based on these findings, the US Fish and Wildlife Service began investigating potential sites for translocation and reintroduction, and established a new population on Midway Island (an island now free of rats) in 2004 (US Geological Survey 2005). This population began breeding in 2005, and additional sites for reintroduction are now under consideration.

Several other studies have harnessed the power of aDNA to assess prior range. Leonard *et al.* (2005) showed that DNA sequences from old wolf specimens in the southern part of their historical range clustered with the Mexican gray wolf clade, suggesting that this taxon be reintroduced to a wider geographic region. Glenn *et al.* (1999) used control region sequences to assess variation across the historical range of the endangered whooping crane (*Grus americana*), and found little evidence of geographic structuring, even between an extinct non-migratory Louisiana population and extant migratory ones. They felt this supported conservation efforts to recreate the resident Louisiana population. Similarly, Hofkin *et al.* (2003) were able to justify reintroduction of Galápagos land iguanas (*Conolophus subcristatus*) to Isla Baltra based on analyses of century-old museum specimens. Isla Baltra populations had gone extinct in the 1940s, but not prior to a small translocation to the nearby islet of Seymour Norte during the 1930s. Hofkin *et al.* sequenced a small section of the cytochrome *b* gene from 12 Isla Baltra museum specimens, and found that one haplotype that currently occurs on Seymour Norte was limited in distribution to that island and Isla Baltra. Thus, the authors recommended that individuals of this haplotype on Seymour Norte should be used for reintroduction to Isla Baltra.

In a twist to the theme above, sometimes aDNA can be used to identify whether populations currently occupying a locality are actually native to that region. An interesting study using aDNA and radiocarbon dating methods considered giant tortoises of the Seychelles and Aldabra (genus *Dipsochelys*; Karanth *et al.* 2005). Previous work had indicated that the modern populations of tortoises of these islands were genetically similar, and that the Seychelles populations may have been derived from early European introductions from Aldabra. Sequences obtained from two putative subfossil bones from the Seychelles matched those of modern Seychelles and Aldabra tortoises and, thus, provided no evidence to dismiss modern Seychelles tortoises as the product of historical translocations. However, as a cautionary tale highlighting the value of accurate dating of specimens used in aDNA studies, radiocarbon dating of these bones indicated that they actually postdated European colonization and were therefore not useful in answering the question of prior range. Determination of whether modern Seychelles tortoises are representative of the native lineage will have to await future genetic characterization of a 2000-year-old bone, which was recovered from the Seychelles and at least provides evidence that tortoises existed in the Seychelles prior to human colonization.

Systematics and forensics

Throughout the application of aDNA, researchers have targeted extinct but often charismatic taxa such as the marsupial wolf (*Thylacinus cynocephalus*; Thomas *et al.* 1989; Krajewski *et al.* 1997), dodo (*Raphus cucullatus*; Shapiro *et al.* 2002), Irish elk (*Megaloceros giganteus*; Lister *et al.* 2005) and sabretooth cats (Barnett *et al.* 2005) for inclusion in systematic studies. Although there would appear to be little value for conservation in determining the evolutionary relationships of already-extinct taxa, studies of this nature have provided a more accurate and thorough description of the biodiversity that has been lost. The inclusion of extinct species in phylogenetic reconstructions might also yield clues as to those groups or lineages that might be most susceptible to future extinction. Most importantly, though, these studies can sometimes provide insight into the still-salvageable processes (e.g. dispersal) that affect biodiversity. For example, while skeletal features of bones from an extinct Hawaiian eagle (*Haliaeetus* spp.) could not distinguish them from putative Old World and New World ancestors, phylogenetic analysis of aDNA indicated that the Hawaiian form most likely arose from an ancestor of the white-tailed eagle, which colonized Hawaii from the Old World (Fleischer *et al.* 2000). Minimal genetic divergence between the Hawaiian form and its Old World relative, combined with the absence of this species from older fossil deposits, suggested that this eagle had colonized the islands relatively recently. In this example, therefore, systematic placement of an extinct species was useful in identifying an ancient source of biodiversity and helping to calibrate the tempo of colonization.

A more detailed understanding of the processes giving rise to biodiversity (e.g. lineage diversification, morphological diversification and speciation) can be obtained when ancient sources of DNA are relatively abundant. Moas, which formed a diverse radiation of flightless New Zealand birds that went extinct approximately 100 years after human contact, are well represented in the subfossil record. Our understanding of the species diversity encompassed by this radiation was, however, hampered until recently by extreme variation in bone morphology. Analysis of single-locus sex-linked nuclear DNA from subfossil remains has helped to resolve taxonomic issues by revealing extreme size dimorphism between sexes that had previously been attributed to differences between species (Bunce *et al.* 2003; Huynen *et al.* 2003). Furthermore, by molecular dating of the splits between extinct taxa distributed across the North and South Islands, a clearer picture has emerged of the influence of geology, ecological specialization, and geographic fragmentation on speciation (Baker *et al.* 2005). This work and similar studies on the extinct giant goose (Paxinos *et al.* 2002b) and moa-nalos of Hawaii (Sorenson *et al.* 1999) can help us to understand the forces that generate and constrain species diversity and may allow us to refine our expectations for the levels of species diversity that can be expected to persist in the future.

Systematic studies using ancient specimens of extant species can also be valuable to conservation. Here, the use of ancient specimens allows for the study of organisms that would otherwise be inaccessible due to their rarity in the wild (Roosevelt's muntjac, Amato *et al.* 1999; large-billed reed

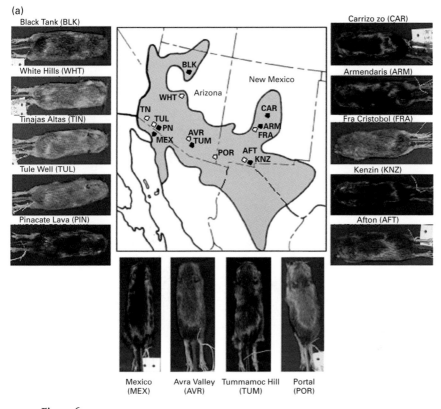

(a)

Black Tank (BLK)

White Hills (WHT)

Tinajas Altas (TIN)

Tule Well (TUL)

Pinacate Lava (PIN)

Carrizo zo (CAR)

Armendaris (ARM)

Fra Cristobol (FRA)

Kenzin (KNZ)

Afton (AFT)

New Mexico

Arizona

BLK

WHT

TN
TUL
PN
MEX
AVR
TUM
POR
AFT
KNZ
CAR
ARM
FRA

Mexico (MEX) Avra Valley (AVR) Tummamoc Hill (TUM) Portal (POR)

Figure 6.3a

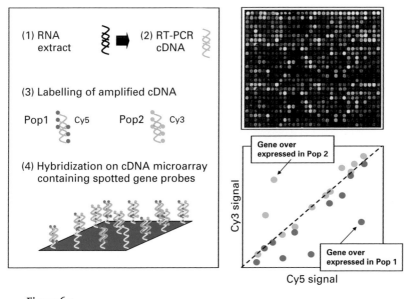

(1) RNA extract ➡ (2) RT-PCR cDNA

(3) Labelling of amplified cDNA

Pop1 Cy5   Pop2 Cy3

(4) Hybridization on cDNA microarray containing spotted gene probes

Gene over expressed in Pop 2

Gene over expressed in Pop 1

Cy3 signal

Cy5 signal

Figure 6.5

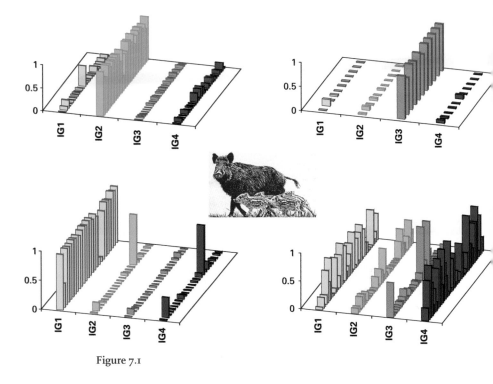

Figure 7.1

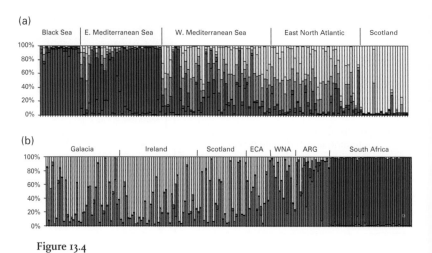

(a)

Black Sea | E. Mediterranean Sea | W. Mediterranean Sea | East North Atlantic | Scotland

(b)

Galacia | Ireland | Scotland | ECA | WNA | ARG | South Africa

Figure 13.4

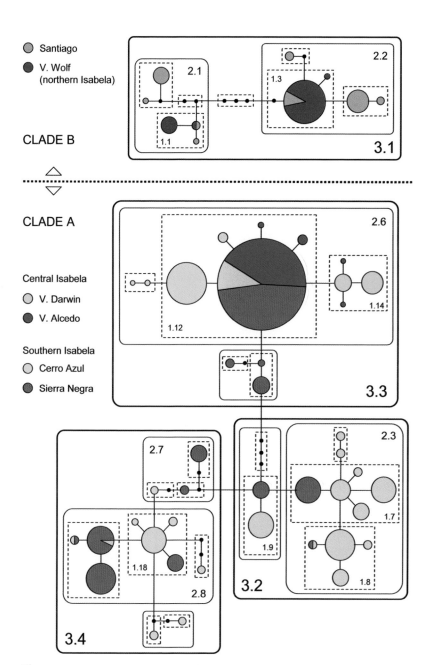

Figure 12.4

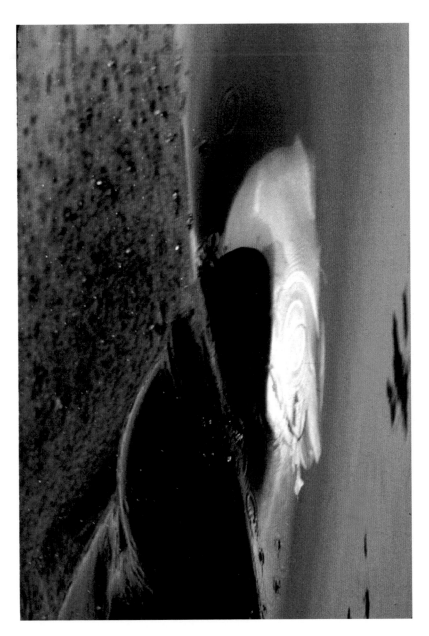

Figure 13.2

warbler (*Acrocephalus orinus*), Bensch *et al.* 2002), due to their localization in politically sensitive parts of the globe (Ethiopian wolf (*Canis simensis*), Roy *et al.* 1994) or, in the case of faked (snake-eating cow, Olson and Hassanin 2003) or mis-identified specimens (Mascarene starling, Olson *et al.* 2005), due to the fact that they never existed at all! Molecular confirmation of taxonomic affiliations may obviate the need for costly field operations and can help set conservation priorities. In addition, if there is value in conserving evolutionary history or, in other words, preserving lineages that represent long distinct branches on an evolutionary tree (Nee and May 1997; Moritz and Faith 1998; Faith *et al.* 2004), then aDNA may provide the only means of quantifying the phylogenetic uniqueness of rare or endangered species which would be hard to obtain from the wild (po'ouli (*Melamprosops phaeosoma*), Fleischer *et al.* 2001; Indian wolves (*Canis* spp.), Sharma *et al.* 2004; Yunnan box turtle (*Cuora yunnanensis*), Parham *et al.* 2004).

Perhaps the field of conservation in which systematic knowledge derived from aDNA has had the most immediate impact is that of wildlife forensics. The ability of law enforcement agencies to identify the specific taxonomic source of plant or wildlife products (e.g. caviar, DeSalle and Birstein 1996; pinniped penises, Malik *et al.* 1997; oak, Deguilloux *et al.* 2002; ivory, Comstock *et al.* 2003), which often bear little resemblance to the organism from which they were derived, has huge potential to discourage illegal hunting, trafficking in endangered species, and the sale of products made from protected species. Shark fins are a good example of wildlife products that are in high demand and that can derive from both legal and protected sources. Fins often lack taxonomically distinct characters and are traditionally sold based on categories unrelated to their taxonomic origin; thus, the species composition of the trade has gone largely unmonitored. Given that increasing demand for shark fins may be creating unsustainable fisheries and encouraging the take of protected species, knowledge of the particular species involved is essential. The use of aDNA obtained from fins (Clarke *et al.* 2006), as well as other shark products including cartilage pills and soup (Hoelzel 2001), should improve monitoring of the shark trade and has already led to reappraisal of the extent to which a protected species (great white shark, *Carcharodon carcharias*) has entered the fin trade (Shivji *et al.* 2005). Similar application of aDNA to taxonomic typing of commercially available whale meat in Japan and Korea has revealed the presence of several protected species (e.g. sei, humpback, fin and blue whales) and has called into question the effectiveness of international hunting moratoriums (Baker *et al.* 1996, 2000).

An emerging extension of forensics is the field of paleomicrobiology (Drancourt and Raoult 2005). Given the importance of infectious disease in driving population dynamics, the application of aDNA techniques to the identification of bacteria, parasites and viruses in old specimens will be a valuable tool for conservationists. With aDNA, it is now possible to diagnose the particular strain of infectious agent responsible for past declines, to assess both the geographical and temporal distribution of disease, and to even monitor the evolution of particular genes responsible for virulence. The vast majority of studies to date have targeted diseases important to the conservation of our own species (e.g. 1918 influenza, Reid *et al.* 1999; tuberculosis, Taylor *et al.* 2005), and much work remains to understand the historical role of disease in endangered species. Even in Hawaii, where avian pox and malaria are known to have contributed to the decline and extinction of much of the native avifauna (van Riper *et al.* 1986), studies of historical specimens are only now beginning to shed light on the temporal composition of vectors (Fonseca *et al.* 2006), timing of disease introduction (R.C. Fleischer *et al.* unpubl.) and the origin and diversity of the parasites responsible for the disease (Beadell *et al.* 2006). Disease information gleaned from aDNA presents an exciting means by which we can uncover the mechanism driving historical declines.

Genetic diversity in historical context

Loss of genetic diversity is of conservation concern because it may be associated with fitness costs due to increased inbreeding and reduced adaptability to such perturbations as disease outbreaks or rapid environmental change. When analyses of modern populations reveal reduced genetic variability, the natural questions that arise are (1) has the population experienced a bottleneck?, (2) how much variation was lost?, and (3) on what time scale was the variation lost? To answer the first two questions, many studies have compared genetic variation within the putatively bottlenecked population to conspecific populations or sister species known to have escaped the bottleneck (Packer *et al.* 1991; Hoelzel *et al.* 1993; Hedrick 1995; Schaeff *et al.* 1997). This comparison, of course, assumes that groups serving as surrogates for the pre-bottleneck population mirror the target population in all facets of their demographic and genetic history except the bottleneck. Studying DNA from the same population pre- and post-bottleneck, however, provides a more definitive means of discovering not only how much genetic variation was lost (if any), but also when the loss occurred.

Direct examination of the loss of genetic diversity in populations that have experienced well-documented bottlenecks has often confirmed the theoretical expectation of lost alleles, and in extreme cases, reduced heterozygosity (Wright 1931; Nei et al. 1975). Using microsatellites, Bouzat et al. (1998) were among the first to provide direct evidence for the loss of alleles in a study of greater prairie chickens (*Tympanuchus cupido*) in Illinois. Similar evidence has been found in bottlenecked whooping cranes (*Grus americana*; Glenn et al. 1999), northern elephant seals (*Mirounga angustirostris*; Weber et al. 2000; Hoelzel et al. 2002), sea otters (*Enhydra lutris*; Larson et al. 2002) and bearded vultures (*Gypaetus barbatus*; Godoy et al. 2004). Importantly, in the case of elephant seals, the loss of genetic diversity due to over-hunting was linked to a decrease in fitness (Hoelzel et al. 2002). This was manifested by a post-bottleneck increase in fluctuating asymmetry, which may be indicative of developmental instability.

In an increasing number of cases, however, changes in the genetic diversity of organisms over the last several hundred years have not met the predictions of a recent population bottleneck and have prompted a reexamination of assumptions regarding the extent and timing of human impact, and the influence of broad-scale environmental change (e.g. Nielsen et al. 1999). In the case of the European otter (*Lutra lutra*), comparison of heterozygosity and allelic diversity in microsatellites among populations collected from the 1880s to present provided little evidence of a recent bottleneck (Pertoldi et al. 2001). Interestingly, though, the distribution of microsatellite alleles did suggest that the population had begun a substantial decline approximately 2000 years ago, possibly in conjunction with environmental changes wrought by early human inhabitants. Similarly, examination of museum specimens of northern right whales (*Eubalaena glacialis*) indicated that an extreme bottleneck had not occurred within the last 150 years but that low modern genetic variation can probably be attributed to intensive hunting efforts which began several hundred years earlier (Rosenbaum et al. 2000).

Another interesting case is that of the nene (*Branta sandvicensis*), an endemic Hawaiian goose and a close relative of the Canada goose (*Branta canadensis*) that was recently limited to the island of Hawaii, but prehistorically occurred on most of the main Hawaiian Islands. The nene population on Hawaii suffered a major decline during the nineteenth and early twentieth centuries, and may have reached a low of 30 or fewer individuals by the 1950s (Paxinos et al. 2002a). Genetic variation in modern nene populations, as would be predicted from this documented bottleneck, is extremely low (Rave et al. 1994), and only a single haplotype was found among mtDNA control region sequences from 26 modern birds, broadly sampled (Paxinos *et al.* 2002a). Oddly though, little or no mitochondrial DNA sequence variation was evident in 14 museum specimens collected between 1833 and 1928, nor in 16 subfossil bones found in archaeological middens radiocarbon dated to 160 to 500 ybp. Although these data alone would support the view that island taxa are naturally genetically depauperate, surprisingly, analysis of 14 older bones dating from 850 to 2540 years old revealed six additional mtDNA haplotypes as well as significantly higher mtDNA sequence variation (see Fig. 9.2) typical of continental geese. Assuming this temporal pattern in genetic variation was not the result of a selective sweep (which would have required a selection coefficient of at

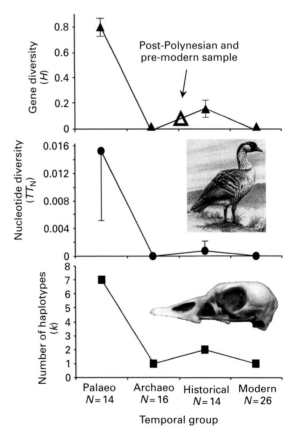

Figure 9.2. Measures of genetic diversity in the Hawaiian nene (pictured) sampled over four time periods suggest a severe bottleneck between 500 and 850 ybp. (Data from Paxinos *et al.* 2002a.)

least 0.1), it appears that a second severe population bottleneck (simulations suggest a decline to 20 to 270 individuals; Paxinos *et al.* 2002a) occurred 500–850 years before recent declines, coincident with the period of greatest anthropogenic extinction of large, flightless birds in Hawaii (Olson and James 1991). Whether the subsequent recovery of nene populations can be attributed to semi-domestication by Hawaiians or imposition of a kapu (taboo; Kirch 1985), these results suggest that loss of genetic variation and serial demographic bottlenecks are not always fatal and that low genetic diversity is not necessarily characteristic of endemic island fauna. Moreover, the results of this study, as with those described before, highlight the utility of aDNA in pinpointing the timing of historical bottlenecks, and by extension, in circumscribing the role that humans have played in wildlife declines.

Response to long-term environmental change

Throughout the Earth's history, plants and animals have been forced to contend with dramatic environmental change. Future changes associated with a predicted global temperature increase of 1.4 to 5 °C by the end of the twenty-first century (Houghton *et al.* 2001), combined with extensive human-mediated habitat fragmentation and destruction, necessitate an understanding of how species respond to environmental change across broad temporal scales, so that we may make informed predictions about future response (Jablonski and Sepkoski 1996). Species respond to environmental change in four main ways: range alteration, phenotypic and genetic adaptation, demographic fluctuation, and if the change is too severe, extinction (Root *et al.* 2003). How a particular species will respond may be a result of traits such as dispersal ability and phenotypic plasticity. Studies using aDNA offer a unique window into the changes that species have undergone in response to past environmental change. For example, amplification of material from coprolites (Poinar *et al.* 1998; Hofreiter *et al.* 2003) and soil cores (Willerslev *et al.* 2003) provides a direct means of monitoring ancient diets in the context of the flora and fauna available at different climatic periods. This knowledge could be important to the conservation of modern species if it can be ascertained that the historical diet of a species, which corresponded to its success, is no longer available in that species' current range. In addition, aDNA provides a way to directly observe genetic changes (and implied changes in effective population size) in response to past environmental perturbations (Orlando *et al.* 2002; Pergams *et al.* 2003; Shepherd *et al.* 2005) and, coupled with a history of demographic and morphological changes independently reconstructed from the fossil record, can provide a more integrated understanding of how species adapt, migrate and persist in response to environmental change.

Hadly and colleagues have provided several nice examples of this type of work. One study compared population histories of two rodent species with different levels of population density, dispersal and substructure in order to understand their individualistic response to climatic change. DNA was recovered from montane voles (*Microtus montanus*) and pocket gophers (*Thomomys talpoides*) from a single locality that traced the mammalian community over 3000 years. Both species prefer wet habitats and a palaeontological study found a 40–50% decline in relative abundance during the medieval warm period (1150–650 ybp). In the case of the pocket gopher, a subterranean rodent with low dispersal ability, its population remained closed to immigration for 3000 years despite a decline in relative abundance and significant climatic change (Hadly *et al.* 1998). During this period, however, the population changed developmentally, as animals became smaller with warmer climates (Hadly *et al.* 1998). In contrast, the montane vole, a high-dispersal species, showed an increase in gene flow during a time of environmental change (Hadly *et al.* 2004).

In a second study, genetic surveys of modern populations of an endemic threatened South American tuco-tuco (*Ctenomys sociabilis*) indicated that this species had experienced a relatively recent bottleneck. However, analysis of aDNA extracted from teeth found in several strata of a late-Holocene raptor roost revealed that this species had persisted for at least 1000 years with little to no genetic variation, as evidenced by mitochondrial haplotype diversity (Hadly *et al.* 2003). Examination of fossils dating back to 10 000 ybp indicated that the major decline in population size probably occurred from 8200 to 3000 ybp as grassland habitat was declining and competition with a congener was increasing, though eruption of a nearby volcano may have also contributed to the population decline (Chan *et al.* 2005). In this case, low modern genetic diversity, which might be viewed as a cause for conservation concern, may be better interpreted in the historical context of long-term environmentally driven change provided by aDNA.

Recently, aDNA obtained from large Beringian mammals has opened a window on numerous natural experiments associated with the environmental changes of the late Pleistocene and has forced us to reinterpret the relative importance of major anthropological, geological, and climatological influences on population histories. One of the most debated topics is the role of early humans in the extinction of North

American megafauna at the end of the Pleistocene. Several studies using the vast warehouse of permafrost-preserved bones (Shapiro and Cooper 2003) have now uncovered genetic evidence of declines in large mammals that predate these extinctions. Analyses of ancient brown bear (*Ursus arctos*) specimens sampled over approximately 50 000 years suggest that genetic diversity had already declined by 15 000 ybp (Leonard *et al.* 2000) and that the most dramatic phylogeographic changes had occurred between 35 000 and 21 000 ybp, possibly due to competition from the short-faced bear (*Arctodus simus*; Barnes *et al.* 2002). Similarly, Shapiro *et al.* (2004) found genetic evidence for a population decline in Beringian bison (*Bison* cf. *priscus*) beginning around 37 000 ybp, substantially predating the presence of large human populations in North America. This decline also predates the last glacial maximum and therefore may be attributable to more subtle ecological changes associated with increased tree cover and increasingly cold and arid climate. In this case, however, humans have not been completely absolved as further analysis of the same data set has suggested that the decline in population became even more precipitous following human colonization (Drummond *et al.* 2005).

CONCLUSIONS

As evidenced by the studies described above, aDNA analysis has exposed a richness of detail that would be difficult to reconstruct from modern DNA data alone. Most notably, aDNA has revealed that the population size and substructure of species is extremely mutable over large time-frames and that modern genetic diversity may shroud more complex population histories. In an age of increasing habitat fragmentation, aDNA will be particularly useful in assessing historical linkage among populations of conservation concern. In addition, studies of aDNA can help us to calibrate modern observed levels of genetic diversity by revealing the genetic diversity with which species have persisted for hundreds or thousands of years. Finally, aDNA studies can help to link changes in genetic variation and changes in population size to the ecological conditions and environmental perturbations that lead to the success or decline of species. Given the level of resolution with which we can now reconstruct historical population parameters and species assemblages, access to aDNA may have made the task of conservation more difficult by widening the time frame on which we can consider the critical question: to which historical state do we target our conservation efforts?

218 | Jon Beadell *et al.*

REFERENCES

Amato, G., Egan, M. G., Schaller, G. B. *et al.*. (1999). Rediscovery of Roosevelt's barking deer (*Muntiacus rooseveltorum*). *Journal of Mammalogy*, **80**, 639–643.

Anderson, C. N. K., Ramakrishnan, U., Chan, Y. L. and Hadly, E. A. (2005). Serial SimCoal: a population genetic model for data from multiple populations and points in time. *Bioinformatics*, **21**, 1733–1734.

Baker, A. J., Huynen, L. J., Haddrath, O., Millar, C. D. and Lambert, D. M. (2005). Reconstructing the tempo and mode of evolution in an extinct clade of birds with ancient DNA: the giant moas of New Zealand. *Proceedings of the National Academy of Sciences USA*, **102**, 8257–8262.

Baker, C. S., Cipriano, F. and Palumbi, S. R. (1996). Molecular genetic identification of whale and dolphin products from commercial markets in Korea and Japan. *Molecular Ecology*, **5**, 671–685.

Baker, C. S., Lento, G. M., Cipriano, F. and Palumbi, S. R. (2000). Predicted decline of protected whales based on molecular genetic monitoring of Japanese and Korean markets. *Proceedings of the Royal Society Series B*, **267**, 1191–1199.

Barnes, I., Matheus, P., Shapiro, B., Jensen, D. and Cooper, A. (2002). Dynamics of Pleistocene population extinctions in Beringian brown bears. *Science*, **295**, 2267–2270.

Barnett, R., Barnes, I., Phillips, M. J. *et al.* (2005). Evolution of the extinct Sabretooths and the American cheetah-like cat. *Current Biology*, **15**, R589–R590.

Beadell, J. S., Ishtiaq, F., Covas, R. *et al.* (2006). Global phylogeographic limits of Hawaii's avian malaria. *Proceedings of the Royal Society Series B*, **273**, 2935–2944.

Bensch, S. and Pearson, D. (2002). The large-billed reed warbler *Acrocephalus orinus* revisited. *Ibis*, **144**, 259–267.

Bouzat, J. L., Lewin, H. A. and Paige, K. N. (1998). The ghost of genetic diversity past: historical DNA analysis of the greater prairie chicken. *American Naturalist*, **152**, 1–6.

Bunce, M., Worthy, T. H., Ford, T., *et al.* (2003). Extreme reversed sexual size dimorphism in the extinct New Zealand moa *Dinornis*. *Nature*, **425**, 172–175.

Chan, Y. L., Lacey, E. A., Pearson, O. P. and Hadly, E. A. (2005). Ancient DNA reveals Holocene loss of genetic diversity in a South American rodent. *Biology Letters*, **1**, 423–426.

Chan, Y. L., Anderson, C. N. K. and Hadly, E. A. (2006). Bayesian estimation of the timing and severity of a population bottleneck from ancient DNA. *PLoS Genetics*, **2**, e59.

Clarke, S. C., Magnussen, J. E., Abercrombie, D. L., McAllister, M. K. and Shivji, M. S. (2006). Identification of shark species composition and proportion in the Hong Kong shark fin market based on molecular genetics and trade records. *Conservation Biology*, **20**, 201–211.

Comstock, K. E., Ostrander, E. A. and Wasser, S. K. (2003). Amplifying nuclear and mitochondrial DNA from African elephant ivory: a tool for monitoring the ivory trade. *Conservation Biology*, **17**, 1840–1843.

Cooper, A. and Poinar, H. (2000). Ancient DNA: do it right or not at all. *Science*, **289**, 1139.

Cooper, A., Rhymer, J., James, H. F. *et al.* (1996). Ancient DNA and island endemics. *Nature*, **381**, 484.

Deguilloux, M. F., Pemonge, M. H. and Petit, R. J. (2002). Novel perspectives in wood certification and forensics: dry wood as a source of DNA. *Proceedings of the Royal Society Series B*, **269**, 1039–1046.

DeSalle, R. and Amato, G. (2004). The expansion of conservation genetics. *Nature Reviews of Genetics*, **5**, 702–712.

DeSalle, R. and Birstein, V. J. (1996). PCR identification of black caviar. *Nature*, **381**, 197–198.

Drancourt, M. and Raoult, D. (2005). Palaeomicrobiology: current issues and perspectives. *Nature Reviews*, **3**, 23–35.

Drummond, A. J. and Rambaut, A. (2003). BEAST VI.O. Available from http://evolve. zoo.ox.ac.uk/beast/.

Drummond, A. J., Pybus, O. G., Rambaut, A., Forsberg, R. and Rodrigo, A. G. (2003). Measurably evolving populations. *Trends in Ecology and Evolution*, **18**, 481–488.

Drummond, A. J., Rambaut, A., Shapiro, B. and Pybus, O. G. (2005). Bayesian coalescent inference of past population dynamics from molecular sequences. *Molecular Biology and Evolution*, **22**, 1185–1192.

Faith, M. S., Reid, C. A. M. and Hunter, J. (2004). Integrating phylogenetic diversity, complementarity, and endemism for conservation assessment. *Conservation Biology*, **18**, 255–261.

Fleischer, R. C. (1998). Genetics and avian conservation. In *Avian Conservation: Research and Management*, ed. J. Marzluff and R. Sallabanks. Washington, DC: Island Press, pp. 29–48.

Fleischer, R. C., Olson, S. L., James, H. E. and Cooper, A. C. (2000). Identification of the extinct Hawaiian eagle (*Haliaeetus*) by mtDNA sequence analysis. *Auk*, **117**, 1051–1056.

Fleischer, R. C., Tarr, C. L., James, H. E., Slikas, B. and McIntosh, C. E. (2001). Phylogenetic placement of the po'ouli, *Melamprosops phaeosoma*, based on mitochondrial DNA sequence and osteological characters. *Studies in Avian Biology*, **22**, 98–103.

Fonseca, D. M., Smith, J. L., Wilkerson, R. C. and Fleischer, R. C. (2006). Pathways of expansion and multiple introductions illustrated by large genetic differentiation among worldwide populations of the southern house mosquito. *American Journal of Tropical Medicine and Hygiene*, **74**, 284–289.

Glenn, T. C., Wolfgang, S. and Braun, M. J. (1999). Effects of a population bottleneck on whooping crane mitochondrial DNA variation. *Conservation Biology*, **13**, 1097–1107.

Godoy, J. A., Negro, J. J., Hiraldo, F. and Donázar, J. A. (2004). Phylogeography, genetic structure and diversity in the endangered bearded vulture (*Gypaetus barbatus*, L.) as revealed by mitochondrial DNA. *Molecular Ecology*, **13**, 371–390.

Goldstein, P. Z. and DeSalle, R. (2003). Calibrating phylogenetic species formation in a threatened insect using DNA from historical specimens. *Molecular Ecology*, **12**, 1993–1998.

Hadly, E. A., Kohn, M. H., Leonard, J. A. and Wayne, R. K. (1998). A genetic record of population isolation in pocket gophers during Holocene climatic change. *Proceedings of the National Academy of Sciences USA*, **95**, 6893–6896.

Hadly, E. A., van Tuinen, M., Chan, Y. and Heiman, K. (2003). Ancient DNA evidence of prolonged population persistence with negligible genetic diversity in an endemic tuco-tuco (*Ctenomys sociabilis*). *Journal of Mammalogy*, **84**, 403–417.

Hadly, E. A., Ramakrishnan, U., Chan, Y. L. *et al.* (2004). Genetic response to climatic change: insights from ancient DNA and phylochronology. *PLoS Biology*, **2**, 1600–1609.

Haig, S. M. (1998). Molecular contributions to conservation. *Ecology*, **79**, 413–425.

Hedrick, P. W. (1995). Elephant seals and the estimation of a population bottleneck. *Journal of Heredity*, **86**, 232–235.

Hedrick, P. W. (2001). Conservation genetics: where are we now? *Trends in Ecology and Evolution*, **16**, 629–636.

Higuchi, R., Bowman, B., Freiberger, M., Ryder, O. A. and Wilson, A. C. (1984). DNA sequences from the quagga, an extinct member of the horse family. *Nature*, **312**, 282–284.

Hoelzel, A. R. (2001). Shark fishing in fin soup. *Conservation Genetics*, **2**, 69–72.

Hoelzel, A. R., Halley, J., O'Brien, S. J. *et al.* (1993). Elephant seal genetic variation and the use of simulation-models to investigate historical population bottlenecks. *Journal of Heredity*, **84**, 443–449.

Hoelzel, A. R., Fleischer, R. C., Campagna, C., le Boeuf, B. J. and Alvord, G. (2002). Impact of a population bottleneck on symmetry and genetic diversity in the northern elephant seal. *Journal of Evolutionary Biology*, **15**, 567–575.

Hofkin, B. V., Wright, A., Altenbach, J. *et al.* (2003). Ancient DNA gives green light to Galápagos Land Iguana repatriation. *Conservation Genetics*, **4**, 105–108.

Hofreiter, M., Jaenicke, V., Serre, D., von Haeseler, A. and Pääbo, S. (2001a). DNA sequences from multiple amplifications reveal artifacts induced by cytosine deamination in ancient DNA. *Nucleic Acids Research*, **29**, 4793–4799.

Hofreiter, M., Serre, D., Poinar, H. N., Kuch, M. and Pääbo, S. (2001b). Ancient DNA. *Nature Genetics*, **2**, 353–359.

Hofreiter, M., Betancourt, J. L., Sbriller, A. P., Markgraf, V. and McDonald, H. G. (2003). Phylogeny, diet, and habitat of an extinct ground sloth from Cuchillo Curá, Neuquén Province, southwest Argentina. *Quarternary Research*, **59**, 364–378.

Hofreiter, M., Serre, D., Rohland, N. *et al.* (2004). Lack of phylogeography in European mammals before the last glaciation. *Proceedings of the National Academy of Sciences USA*, **101**, 12963–12968.

Houghton, J. T., Ding, Y., Griggs, D. J. *et al.* (2001). *Climate Change 2001: The Scientific Basis*, Intergovernmental Panel on Climate Change. Cambridge: Cambridge University Press.

Huynen, L., Millar, C. D., Scofield, R. P. and Lambert, D. M. (2003). Nuclear DNA sequences detect species limits in ancient moa. *Nature*, **425**, 175–178.

Jablonski, D. and Sepkoski, J. J. Jr (1996). Paleobiology, community ecology, and scales of ecological pattern. *Ecology*, **77**, 1367–1378.

Jaenicke-Després, V., Buckler, E. S., Smith, B. D. *et al.* (2003). Early allelic selection in maize as revealed by ancient DNA. *Science*, **302**, 1206–1208.

Karanth, K. P., Palkovacs, E., Gerlach, J., *et al.* (2005). Native Seychelles tortoises or Aldabran imports? The importance of radiocarbon dating for ancient DNA studies. *Amphibian-Reptilia*, **26**, 116–121.

Kirch, P. V. (1985). *Feathered Gods and Fishhooks: An Introduction to Hawaiian Archaeology and Prehistory.* Honolulu: University of Hawaii Press.

Krajewski, C., Buckley, L. and Westerman, M. (1997). DNA phylogeny of the marsupial wolf resolved. *Proceedings of the Royal Society Series B,* **264**, 911–917.

Krause, J., Dear, P. H., Pollack, J. L. *et al.* (2006). Multiplex amplification of the mammoth mitochondrial genome and the evolution of Elephantidae. *Nature,* **439**, 724–727.

Larson, S., Jameson, R., Etnier, M., Fleming, M. and Bentzen, P. (2002). Loss of genetic diversity in sea otters (*Enhydra lutris*) associated with the fur trade of the 18th and 19th centuries. *Molecular Ecology,* **11**, 1899–1903.

Leonard, J. A., Wayne, R. K. and Cooper, A. (2000). Population genetics of Ice Age brown bears. *Proceedings of the National Academy of Sciences USA,* **97**, 1651–1654.

Leonard, J. A., Vilà, C. and Wayne, R. K. (2005). Legacy lost: genetic variability and population size of extirpated US grey wolves (*Canis lupus*). *Molecular Ecology,* **14**, 9–17.

Leonard, J. A., Shanks, O., Hofreiter, M., *et al.* (2006). Animal DNA in PCR reagents plagues ancient DNA research. *Journal of Archaeological Science,* **34**, 1361–1366.

Lindahl, T. (1993). Instability and decay of the primary structure of DNA. *Nature,* **362**, 709–715.

Lister, A. M., Edwards, C. J., Nock, D. A. W. *et al.* (2005). The phylogenetic position of the 'giant deer' *Megaloceros giganteus. Nature,* **438**, 850–853.

Malik, S., Wilson, P. J., Smith, R. J., Lavigne, D. M. and White, B. N. (1997). Pinniped penises in trade: a molecular-genetic investigation. *Conservation Biology,* **11**, 1365–1374.

Moritz, C. (2002). Strategies to protect biological diversity and the evolutionary processes that sustain it. *Systematic Biology,* **51**, 238–254.

Moritz, C. and Faith, D. P. (1998). Comparative phylogeography and the identification of genetically divergent areas for conservation. *Molecular Ecology,* **7**, 419–429.

Nee, S. and May, R. M. (1997). Extinction and the loss of evolutionary history. *Science,* **278**, 692–694.

Nei, M., Maruyama, T. and R. Chakraborty. (1975). The bottleneck effect and genetic variability in populations. *Evolution,* **29**, 1–10.

Nielsen, E. E., Hansen, M. M. and Loeschcke, V. (1999). Genetic variation in time and space: microsatellite analysis of extinct and extant populations of Atlantic salmon. *Evolution,* **53**, 261–268.

Nixon, K. C. and Wheeler, Q. C. (1991). An amplification of the phylogenetic species concept. *Cladistics,* **6**, 211–223.

Noonan, J. P., Hofreiter, M., Smith, D. *et al.* (2005). Genomic sequencing of Pleistocene cave bears. *Science,* **309**, 597–600.

Noonan, J. P., Coop, G., Kudaravalli, S. *et al.* (2006). Sequencing and analysis of Neanderthal genomic DNA. *Science,* **314**, 1113–1118.

Olson, L. E. and Hassanin, A. (2003). Contamination and chimerism are perpetuating the legend of the snake-eating cow with twisted horns (*Pseudonovibos spiralis*): a case study of the pitfalls of ancient DNA. *Molecular Phylogenetics and Evolution,* **27**, 545–548.

Olson, S. L. and James, H. F. (1991). Descriptions of 32 new species of birds from the Hawaiian Islands. *Ornithological Monographs,* **45**, 1–88.

Olson, S. L., Fleischer, R. C., Fisher, C. T. and Bermingham, E. (2005). Expunging the 'Mascarene starling' *Necrospar leguati*: archives, morphology and molecules topple a myth. *Bulletin of the British Ornithologists' Club*, 125, 31–42.

Orlando, L., Bonjean, D., Bocherens, H. *et al.* (2002). Ancient DNA and the population genetics of cave bears (*Ursus spelaeus*) through space and time. *Molecular Biology and Evolution*, 19, 1920–1933.

Pääbo, S., Poinar, H., Serre, D. *et al.* (2004). Genetic analyses from ancient DNA. *Annual Review of Genetics*, 38, 645–679.

Packer, C., Pusey, A. E., Rowley, H., Gilbert, D. A., Martenson, J. and O'Brien, S. J. (1991). Case-study of a population bottleneck: lions of the Ngorongoro crater. *Conservation Biology*, 5, 219–230.

Parham, J. F., Stuart, B. L., Bour, R. and Fritz, U. (2004). Evolutionary distinctiveness of the extinct Yunnan box turtle (*Cuora yunnanensis*) revealed by DNA from an old museum specimen. *Proceedings of the Royal Society Series B*, 271, S391–S394.

Paxinos, E. E., James, H. F., Olson, S. L., *et al.* (2002a). Prehistoric decline of genetic diversity in the nene. *Science*, 296, 1827.

Paxinos, E. E., James, H. F., Olson, S. L. *et al.* (2002b). mtDNA from fossils reveals a radiation of Hawaiian geese recently derived from the Canada goose (*Branta canadensis*). *Proceedings of the National Academy of Sciences USA*, 99, 1399–1404.

Pergams, O. R. W., Barnes, W. M. and Nyberg, D. (2003). Rapid change in mouse mitochondrial DNA. *Nature*, 423, 397.

Perry, W. L., Lodge, D. M. and Feder, J. L. (2002). Importance of hybridization between indigenous and nonindigenous freshwater species: an overlooked threat to North American biodiversity. *Systematic Biology*, 51, 255–275.

Pertoldi, C., Hansen, M. M., Loeschcke, V. *et al.* (2001). Genetic consequences of population decline in the European otter (*Lutra lutra*): an assessment of microsatellite DNA variation in Danish otters from 1883 to 1993. *Proceedings of the Royal Society Series B*, 268, 1775–1781.

Poinar, H. N., Hofreiter, M., Spaulding, W. G. *et al.* (1998). Molecular coproscopy: dung and diet of the extinct ground sloth *Nothrotheriops shastensis*. *Science*, 281, 402–406.

Poinar, H. N., Schwarz, C., Qi, J. *et al.* (2006). Metagenomics to paleogenomics: large-scale sequencing of mammoth DNA. *Science*, 311, 392–394.

Rave, E. H., Fleischer, R. C., Duvall, F. and Black, J. M. (1994). Genetic analyses through DNA-fingerprinting of captive populations of Hawaiian geese. *Conservation Biology*, 8, 744–751.

Ray, D. A., Dever, J. A., Platt, S. G. *et al.* (2004). Low levels of nucleotide diversity in *Crocodylus moreletii* and evidence of hybridization with *C. acutus*. *Conservation Genetics*, 5, 449–462.

Reid, A. H., Fanning, T. G., Hultin, J. V. and Taubenberger, J. K. (1999). Origin and evolution of the 1918 'Spanish' influenza virus hemagglutinin gene. *Proceedings of the National Academy of Sciences USA*, 96, 1651–1656.

Rhymer, J. M. (2001). Evolutionary relationships and conservation of the Hawaiian anatids. *Studies in Avian Biology*, 22, 61–67.

Rodrigo, A. G. and Felsenstein, J. (1999). Coalescent approaches to HIV population genetics. In *The Evolution of HIV*, ed. K. A. Crandall. Baltimore: Johns Hopkins University Press, pp. 233–272.

Root, T., Price, J. T., Hall, K. R., Schneider, S. H., Rosenzweig, C. and Poinds, A. J. (2003). Fingerprints of global warming on wild animals and plants. *Nature*, **421**, 57–60.

Rosenbaum, H. C., Egan, M. G., Clapham, P. J. *et al.* (2000). Utility of North Atlantic right whale museum specimens for assessing changes in genetic diversity. *Conservation Biology*, **14**, 1837–1842.

Rosenberg, N. A. and Nordborg, M. (2002). Genealogical trees, coalescent theory and the analysis of genetic polymorphisms. *Nature Reviews Genetics*, **3**, 380–390.

Roy, M. S., Girman, D. J. and Wayne, R. K. (1994). The use of museum specimens to reconstruct the genetic variability and relationships of extinct populations. *Experientia*, **50**, 551–557.

Roy, M. S., Geffen, E., Smith, D. and Wayne, R. K. (1996). Molecular genetics of pre-1940 red wolves. *Conservation Biology*, **10**, 1413–1424.

Schaeff, C. M., Kraus, S. D., Brown, M. W. *et al.* (1997). Comparison of genetic variability of North and South Atlantic right whales (*Eubalena*), using DNA fingerprinting. *Canadian Journal of Zoology*, **75**, 1073–1080.

Shapiro, B. and Cooper, A. (2003). Beringia as an Ice Age genetic museum. *Quaternary Research*, **60**, 94–100.

Shapiro, B., Sibthorpe, D., Rambaut, A. *et al.* (2002). Flight of the dodo. *Science*, **295**, 1683.

Shapiro, B., Drummond, A. J., Rambaut, A. *et al.* (2004). Rise and fall of the Beringian steppe bison. *Science*, **306**, 1561–1565.

Sharma, D. K., Maldonado, J. E., Jhala, Y. V. and Fleischer, R. C. (2004). Ancient wolf lineages in India. *Proceedings of the Royal Society Series B*, **271**, S1–S4.

Shepherd, L. D., Millar, C. D., Ballard, G. *et al.* (2005). Microevolution and mega-icebergs in the Antarctic. *Proceedings of the National Academy of Sciences USA*, **102**, 16 717–16 722.

Shivji, M. S., Chapman, D. D., Pikitch, E. K. and Raymond, P. W. (2005). Genetic profiling reveals illegal international trade in fins of the great white shark, *Carcharodon carcharias*. *Conservation Genetics*, **6**, 1035–1039.

Sorenson, M. D., Cooper, A., Paxinos, E. E. *et al.* (1999). Relationships of the extinct moa-nalos, flightless Hawaiian waterfowl, based on ancient DNA. *Proceedings of the Royal Society Series B*, **266**, 2187–2194.

Stiller, M., Green, R. E., Ronan, M. *et al.* (2006). Patterns of nucleotide misincorporations during enzymatic amplification and direct large-scale sequencing of ancient DNA. *Proceedings of the National Academy of Sciences USA*, **103**, 13 578–13 584.

Taylor, G. M., Young, D. B. and Mays, S. A. (2005). Genotypic analysis of the earliest known prehistoric case of tuberculosis in Britain. *Journal of Clinical Microbiology*, **43**, 2236–2240.

Thomas, R. H., Schaffner, W., Wilson, A. C. and Pääbo, S. (1989). DNA phylogeny of the extinct marsupial wolf. *Nature*, **340**, 465–467.

US Geological Survey (2005). *Translocation of Endangered Laysan Ducks to Midway Atoll National Wildlife Refuge (2004–5)*, US Geological Survey Fact Sheet 2005–3128. Washington, DC: US Government Printing Office.

van Riper III, C., van Riper, S. G., Goff, M. L. and Laird, M. (1986). The epizootiology and ecological significance of malaria in Hawaiian land birds. *Ecological Monographs*, **56**, 327–344.

van Tuinen, M., Ramakrishnan, U. and Hadly, E. (2004). Studying the effect of environmental change on biotic evolution: past genetic contributions, current work and future directions. *Philosophical Transactions of the Royal Society Series A*, **362**, 2795–2820.

Wayne, R. K. and Jenks, S. M. (1991). Mitochondrial DNA analysis implying extensive hybridization of the endangered red wolf *Canis rufus*. *Nature*, **351**, 565–568.

Wayne, R. K. and Morin, P. A. (2004). Conservation genetics in the new molecular age. *Frontiers in Ecology and the Environment*, **2**, 89–97.

Wayne, R. K., Leonard, J. A. and Cooper, A. (1999). Full of sound and fury: the recent history of ancient DNA. *Annual Review of Ecology and Systematics*, **30**, 457–477.

Weber, D. S., Stewart, B. S., Garza, J. C. and Lehman, N. (2000). An empirical genetic assessment of the severity of the northern elephant seal population bottleneck. *Current Biology*, **10**, 1287–1290.

Willerslev, E. and Cooper, A. (2005). Ancient DNA. *Proceedings of the Royal Society Series B*, **272**, 3–16.

Willerslev, E., Hansen, A. J., Binladen, J. *et al.* (2003). Diverse plant and animal genetic records from Holocene and Pleistocene sediments. *Science*, **300**, 791–795.

Wright, S. (1931). Evolution in Mendelian populations. *Genetics*, **16**, 97–159.

From genetic data to practical management:
issues and case studies

Future-proofing genetic units for conservation: time's up for subspecies as the debate gets out of neutral!

MICHAEL W. BRUFORD

INTRODUCTION

Conservation genetics is a maturing discipline. Since the millennium and with an established field-specific journal, large numbers of papers are published in the field and many relevant issues are debated widely in the literature. Perhaps still the most active of these debates centres on the longstanding issue of identifying and diagnosing units for conserva-tion. The purpose of this short essay is to take up a few threads from this debate, dating back to 2000 when Crandall et al. (2000) published an article in *Trends in Ecology and Evolution (TREE)*, suggesting the use of ecological and genetic exchangeability as criteria for diagnosing con-servation units, an article that was pivotal in encouraging the debate towards adaptive variation in conservation, the general theme of this essay. I do not intend to exhaustively review earlier discussions on the issue, which most observers would agree dates back to Ryder's paper in the first issue of *TREE* in 1986. Here I will briefly overview recent opinions on diagnosing units for conservation and the role of neutral and adaptive genetic variation in this. I will focus a little on one con-troversial example, which serves to shed light on where some of the current debate is focused. I will then discuss perhaps the major recent development in conservation unit designation: the use of adaptive genetic markers, the concept of exchangeability and a recent proposal for a 'Population Adaptive Index'. Finally, I will briefly discuss the issue of predictive conservation genetics and how geneticists and wildlife managers might coalesce around using new tools to take present-day molecular data and evaluate the likely changes in diversity under differ-ent management approaches that may be applied.

Population Genetics for Animal Conservation, eds. G. Bertorelle, M. W. Bruford, H. C. Hauffe, A. Rizzoli and C. Vernesi. Published by Cambridge University Press. © Cambridge University Press 2009.

CURRENT SITUATION

One of the key elements of the current debate about how and whether adaptive genetic variation should be assayed when assessing conservation units is to what extent neutral genetic diversity can or should be used as a proxy for adaptive diversity or adaptive potential. Such an assumption has run deep through the literature over the last twenty years but was directly questioned by Hedrick in 2001. Hedrick compared neutral, detrimental and adaptive variation, highlighting how each has been studied in conservation genetics. Continuing on previous themes he pointed out that statistical anomalies can result in high-precision neutral markers 'over-diagnosing' conservation units, that statistical significance in such studies might be biologically misleading (and vice versa) and that certain simple demographic processes (such as those leading to genetic drift) might also result in over-diagnosis (see also Bruford 2002). Hedrick also discussed adaptive variation directly and pointed out that selection experiments on many endangered species are both logistically and ethically problematic, leading to the more common use of candidate genes (e.g. the major histocompatibility complex (MHC); Hedrick and Parker 1998) or to an assessment of the population genetic behaviour of neutral markers linked to genomic regions under selection (e.g. MHC-linked microsatellites, which can show very unusual patterns of variation, often as a result of balancing or allele-specific selection, e.g. Aguilar *et al.* 2004).

Shortly afterwards, Moritz (2002) further developed the interplay between adaptive and neutral variation and how conservation management decisions to favour one kind of variation may not be in the best interests of the other. He firmly advocated protecting evolutionarily significant units (ESUs) as evolutionary lineages which cannot be recovered, but stated also that adaptive trait variation was in principle more difficult to maintain due to the complexities of understanding genomic architecture in isolated populations and might involve prioritizing areas for ESUs which incorporate the most environmental heterogeneity in the expectation that this would lead to the maintenance of different adaptive variants. Around this time it was becoming clear that genome-scale scans for genetic diversity in non-model organisms were becoming technically feasible and DeSalle and Amato (2004) emphasized the future use of genome screening technologies and their likely interface with landscape-scale questions (landscape genetics and genomics) in a conservation context. These are both issues being continually expanded at the present time (see Kohn *et al.* 2006), now that the data-handling issues are becoming resolved. For example, Joost *et al.* (2007) have recently carried out spatial analysis using a landscape genomics approach (using spatial coincidence analysis and logistic regression) to identify candidate adaptive loci using very large numbers of markers and environmental variables at the same time. This area is likely to develop rapidly.

DeSalle and Amato (2004) following on from Goldstein *et al.* (2000) continue to advocate the use of character-based diagnosis for conservation as opposed to frequency-based or genetic-distance-based methods. Although their approach is essentially typological in its perspective, they contend that adhering to such an approach avoids confusion and misinterpretation of frequency and distance data in an evolutionary context: a problem that has long bedevilled conservation genetics (see Hedrick 2001). However, with the ever-expanding and highly informative data sets available in conservation genetics, the use of sometimes single, fixed evolutionary characters to describe units for conservation may be even more likely to over-diagnose in the same way as described above, and this is an issue that needs to be addressed. Very recently Byrd Davis *et al.* (2008) have proposed using evolutionary 'hot spots' of neo-endemism as an approach to identifying regions of ongoing diversification (biotic and abiotic); this approach is designed to highlight the presence of narrowly endemic, recently evolved taxa which have arisen in habitats with ongoing diversification. Such an approach promises to bridge the gap between phylogenetic approaches to conservation and landscape analysis. Other approaches to linking phylogenetic diversification with landscapes are becoming more widely used. For example, Moodley and Bruford (2007) recently analysed the evolutionary relationships between populations of a sub-Saharan generalist ungulate species to refine ecoregion designation and incorporate an evolutionary element to habitat pattern analysis.

One aspect of the debate on conservation units which has developed quite extensively in the literature is the relevance of subspecies in conservation, particularly with relevance to the US Endangered Species Act (ESA). Ryder (1986) thought of ESUs and subspecies as essentially synonymous in most cases, however it is becoming clear that this simply is not holding up to detailed scrutiny. Zink *et al.* (2000) with the case of the Californian gnatcatcher (*Polioptila californica*) and Zink (2004) provided some interesting insight into this question from the perspective of (admittedly vagile) avian species and their conservation, questioning particularly the use of mtDNA phylogenies to define units for conservation and how monophyletic currently described avian subspecies have proved to be, using mtDNA. To reinforce this, a further clear example recently appeared in the literature

(Johnson *et al.* 2005) which showed that both extant and ancient DNA specimens from the Cape Verde kite (*Milvus milvus fasciicauda*), one of the world's rarest raptors and previously described as an endemic subspecies of the red kite, were not monophyletic and instead grouped either as a random sample of continental red kites (ancient samples) or black kites (modern samples). The clear implications of this research is that the Cape Verde kite should be declassified as a distinct evolutionary entity worthy of special conservation efforts and there are a number of examples now in the literature which are showing the same thing.

I will briefly describe a highly controversial example of the above problem, which has appeared in a series of exchanges in the pages of *Animal Conservation* and *Molecular Ecology* over the last few years.

Preble's meadow jumping mouse: an example of the problem

A paper appeared in *Animal Conservation* in 2005 (Ramey *et al.* 2005), which described an ostensibly quite comprehensive analysis of morphological data and mitochondrial and microsatellite DNA in the threatened, ESA-listed Preble's meadow jumping mouse (*Zapus hudsonius preblei*). Meadow jumping mice are widely and mostly continuously distributed throughout North America, but this particular subspecies is isolated, mostly in Colorado, from the nearest two populations of other subspecies by a few hundred kilometres and had previously been described as distinct based on cranial characters and pelage characteristics. The study re-examined these analyses and added analysis of a segment of the mitochondrial control region and five polymorphic microsatellites, finding (1) that principal component analysis of the morphometric data placed this taxon within the range of the two closest subspecies with discriminant function analysis only correctly assigning 42% of individuals, (2) mtDNA was not reciprocally monophyletic, with all mtDNA haplotypes also being found in one of the two closest subspecies and (3) microsatellites revealed low genetic structure, with no taxon-relevant groups emerging from the data using Bayesian clustering, and there was evidence for recent gene flow. In line with the results presented, and continuing the theme above, the authors suggested a delisting for this taxon to prevent further misallocation of funds. Important to note was that the authors presented this analysis in the kind of hypothesis-testing context encouraged by Crandall *et al.* (2000) (as will be explained later) and invoked genetic and ecological exchangeability as a key element of their hypothesis-testing approach, citing an examination of relevant literature for ecological exchangeability and examining their own data for evidence of genetic exchangeability.

The paper provoked much controversy. Strong rebuttals and responses were published in an issue of *Animal Conservation* the following year (Crandall 2006; Martin 2006; Ramey *et al.* 2006; Vignieri *et al.* 2006) where accusations of selective interpretation of the data, biased advocacy and poor handling of the paper were variously put forward. Elements of this argument are summarized by Cronin (2007) largely in the context of the subjectivity of subspecific taxonomy, but it is worth re-evaluating the arguments (below) and placing them in the context of the neutral versus adaptive issues discussed above because they seem to be key to the disagreements between those involved in this case.

Vignieri *et al.* (2006) took issue with most aspects of the study as originally presented. However, in particular they focused on the evidence presented regarding ecological exchangeability and the lack of experimental data to substantiate this. In fairness to Ramey *et al.* (2005), it is possible to interpret Crandall *et al.*'s (2000) recommendations for inferring ecological exchangeability as not necessarily requiring experimentation: this recommendation is only fully developed in a much later paper from Crandall's group (Bader *et al.* 2005). It is nevertheless arguable that the evidence presented by Ramey *et al.* (2005) and the way in which ecological exchangeability is framed within a hypothesis-testing context, was perhaps overemphasized given the strength of evidence available. Concerns were also expressed over the molecular data, especially the low number of microsatellite markers used and whether the expectation of complete lineage sorting for mtDNA haplotypes in these recently diverged taxa was a realistic rejection criterion to use. This critique and Ramey *et al.*'s robust response was accompanied by a serious attack on the authors by Martin (2006). In his paper, Martin accused Ramey of 'advocacy' and a general lack of objectivity over the ESA, the way it is enacted and the way that funds are distributed based on genetic or systematic evidence. Martin, importantly, also called for a retraction and independent verification of the analysis before any action was taken at the management level. I find that the description of the peer review process given in Crandall's (2006) response seems familiar, plausible and fair. The issue of bias and advocacy with respect to the ESA has been a strong feature of this case, but also it seems clear that, on the basis of the original paper and the ensuing discussion, the merits of the science could have been debated endlessly.

Following on from Martin's (2006) paper, in December 2006 King *et al.* published a comprehensive genetic reanalysis of Preble's mouse. Although King *et al.* did not address the issue of morphological distinctiveness or ecological exchangeability, their study was impressive in its thoroughness: they increased the number of microsatellites used to 21, sampled much more extensively (although using fewer geographic sites) and analysed approximately 1300 bp of mitochondrial DNA including 1000 bp of the cytochrome *b* gene. On the face of it, their results could not have been more contrasting to Ramey's. They found strong evidence for genetic structure within *Z. h. preblei* and between it and the other subspecies compared, which they attributed largely to the increased number and power of the microsatellites used and, surprisingly, they also found no mitochondrial haplotype sharing between *preblei* and the other subspecies. They also cited difficulties in obtaining common reference samples for a direct comparison with the previous study and different control region sequences from the same museum samples they were able to access and test. They concluded that their data supported the evolutionary distinctiveness of *Z. h. preblei* and its protection under the ESA and the evidence from their data (at least to this author) seems quite compelling. Subsequently Ramey *et al.* (2007) have responded to this paper and the central tenet of their argument is that although *preblei* could be shown by King *et al.* to be distinct from the other subspecies, the levels of distinctiveness are more equivalent to that of a management unit (MU, sensu Moritz 1994b) than an ESU or a subspecies, although as Cronin (2007) points out, despite Ryder's (1986) original thinking, the lines between all these units can be blurred and subjective. They also respond that the lack of morphometric re-analysis and ecological hypothesis testing renders the study inconclusive with respect to their original question (alongside counter-claims of sample mishandling by King *et al.* and of changing the paper post-review in proof!). It seems likely that the story will run and run, and the issue of the eventual delisting (or not) of this taxon is going to be a litmus test for the power of genetic data in the ESA.

All of this could be viewed as an extremely long and unfortunate saga and bad news for integrity of conservation genetics and its relevance to the ESA. However, of equal importance is the pivotal role that more analysis of adaptive variation could and probably should have played in the debate. Essentially the original neutral genetic data, regardless of their accuracy or otherwise, could be interpreted in a number of ways. King *et al.*'s (2006) molecular data seem more comprehensive and compelling, but the lack of adaptive variation analysis in that study is a problem: ecological exchange-ability experiments remain to be done and given the controversy would perhaps be a good idea. However, it is also clear that the requirement for such experiments *in all cases* could potentially impede the listing/delisting process due to their protracted nature. Therefore, a decision tool used to assess the strength of a priori evidence for a lack of exchangeability would seem to be a necessity to avoid planning blight and allow timely decisions to be taken. Perhaps one positive thing that will emerge from this episode is that new standards of analysis will be required for ESA listing/delisting and hopefully this could be adopted globally. It seems that a serious attempt to analyse adaptive variation should be included in such studies where possible, which will hopefully allow the distinction between statistical and biological significance in designating units for conservation.

GETTING OUT OF 'NEUTRAL'

Crandall *et al.*'s (2000) paper has already wielded considerable influence on conservation genetics in the new millennium. During the 1990s a largely phylogenetic approach to conservation unit definition began to predominate and theirs was an attempt to go back to some of the original meaning in Ryder's paper (Ryder 1986) and address the rising concern in both the genetics and conservation worlds that there was little good evidence that neutral diversity and 'meaningful' adaptive diversity would be always correlated. Specifically they mention the tension inherent in maintaining isolation among monophyletic ESUs and the goal of maintaining adaptive (and other) genomic variation. They also argued (as has Zink) that some of the genetic criteria for ESUs and MUs advanced, for example by Moritz (1994a, b), are not necessarily expected to hold for the most vagile of species (e.g. birds and carnivores) and the fact that 'over-diagnosis' was an almost inevitable consequence of the distorted demographic profiles of isolated populations of endangered species, regardless of how evolutionarily meaningful this may (or may not) be. Species with large historical distributions and an isolation-by-distance pattern of genetic structure in pristine environments are likely to be particularly problematic in this sense, for example.

Crandall *et al.* (2000) therefore advanced the idea of using evidence for genetic and ecological exchangeability over short and longer timescales as one way to add a hypothesis-testing component to unit designation, where full exchangeability would act as a null hypothesis, enabling ecological data to be overlaid onto the genealogy of the taxa being tested. They argued that ecological exchangeability arises largely through 'shared fundamental adaptations' and that the characters used for such an analysis should be demonstrably heritable. Their protocol involved an appraisal of historical data (e.g. ancient DNA analysis, pollen data, museum information) for the long-term evidence, the use of genetic data of differing evolutionary rates to assess both short- and long-term genetic exchange and a phylogenetic (e.g. network-based) analysis incorporating different timescales. They proposed a cross-hair analysis of the strength of evidence for adaptive distinctiveness on short- and long-term information on genetic and ecological exchange-ability (+/− categorization) (see Fig. 10.1, taken from Crandall *et al.* 2000). The cases described in Fig. 10.1 detail a range of possible scenarios, but when the authors surveyed 98 relevant case studies from the literature, they found that the large majority fell into Case 8 (treat as a single population) mainly only rejecting recent exchange or failing to reject any exchangeability. However, the problem with this approach is the evidence base. It remains the case that although genetic analysis (provided it is carried out properly) can often shed light on short- and longer-term genetic exchange, the ecological evidence is in many cases lacking (as could be argued for short-term ecological exchange in Preble's jumping mouse, for example).

It was implicitly interpreted by many (e.g. Vignieri *et al.* 2006, when responding to Ramey *et al.* 2005) but not all, that the 'acid test' for ecological exchangeability could only be through direct manipulative experimentation. Bader *et al.* (2005) acknowledged the 'considerable confusion' surrounding this issue and discussed both the fact that ecological experiments could be difficult over realistic timescales (including being statistically challenging) but also the commonly acknowledged fact that for many species which are the focus of conservation measures, experimentation is simply impossible or unethical. Of particular concern in exchangeability experiments is the confounding effect of phenotypic plasticity where the same genes are differentially expressed to produce divergent traits in different environments (in other words genetically controlled 'versatility'). It is potentially very challenging to study such effects in many species and in heterogeneous environmental conditions experienced in the wild. It is worth noting, however, that exchangeability experiments do not always have to be reciprocal: this will depend on the experimental design and the management approach being tested. For instance, in the face of climate change, non-reciprocal exchangeability experiments will be required to assess the likely effects of moving populations to different latitudes.

It is important to state that some of these ideas are quite controversial: exchangeability studies are not without their opponents and experiments like the ones proposed by Bader *et al.* (2005) could have ethical problems and ill-advised translocations have had a strong and negative impact on biodiversity throughout the world. Further, the advisability of such experimentation is likely to be context-dependent. Zink (2007) has again raised concerns about widely distributed species, citing the example of the great

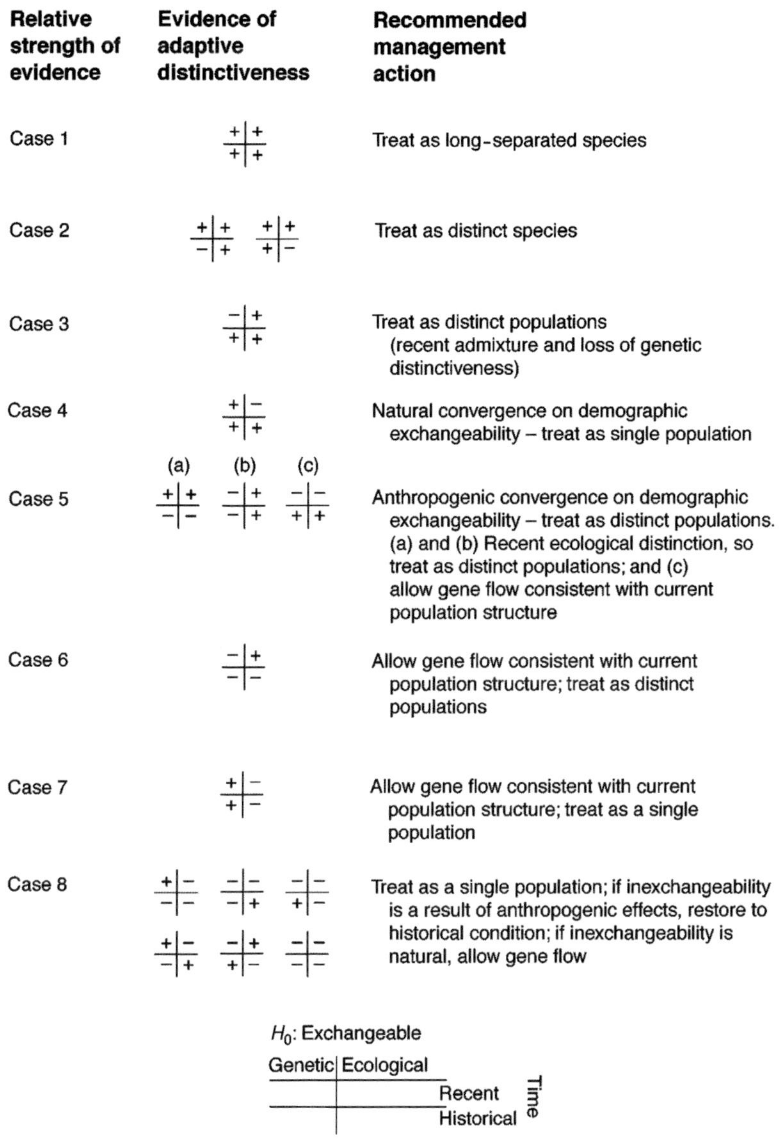

Relative strength of evidence	Evidence of adaptive distinctiveness	Recommended management action
Case 1	$\frac{+\mid+}{+\mid+}$	Treat as long-separated species
Case 2	$\frac{+\mid+}{-\mid+}$ $\frac{+\mid+}{+\mid-}$	Treat as distinct species
Case 3	$\frac{-\mid+}{+\mid+}$	Treat as distinct populations (recent admixture and loss of genetic distinctiveness)
Case 4	$\frac{+\mid-}{+\mid+}$	Natural convergence on demographic exchangeability – treat as single population
Case 5	(a) $\frac{+\mid+}{-\mid-}$ (b) $\frac{-\mid+}{-\mid+}$ (c) $\frac{-\mid-}{+\mid+}$	Anthropogenic convergence on demographic exchangeability – treat as distinct populations. (a) and (b) Recent ecological distinction, so treat as distinct populations; and (c) allow gene flow consistent with current population structure
Case 6	$\frac{-\mid+}{-\mid-}$	Allow gene flow consistent with current population structure; treat as distinct populations
Case 7	$\frac{+\mid-}{+\mid-}$	Allow gene flow consistent with current population structure; treat as a single population
Case 8	$\frac{+\mid-}{-\mid-}$ $\frac{-\mid-}{-\mid+}$ $\frac{-\mid-}{+\mid-}$ $\frac{+\mid-}{-\mid+}$ $\frac{-\mid+}{+\mid-}$ $\frac{-\mid-}{-\mid-}$	Treat as a single population; if inexchangeability is a result of anthropogenic effects, restore to historical condition; if inexchangeability is natural, allow gene flow

H_0: Exchangeable

Genetic | Ecological

Recent

Historical

Time

Figure 10.1. Categories of population distinctiveness based on rejection (+) or failure to reject (–) the null hypothesis (H_0) of genetic and ecological exchangeability, for both recent and historical time frames. As the case numbers increase (from Case 1 to Case 8), there is decreasing evidence for significant population differentiation. (From Crandall et al. 2000.)

tit, a passerine bird distributed across Eurasia which has been the focus for numerous quantitative genetic studies in the past and which has been shown to show small-scale local differences in crucial life-history characters. In contrast however, several studies have shown that this cosmopolitan species shows remarkably little mtDNA structure throughout its geographic range and there are a number of such examples for different avian species in the literature. Such extreme incongruity between ecological and genetic patterns makes strict application of the cross-hair approach of Crandall and colleagues unlikely to achieve anything other than a 'Case 8' status, so should be treated with caution. Interestingly, past accidental and uncontrolled translocation events potentially provide valuable examples of this kind of ecological experiment and should perhaps be studied more closely and collectively to assess if any general results emerge applicable to species otherwise unsuitable for such experiments (see chapter by Bertorelle *et al.*, this volume).

Conservation genetics is routinely incorporating a few candidate genes in analysis of genetic diversity and differentiation, especially the MHC (see Hedrick 2001). However, it is becoming clear that the candidate gene approach will be a long and laborious effort, enhanced by genome projects but constrained by environmental heterogeneity and phylogenetic effects. Although using neutral markers as a proxy for adaptive variation is fraught with problems, methods have been developed to detect selection using neutral markers (e.g. Luikart *et al.* 2003; Joost *et al.* 2007) and these are now being applied. Most recently, the approach to detect genetic outlier markers described by Beaumont and Nichols (1996) and Luikart *et al.* (2003) has been adapted by Bonin *et al.* (2007; see also chapter by Bonin and Bernatchez, this volume) to develop a 'population adaptation index' (PAI).

This is the first time that a method has been proposed to link genome scan data to the study of conservation units. Bonin *et al.* (2007) used amplified fragment length polymorphism (AFLP) and outlier analysis to study populations of two species, an amphibian and a plant (the common frog and Austrian dragonhead). The process uses the principle of complementarity to maximize diversity in conservation programmes (very similar to the method commonly used in domestic livestock diversity originally proposed by Weitzmann in 1992 and also advanced by Moritz (2002)). In practice, AFLP was used to generate potentially adaptive (i.e. divergent outlier) markers and 'neutral' diversity and genetic differentiation indices computed from the neutral data, with the outlier loci being analysed separately. The PAI is simply computed as the percentage of the adaptive (outlier) loci possessing significantly different band frequencies in two or more

Table 10.1 *Levels of neutral and adaptive genetic diversity, measured through the population adaptive index (PAI) in common frog and Austrian dragonhead populations*

Population	Proportion of polymorphic loci	Nei's gene diversity	PAI
Common frog			
AI	0.52	0.18	0.07
CO	0.69	0.23	0.21
PP	0.74	0.25	0.14
RM	0.68	0.23	0.43
TE	0.68	0.23	0.29
TI	0.66	0.22	0.29
Austrian dragonhead			
BE	0.58	0.21	0.43
CH	0.75	0.25	0.29
ES	0.61	0.22	0.21
FO	0.58	0.20	0.14
LA	0.60	0.21	0.14
RE	0.64	0.22	0.14
VA	0.69	0.21	0.21

Source: Bonin *et al.* (2007).

population comparisons. This approach should identify the populations with rare or unique adaptive loci, while eliminating false positives, and hence allow population sets to be selected which are complementary in their adaptations, allowing the maximum adaptive and neutral diversity to be conserved. Unsurprisingly, the authors found that neutral and adaptive diversity indices did not correlate in all cases, but that the complementarity approach was efficient at maximizing diversity for both kinds of variation. They suggested that populations providing conflicting evidence for neutral and adaptive diversity should make use of additional information (e.g. demographic history) to help understand further the potential source of the incongruity. Table 10.1 shows the range of values for populations of common frog and Austrian dragonhead detected using neutral (first two columns) and adaptive (third column) markers.

This method is a potentially fruitful approach to incorporating anonymous markers under selection into conservation programmes but the link between these markers and phenotype remains an open question; a parallel study using this approach and phenotypic data would be a good way to extend the method. It is clear that the nature of the adaptive variation being conserved (e.g. to pristine or anthropogenically altered habitats) is an important consideration in this and other approaches attempting to assay adaptive variation in threatened taxa: genome scans on their own are unlikely to provide conservation managers with the information they need (see also chapter by Beaumont, this volume).

FUTURE-PROOFING

All the methods described above are designed to adhere to the precautionary principle. That is they aim to conserve diversity (either neutral or adaptive) under the assumption that the presence of that diversity, i.e. a set of alleles, will future-proof the populations concerned against events which may alter their future environment. Although conservation managers may view this as insurance against the effects of different management options in the future, a question frequently asked is what might be the consequences on neutral or adaptive diversity for different management decisions they may take. This is where conservation genetics is still in its infancy, yet the need seems great. Stochastic simulation tools (such as the software VORTEX: Lacy *et al.* 2005) are fortunately becoming more sophisticated in this regard, with recent versions allowing incorporation of marker allele frequencies from molecular data sets, and even allowing allelic characters to be associated with life-history parameters. Although examples of such approaches have yet to reach the literature, they should do so in the near future. Only when such approaches are commonplace, and allow conservation managers to assess the genetic diversity consequences of their actions will they become routinely used in action plans and will they become to be used to guide specific management decisions. Bowen and Roman (2005) cite future components of biodiversity planning as central to integrated conservation management and in the current climate of global environmental change this has never been more apposite. Population and evolutionary geneticists need to take note of this imperative and develop better ways to simulate genetic diversity responses in endangered populations if we are to make the conceptual leap from precautionary genetic resource management to a more proactive paradigm.

REFERENCES

Aguilar, A., Roemer, G., Debenham, S. *et al.* (2004). High MHC diversity maintained by balancing selection in an otherwise genetically monomorphic mammal. *Proceedings of the National Academy of Sciences USA,* **101**, 3490–3494.

Bader, R. B., Belk, M. C., Shiozawa, D. K. and Crandall, K. A. (2005). Empirical tests for ecological exchangeability. *Animal Conservation,* **8**, 239–247.

Beaumont, M. A. and Nichols, R. A. (1996). Evaluating loci for use in the genetic analysis of population structure. *Proceedings of the Royal Society Series B*, **263**, 1619–1626.

Bonin A., Nicole, F., Pompanon, F., Miaud, C. and Taberlet, P. (2007). Population Adaptive Index: a new method to help measure intraspecific genetic diversity and prioritize populations for conservation. *Conservation Biology*, **3**, 697–708.

Bowen, B. W. and Roman, J. (2005). Gaia's handmaidens: the Orlog model for conservation biology. *Conservation Biology*, **19**, 1037–1043.

Bruford, M. W. (2002). Biodiversity: evolution, species, genes. In *Conserving Bird Biodiversity: General Principles and their Application*, ed. K. Norris and D. J. Pain. Cambridge: Cambridge University Press, pp. 1–19.

Byrd Davis, E., Koo, M. S., Conroy, C. J. L. and Moritz, C. (2008). The California Hotspots Project: identifying regions of rapid diversification of mammals. *Molecular Ecology*, **17**, 120–138.

Crandall, K. A. (2006). Advocacy dressed up as scientific critique. *Animal Conservation*, **9**, 250–251.

Crandall, K. A., Bininda-Emonds, O. R. P., Mace, G. M. and Wayne, R. K. (2000). Considering evolutionary processes in conservation biology. *Trends in Ecology and Evolution*, **15**, 290–295.

Cronin, M. A. (2007). The Preble's meadow jumping mouse: subjective subspecies, advocacy and management. *Animal Conservation*, **10**, 159–161.

DeSalle, R. and Amato, G. (2004). The expansion of conservation genetics. *Nature Reviews Genetics*, **5**, 702–712.

Goldstein, P. Z., DeSalle, R., Amato, G. and Vogler, A. P. (2000). Conservation genetics at the species boundary. *Conservation Biology*, **14**, 120–131.

Hedrick, P. W. (2001). Conservation genetics: where are we now? *Trends in Ecology and Evolution*, **16**, 629–636.

Hedrick, P. W. and Parker, K. M. (1998). MHC variation in the endangered Gila topminnow. *Evolution*, **52**, 194–199.

Johnson, J. A., Watson, R. T. and Minde, D. P. (2005). Prioritizing species conservation: does the Cape Verde kite exist? *Proceedings of the Royal Society Series B*, **272**, 1365–1371.

Joost, S., Bonin, A, Bruford, M. W. *et al.* (2007). A spatial analysis method (SAM) to detect candidate loci for selection: towards a landscape genomics approach to adaptation. *Molecular Ecology*, **16**, 3955–3969.

King, T. L., Switzer, J. F., Morrison, C. L. *et al.* (2006). Comprehensive genetic analyses reveal evolutionary distinction of a mouse (*Zapus hudsonius preblei*) proposed for delisting from the US Endangered Species Act. *Molecular Ecology*, **15**, 4331–4359.

Kohn, M. H., Murphy, W. J., Ostrander, E. A. and Wayne, R. K. (2006). Genomics and conservation genetics. *Trends in Ecology and Evolution*, **21**, 629–637.

Lacy, R. C., Borbat, M. and Pollak, J. P. (2005). *VORTEX: A Stochastic Simulation of the Extinction Process*, v. 9.50. Brookfield, IL: Chicago Zoological Society.

Luikart, G., England, P. R., Tallmon, D., Jordan, S. and Taberlet, P. (2003). The power and promise of population genomics: from genotyping to genome typing. *Nature Reviews Genetics*, **4**, 981–994.

Martin, A. (2006). Advocacy dressed up as science: response to Ramey *et al.* (2005). *Animal Conservation*, **9**, 248–249.

Moodley, Y. and Bruford, M. W. (2007). Molecular biogeography: towards an integrated framework for conserving pan-African biodiversity. *PLoS ONE*, Issue 5, e454.

Moritz, C. (1994a). Applications of mitochondrial DNA analysis in conservation: a critical review. *Molecular Ecology*, **3**, 401–411.

Moritz, C. (1994b). Defining evolutionarily significant units for conservation. *Trends in Ecology and Evolution*, **9**, 373–375.

Moritz, C. (2002). Strategies to protect biological diversity and the evolutionary processes that sustain it. *Systematic Biology*, **51**, 238–254.

Ramey, II R. R., Liu, H. P., Epps, C. W., Carpenter, L. M. and Wehausen, J. D. (2005). Genetic relatedness of Preble's jumping mouse (*Zapus hudsonius prebei*) to nearby species of *Z. hudsonius* as inferred from variation in cranial morphology, mitochondrial DNA and microsatellite DNA: implications for taxonomy and conservation. *Animal Conservation*, **8**, 329–346.

Ramey, II R. R., Wehausen, J. D., Liu, H. P., Epps, C. W. and Carpenter, L. H. (2006). Response to Vignieri *et al.* (2006): should hypothesis testing or post hoc interpretation of results guide the allocation of conservation effort? *Animal Conservation*, **9**, 244–247.

Ramey, II R. R., Wehausen, J. D., Liu, H. P., Epps, C. W. and Carpenter, L. M. (2007). How King *et al.* (2006) define an 'evolutionary distinction' of a mouse subspecies: a response. *Molecular Ecology*, **16**, 3518–3521.

Ryder, O. A. (1986). Species conservation and systematics: the dilemma of subspecies. *Trends in Ecology and Evolution*, **1**, 9–10.

Vignieri, S. N., Hallerman, E. M. and Bergstrom, B. J., (2006). Mistaken view of taxonomic validity undermines conservation of an evolutionarily distinct mouse: a response to Ramey *et al.* (2005). *Animal Conservation*, **9**, 237–243.

Weitzman, S. (1992). 'On diversity'. *Quarterly Journal of Economics*, **107**, 363–405.

Zink, R. M. (2004). The role of subspecies in obscuring avian biological diversity and misleading conservation policy. *Proceedings of the Royal Society Series B*, **271**, 561–564.

Zink, R. M. (2007). Ecological exchangeability versus neutral molecular markers: the case of the great tit. *Animal Conservation*, **10**, 369–373.

Zink, R. M., Barrowough, G. F., Atwood, J. L. and Blackwell-Rago, R. C. (2000). Genetics, taxonomy and conservation of the threatened California gnatcatcher. *Conservation Biology*, **14**, 1394–1405.

11

Genetic diversity and fitness-related traits in endangered salmonids

KATRIINA TIIRA AND CRAIG R. PRIMMER

INTRODUCTION

An individual's fitness can be largely influenced by its genetic diversity. Low survival and poor fecundity are just few examples of the consequences of the loss of genetic variation (Mitton 1993; Falconer and Mackay 1996). Matings between closely related parents can result in progeny with lowered fitness, a phenomenon generally known as inbreeding depression (Lynch and Walsh 1998). However, genetic drift in small and isolated populations can also result in increased homozygosity, even in the absence of matings between close kin (Shields 1993). On the other hand, matings between genetically differentiated populations or strains often produce individuals with higher fitness during the first generation; a phenomenon called heterosis (Mitton 1993).

The earliest observations of the strong influence of an individual's genetic diversity on its fitness came from domestic and laboratory animals (reviewed by Falconer and Mackay 1996), and from zoos (Ralls and Ballou 1983) where inbreeding was seen to cause harmful effects on survival, fecundity and growth (direct evidence). In wild populations, where pedigrees have been known, the negative effects of inbreeding have also been observed (Kruuk *et al.* 2002; Reid *et al.* 2003). However, pedigree information is rarely available in wild populations. The majority of information on the genetic diversity of natural populations has actually been obtained using molecular markers believed to be generally neutral, namely allozyme and DNA markers (indirect evidence) (Frankham 1995; Haig and Avise 1996). Positive associations have observed between protein heterozygosity as assessed by allozymes and several fitness-related traits such as growth, metabolic efficiency, body size, fecundity and survival (Mitton and Grant 1984; Charlesworth and Charlesworth 1987; Danzmann *et al.* 1987, 1989; Allendorf and Leary 1988; Britten 1996; David 1998; Wang *et al.* 2002).

Population Genetics for Animal Conservation, eds. G. Bertorelle, M. W. Bruford, H. C. Hauffe, A. Rizzoli and C. Vernesi. Published by Cambridge University Press. © Cambridge University Press 2009.

With the development of the polymerase chain reaction (PCR), nuclear DNA molecular markers have become popular in heterozygosity–fitness studies. Microsatellites have been used extensively as genetic markers in a wide range of biological studies, not least in conservation genetics (Beaumont and Bruford 1999). Variations in microsatellites, which are highly variable non-coding sections of DNA, have been found to be positively connected with several fitness-related traits (Coltman et al. 1998; Coulson et al. 1998, 1999; Amos et al. 2001; Hansson et al. 2001; Höglund et al. 2002; Acevedo-Whitehouse et al. 2003; Tiira et al. 2006). However, criticism has been presented against using molecular markers as surrogates for variation in quantitative traits (Karhu et al. 1996; Reed and Frankham 2001). According to a meta-analysis conducted by Reed and Frankham (2001) the correlation between molecular and quantitative measures of genetic variation within populations is very weak, with the major reason for the weak association most likely being differences in selective forces between molecular and quantitative measures of genetic variation. Therefore the relationship between molecular markers and quantitative trait variation remains uncertain and more research is needed to investigate this association with different quantitative traits.

Nevertheless, evidence for positive associations between molecular markers and genetic variation exists. How, then, it is possible to get information on the agent loci (i.e. the loci directly contributing to observed trait variation) using molecular markers that are mainly neutral? The direct effect hypothesis suggests that positive associations between heterozygosity and fitness originate because the scored loci themselves (usually allozymes) influence fitness in an overdominant manner (functional overdominance) (David 1998; Lynch and Walsh 1998; Hansson and Westerberg 2002). The direct effect hypothesis, however, is normally not relevant for microsatellites, as they are usually considered to be selectively neutral. The two main hypotheses, which can explain the positive association between molecular variation at neutral loci and fitness, are the general and local effect hypotheses (Hansson and Westerberg 2002). Populations which experience partial inbreeding can experience correlations between the homozygosity of different loci, also between unlinked loci (Charlesworth 1991). This general effect hypothesis is also known as the identity disequilibrium hypothesis (David 1998; Lynch and Walsh 1998; Hansson and Westerberg 2002) and it has been the main hypothesis proposed to explain the detected positive correlations between heterozygosity and fitness in many studies (Coltman et al. 1998; Coulson et al. 1998; Slate et al. 2000; Höglund et al. 2002). On the other hand, according to the local effect hypothesis the neutral microsatellites can mark large fragments of the chromosome through linkage disequilibrium (LD), and thus co-segregate with selected loci (Bierne *et al.* 1998; Hansson and Westerberg 2002). This can be expected in recently expanded populations. LD has generally been thought to appear rarely in natural populations, however two recent studies suggest that it can be an important factor creating heterozygosity–fitness association in natural populations (Balloux *et al.* 2004; Hansson *et al.* 2004; Hansson and Westerberg 2008).

Among the genetic measures used to describe genetic diversity of an individual, heterozygosity is the most frequently used. A fairly recently developed measure of genetic diversity, mean d^2 (Coulson *et al.*, 1998), invoked an enthusiastic wave in heteozygosity–fitness correlation (HFC) research, where several positive associations were found (Coltman *et al.* 1998; Coulson *et al.* 1999; Hansson *et al.* 2001; Höglund *et al.* 2002). Mean d^2 is the square of the difference in repeat units between the two alleles at a locus, and it assumes a stepwise mutation model (Coulson *et al.* 1998; Estoup and Cornuet 1999). However, the majority of evaluations suggest that heterozygosity may be a more efficient estimator of genetic diversity than mean d^2 in most cases (Hedrick *et al.* 2001; Tsitrone *et al.* 2001; Slate and Pemberton 2002), at least when a sufficient number of loci are analysed (Slate and Pemberton 2002). Mean d^2 tends to have larger between-locus variance in effect sizes compared with heterozygosity (Coltman and Slate 2003), in addition, the microsatellites may not evolve in a stepwise process (Ellegren 2000), which is an important theoretical assumption for using mean d^2. Another measure of genetic diversity frequently used in HFC studies is internal relatedness (IR) (Amos *et al.* 2001). IR is an estimate of parental similarity, which weights the genotype by the frequencies of the alleles involved (Amos *et al.* 2001) whereby shared rare alleles are weighted more heavily compared to shared common alleles, similarly to the relatedness value of Queller and Goodnight (1989). In addition to above-mentioned measures of individual genetic diversity, Ritland's (1996) marker-based method for estimating individual genetic relatedness is occasionally used as a coefficient of individual inbreeding. This method-of-moments estimator (MME) is suggested to be particularly useful for highly polymorphic markers, such as microsatellites (Ritland 1996).

The majority of published studies report a positive association usually of low effect between genetic diversity and fitness (Wang *et al.* 2002), however these associations are not universal (Britten 1996; David 1998). In addition, non-significant results are probably underrepresented due to publication bias. Several factors can influence the appearance of genetic diversity–fitness correlations: the genetic background of the population being studied, the sample size, the number of studied loci and the genetic indices (David 1998; Tsitrone et al. 2001; Knaepkens et al. 2002). In addition, the studied fitness traits are also important; for example life-history traits (e.g. survival) have been found to be more closely associated with genetic variation than morphological traits (Coltman and Slate 2003). Furthermore, the association between genetic diversity and fitness may vary according to age, life stage, reproduction strategy and sex of studied individuals (Liskauskas and Ferguson 1991; David 1998; Altukhov et al. 2000).

Even though the majority of inbreeding studies have been conducted with captive populations (Lacy et al. 1993), and the need for information on the extent of inbreeding depression in the wild is urgent, the studies done with captive populations can still offer important knowledge of relevance for wild populations. For example, the possibility to make controlled experiments with individuals with known levels of genetic diversity can help us to gain information impossible to obtain otherwise.

The two case studies presented in this chapter (study I: Tiira et al. 2003; study II: Primmer et al. 2003), are examples of studying genetic diversity–fitness associations in a captive population of endangered wild species of salmonid using controlled experiments. In both of these studies, we use the method of Primmer et al. (2003) to estimate the genetic variation of the offspring based on the information of their parents' genotypes. This is possible as the genetic variability of an individual is determined by the combination of the genetic material inherited from its parents. This knowledge of the estimated genetic variation of test fish was used in designing the experimental setup. Both of the studies investigate a question rarely addressed in genetic diversity–fitness association studies, namely the possible association between genetic diversity and behaviour.

BACKGROUND OF THE STUDY SPECIES

Endangered salmonid populations

Salmonid fishes are an economically important group of species. Besides their value as a food source, these species are also favoured in recreational fishing. Unfortunately, many salmonid populations are now endangered, over-fishing and environmental degradation being among the salient factors (Allendorf and Waples 1996). Cultivation of salmonid species in hatcheries and fish farms is extremely common all over the world. The two main aims in captive breeding of salmonids are commercial production for food and the conservation of the particular species or populations. Wild populations of salmonids are enhanced by releasing captive-reared fish. The purpose of captive breeding is usually to enlarge the gene pool of the particular wild population and also to assure sufficient numbers of fish to bear the fishing pressure. This is also the case with our study population of Lake Saimaa salmon (*Salmo salar* m. *sebago* Girard). This species originates from Lake Saimaa, which is a glacial lake in Eastern Finland, formed by rapid land upheaval (20 metres per century) following the last ice age. Saimaa is a lake system including a total of 13 710 islands and only 0.5% of the total water area is over 50 m deep (Kuusisto 1999). Before glaciation, Saimaa was a network of rivers rather than a maze of lakes. As waterways to the sea were closed during land upheaval, Lake Saimaa salmon represent a non-anadromous population of the species, i.e. they spend the entire life cycle in fresh water.

The Lake Saimaa salmon has been virtually non-existent since 1971. Despite extensive stocking (50 000–100 000 individuals per year) for over a decade, the effective number of mature individuals returning to native spawning grounds since 1990 has never exceeded 50 per year for either population, and in some years has been fewer than 10 (Pursiainen *et al.* 1998). Despite the intensive stocking, the majority of the stocked fish of salmon are fished in their first year with small-mesh nets used for perch and whitefish.

Our study fish originated from hatchery stocks in Saimaa Fisheries Research and Aquaculture, which are reared for conservation and supplementary stocking purposes. The genetic diversity in Lake Saimaa salmon, measured both with allozymes (Vuorinen 1982) and microsatellites (Primmer *et al.* 2000; Tonteri *et al.* 2005), is very low. According to Vuorinen (1982) the average heterozygosity per individual per locus (33 studied loci) was only 1.1%, and in a recent study using 15 microsatellites, Lake Saimaa salmon had the lowest observed heterozygosity (0.29) among 15 salmon populations from northern Europe (Tonteri *et al.* 2005). This is not unexpected due to the low number of founder individuals used for creating broodstocks (Pursiainen *et al.* 1998), and hence there is also reason to expect that inbreeding may have played an important role in the recent history of the population.

Behaviour of juvenile salmonid fish: the importance of aggression

Salmonid eggs are usually buried in gravel. At first, the fry that hatch from the eggs utilize the nutrients from their yolk sac, and only after few weeks do they start using exogenous food. In the river, juvenile fish are rather stationary on the bottom, and feed on drifting food (Keenleyside and

Yamamoto 1962). As the number of profitable feeding territories or stations (in terms of energetics, number of bypassing food items) in the river may be limited, juvenile fish can experience extensive resource competition, so that a relatively small proportion of fish will survive. For example, migratory brown trout commonly occur at high densities immediately after swimming up from the gravel in the spring, but by autumn the juvenile density can be reduced by up to 80% (Elliott 1986). Hence, the ability to acquire and defend a feeding station is crucial for the future success and survival of the young fish and there is good reason to believe that aggressiveness and competitive ability are important fitness-related traits in salmonids. Fish without a territory are more likely to die due to starvation (Elliot 1994) or predation (Brännäs 1995) than territorial fish. In addition, dominant and more aggressive Atlantic salmon (*Salmo salar*) with higher metabolic rates are also more likely to adopt a faster developmental pathway, where dominants migrate and mature earlier compared with subordinates (Metcalfe et al. 1989; Metcalfe 1991, 1998; Nicieza and Metcalfe 1999). This is because important size-correlated life-history events such as the timing of migration and maturation (Thorpe 1989; Bohlin et al. 1993) are strongly affected by dominance status. In addition, fish with poorer competitive ability exhibit slower growth, later migration age, and possibly also later maturation (Metcalfe et al. 1989; Metcalfe 1991, 1998; Nicieza and Metcalfe 1999). Early territory establishment and high standard metabolic rate are important predictors of high dominance status and therefore also high competitive ability (Metcalfe et al. 1989, 1995). Interestingly, there are also indications that early-feeding salmon fry are more heterozygous (Metcalfe 1998). As early-feeding fish generally have higher standard metabolic rates and are more dominant (Metcalfe et al. 1995; Cutts et al. 1999), we might also expect to find an association between genetic variation and competitive ability at the individual level.

GENETIC DIVERSITY AND COMPETITIVE ABILITY: EXPERIMENT I

Several studies have investigated a multitude of factors associated with competitive ability. The majority of these studies have focused on the effect of morphometric traits like size (Schuett 1997), or traits such as earlier experience (Hsu and Wolf 1999) and prior residence (Cutts et al. 1999) on individual competitive ability. Physiological traits, including metabolic rate (Røskaft et al. 1986; Metcalfe et al. 1995), have also been found to be connected to dominance position. However, studies concerning the effect of genetic variability, measured with genetic markers, on an individual's competitive ability, or behaviour in general were still scarce (Lahti 2001; Höglund et al. 2002). This lack of studies was even more striking given the wide-ranging, although moderate, effects genetic diversity has been observed to have on fitness-related traits in various species (reviewed in Mitton and Grant 1984; David et al. 1995; Falconer and Mackay 1996; Crnokrak and Roff 1999; Wang et al. 2002).

In this study the aim was to determine whether groups of families of juvenile salmon, which are known to differ in genetic diversity, would also have different levels of aggression. In order to create groups of families with differing levels of genetic diversity we used a method first presented by Primmer et al. (2003). With this method we were able to estimate the expected level of offspring genetic diversity based on parental genotype data. These estimated indices of genetic diversity have been shown to accurately predict the average level of genetic diversity in Saimaa salmon (Primmer et al. 2003).

Methods

Estimating offspring genetic diversity

The parents of the test fish, namely 51 females and 49 males, were first-year hatchery generation fish maintained at the Saimaa Fisheries Research and Aquaculture station in Enonkoski, southeastern Finland. These 51 females were offspring of one female mated with three males, whereas the 49 males were obtained from matrix fertilizations by crossing 21 females with 15 males. The DNA from these anaesthetized (buffered tricaine ~150 mg/l) 100 parent fish were taken and analysed using 11 microsatellite markers (the details of DNA extraction, PCR, gel electrophoresis and data analysis are described by Tiira et al. 2003). Females were year class 1990, whereas males were year class 1996. The large difference (6 yrs) in the age of the females and males was chosen to avoid matings between full- and half-siblings.

The following estimators of genetic variability were calculated for each possible parental pair of individuals:

(a) mean $d^2 = \Sigma[(i_a - i_b)^2/n])$, where i_a and i_b are the relative number of repeat units in an individual's microsatellite alleles, and n is the number of loci analysed (Coulson et al. 1998)

(b) mean d^2_{scaled} = the mean d^2 of each locus was standardized with its locus-specific SD, and then averaged over all loci analysed in an individual (Coltman et al. 1998)

(c) H_{OBS} = number of heterozygous loci/the total number of loci.

Multilocus microsatellite genotypes for all possible offspring genotypic combinations in a family were simulated based on the parental genotypes (4^n combinations where potential parents have been analysed for n loci). Then, mean d^2, d^2_{scaled} and H_{OBS} values were calculated for each of the simulated multilocus genotypes and these values then averaged over all 1.05×10^6 genotypic combinations resulting in one value per male–female pair for each estimator. This was done using a computer program which assumed Mendelian inheritance, no mutation and no linkage between loci (R2D2: available from http://users.utu.fi/primmer/; Primmer *et al.* 2003). The genetic indices calculated for each pair represented mean d^2, d^2_{scaled} and estimated heterozygosity, hereafter referred to as mean d^2_{EST}, mean $d^2_{scaled\text{-}EST}$ and H_{EST}. Among these 2136 possible pairs, we then selected 10 pairs with relatively low genetic diversity and 10 pairs with high genetic diversity values (Table 11.1). These groups formed the basis of the experiment and are hereafter referred to as HIGH and LOW genetic diversity groups. The choice was initially made based on $d^2_{scaled\text{-}EST}$ values, as this measure has been suggested to reduce the bias of any single locus on the estimate (Coltman *et al.* 1998) and two of the 11 loci used in this study exhibited considerably larger allele size ranges in potential parents than the 9 remaining loci, which could have potentially biased estimates based on d^2. However, the H_{EST} and mean d^2_{EST} values also differed significantly between the HIGH and LOW groups and hence represent groups generally differing in their level of genetic diversity. Following artificial fertilization, the offspring were reared in family-wise hatching compartments. Our study was conducted in common-garden conditions to assure that the possible observed differences were due to either genetic differences or environmental parental (mostly maternal) effects (Bernardo 1996).

Table 11.1 *Descriptive genetic diversity indices for the* HIGH *and* LOW *genetic diversity experimental groups as estimated from parental genotypes (see text for details)*

	HIGH group			Low group		
	Mean $d^2_{scaled\text{-}EST}$	Mean d^2_{EST}	H_{EST}	Mean $d^2_{scaled\text{-}EST}$	Mean d^2_{EST}	H_{EST}
Max.	1.368	31.178	0.752	0.550	6.645	0.499
Min.	0.706	7.1734	0.365	0.222	0.758	0.203
Median	1.145	10.579	0.637	0.439	3.688	0.298
Mean	1.115	15.766	0.615	0.424	3.345	0.325
Variance	0.026	82.177	0.013	0.010	2.736	0.017

Source: Tiira *et al.* (2003).

Behaviour trials

Aggressive behaviour was monitored for groups of six similarly sized fish originating from the same family. Three replicated groups from each family (10 LOW and 10 HIGH families) were used, except for one family from which we had only two replicated groups. The same fish were not used more than once in the experiment. The fish were observed once a day for a 3-day period, one observation session lasting for 30 min. We recorded the following aggressive behaviours: *charge, chase, lateral display, nip* (Keenleyside and Yamamoto 1962), *approach* (Symons 1968) and *circle* (Johnsson and Åkerman 1998). To further investigate the importance of different forms of aggressive acts, these behaviours were broken down into two groups. *Approach* and *charge*, which were the least costly and risky behaviours, were classified as mild aggressions. *Nip, chase, circle* and *lateral display* were regarded as overt aggressions as they were more costly behaviours (*chase*), required physical contact (*nip*), or took place in an actual fighting situation, where both fish were motivated to fight (*circle, lateral display*). Although lateral displays are not conventionally categorized as overt aggressions, it was the best option in this case as they were seen only in intense fighting situations. The number of all aggressive behaviours was summed into one variable, and the mean aggression rate over the three 30-min observation sessions per aquarium was calculated and used in later statistical analyses:

$$\text{mean total aggression} = (x_1 + x_2 + x_3)/3$$

where the x_i is the number of all (*charge + chase + lateral display + nip + approach + circle*) aggressive acts performed by six fish during one observation period (30 min).

Similarly, we calculated mean overt and mean mild aggression rates for each 30-min observation period.

Results and discussion

Salmon fry estimated to have low level of genetic diversity were less aggressive compared with fry with higher estimated levels of genetic diversity (Figs. 11.1 and 11.2 and Table 11.2). This difference between the groups LOW and HIGH was even more clear in overt aggressions (Fig. 11.1, Table 11.2), which indicates that the effect of genetic diversity was clearer when considering only costly and risky aggression forms. Aggressiveness was also correlated positively with family-wise genetic estimates; estimated $d^2_{\text{scaled}\,-\text{EST}}$ ($r = 0.63$, $P = 0.003$, $N = 20$, Fig. 11.2a) and with H_{EST} ($r = 0.62$,

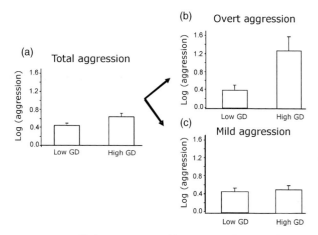

Figure 11.1. (a) Overall, (b) overt or (c) mild aggression (log-transformed + SE) for a 30-min observation period in two groups having either low or high estimated genetic diversity (GD). The number of replicates in both groups was 10. (Adapted from Tiira *et al.* 2003.)

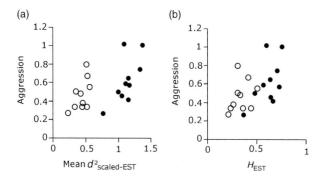

Figure 11.2. Mean aggression (log-transformed) of offspring in the HIGH (black circles) and LOW (white circles) genetic diversity families plotted against family-wise (a) $d^2_{scaled-EST}$ values, (b) H_{EST} values. (From Tiira *et al.* 2003.)

$P = 0.004$, $N = 20$, Fig. 11.2b), where individuals with high genetic diversity behaved more aggressively. These results suggest that large variation in individual competitive ability might partly be explained by individual genetic diversity; fish with less genetic diversity behaved less aggressively than fish with high genetic diversity. The fact that significant positive association was also seen in the HIGH group ($d^2_{scaled-EST}$: $r = 0.69$, $P = 0.029$, H_{EST}: $r = 0.56$, $P = 0.09$) and not just between the groups, gives further support for this hypothesis. Other possible factors causing differences in aggressive behaviour, such as egg-size-related maternal effects

Table 11.2 *Nested ANOVA describing the effects of genetic diversity group and family (nested within group) on total aggressiveness and the number of overt and mild aggressions. Our approach was hierarchical, the prime interest being in the variation of total aggression between groups and families, rather than all variables being of equal importance. Therefore we have not presented corrections for multiple comparisons*

Source	MS	df	F	P^a
Total aggression				
Group	0.585	1	4.466	**0.048**
Family (Group)	0.131	18	0.970	0.510
Error	0.135	39		
Overt aggressions				
Group	1.246	1	4.829	**0.041**
Family (Group)	0.258	18	1.056	0.427
Error	0.244	39		
Mild aggressions				
Group	0.027	1	0.351	0.561
Family (Group)	0.077	18	1.006	0.474
Error	0.076	39		

^{a} Bold figures significant at $P < 0.05$.
Source: Tiira *et al.* (2003).

($t = 0.026$, df $= 18$, $P = 0.979$) or body size (family specific body weight $r = -0.15$, $P = 0.52$), were not associated with aggressiveness.

Recently, there has been an increase in studies investigating the effect of genetic variability on behaviour (Lahti 2001; Höglund *et al.* 2002; Tiira *et al.* 2003; Callardo and Neira 2005; Tiira *et al.* 2005; Hoffman *et al.* 2007; Välimäki *et al.* 2007). Nevertheless, the majority of information on the effects of genetic diversity on behaviour have been obtained using inbred and outbred laboratory strains of mice (Eklund 1996; Meagher *et al.* 2000) and *Drosophila* (Latter and Robertson 1962; Latter and Sved 1994). The low aggressive level of salmon from the LOW group may indicate negative inbreeding effects, as the genetic diversity in this population is very low (microsatellite $H_o = 0.29$: Tonteri *et al.* 2005). Inbreeding has earlier been found to decrease aggressiveness in mice, as male mice from inbred lines show less aggressive behaviour and are poorer competitors compared with males from outbred lines (Eklund 1996; Meagher *et al.* 2000). Alternatively, the high aggressiveness exhibited by fish in the HIGH group can be an indicator of better performance due to heterosis, or a combination of both factors.

However, although the inbreeding hypothesis can explain the low aggressiveness of the fish having less genetic diversity, another, very relevant explanation exists for the observed result. Salmonid fish usually are less aggressive to close relatives than towards unrelated fish (Olsén *et al.* 1998 and references therein; Griffiths and Armstrong 2002; but see Griffiths and Armstrong 2001). The 'phenotype matching hypothesis' (Slater 1994) suggests that individuals either have a learned or genetically dictated recognition template against which individuals compare other conspecifics (Porter *et al.* 1983). The siblings in the LOW group, which shared a higher number of alleles with each other, might also have a more similar recognition template, resulting in enhanced kin recognition, or interpreted in another way, a lowered efficiency of recognising potential competitors. Both phenomena should lead to lowered aggression. This interpretation is supported by studies where reduced aggressiveness and inability to recognize related competitors were found to result from insufficient genetic differentiation due to inbreeding (Nevison *et al.* 2000).

GENETIC VARIATION AND COMPETITIVE ABILITY: EXPERIMENT II

In Experiment I, we found that groups of salmon fry siblings with less genetic variation behaved less aggressively towards each other, compared with groups of siblings having more genetic variation; however, whether this was due to the level of genetic diversity or phenotype matching remained uncertain. In Experiment II, using a slightly different experimental set up, we aimed to solve the question which remained open in Experiment I: does the genetic variation itself have an effect on competitive ability? In this study we also used offspring with estimated genetic diversity; however, this time we used behavioural trials with two fish, one having either low or high estimated amount of genetic variation, and as an opponent for these test fish we used fish having intermediate levels of estimated genetic variation. This was done in order to achieve as equal competition situations as possible. As competitive ability includes not only aggressiveness, but also the ability to obtain food in competitive situations, we included both traits in this study.

Methods

We used the Primmer *et al.* (2003) method to estimate the genetic level of offspring again in this experiment in order to have fish with known levels of genetic variation. This method is described in detail in section 'Estimating offspring genetic diversity', above. The parents of the test fish were from the same age classes as in Experiment I, and these fish were genotyped with 11 microsatellite markers (Tiira *et al.* 2003). We then chose two pairs expected to produce, on average, offspring with low $d^2_{scaled\text{-}EST}$ value (values 0.239, 0.457) and two pairs with high offspring $d^2_{scaled\text{-}EST}$ value (values 0.988, 1.479) from 598 possible combinations of female and male matings. In addition, six pairs with intermediate $d^2_{scaled\text{-}EST}$ values (0.794–0.916) were chosen. Offspring from these intermediate $d^2_{scaled\text{-}EST}$ families were used as opponents for the test fish (with low or high genetic diversity) in the behavioural trials. The chosen pairs were artificially mated in the autumn and 15 eggs from each of the low/high families and 10 eggs from each intermediate family were placed in individual rearing compartments (10 × 5.5 × 9.8 cm), located in standard hatchery troughs. The eggs from each family were randomly placed in the compartments to avoid any effect of the growing environment being manifested only in certain families. After swim-up, the fish were fed with standard hatchery food (nutraG EWOS, diameter 0.6 mm) ad libitum. The study fish were kept in similar conditions, and the hatchery practices were same for all fish starting from the egg phase. The study fish were kept in their individual rearing compartments until they entered the behavioural trials. This is because the early social experience can affect the outcome of encounters in salmonids (Johnsson and Åkerman 1998). As a result of our rearing practice the fish were socially naïve.

Behavioural trials

Competitive ability was measured in trials between two size-matched fish. In each trial, there was one fish (opponent) estimated to have an intermediate mean d^2_{scaled} value together with the test fish (from either low or high mean estimated d^2_{scaled} family). In total, 26 such trials were conducted. We used fish with intermediate estimated mean d^2_{scaled} values as opponents in order to standardize the genetic variation of the opponents and its possible effect on the results. Fish from all the six control families were used as opponents.

Before entering the behavioural trials, the length of the fish was measured and the fish were individually marked with Visible Implant Fluorescent Elastomer tags (VIE: Northwest Marine Technology, Inc., Washington) under anaesthesia (neutralized MS 222). Four different colours were used in the marking: yellow, red, green and orange. A combination of two colours was used for each individual. The size of the marks was on average 1 mm.

The fish were placed in the observation aquaria (16 aquaria, 30 cm × 25 cm, water depth 30 cm) after they had recovered from the anaesthesia, and the observations were started the following day. Fish were observed as outlined in the previous section. The fish were fed with an equal amount of food at the beginning of each session. In addition, the fish were fed once in the evening. Food pellets were provided in a floating plastic frame (diameter 8 cm), which prevented the pellets floating through outlet and ensured that the food was always provided at the same location.

During the observation sessions we recorded aggressions (initiator, receiver and the form of aggression) and feeding (identity of the individual and the number of eaten pellets). Aggressive behaviours were classified as in Experiment I. However, for the statistical analysis all aggression forms were summed up into one variable and mean aggression rate for the three 30-min observation sessions per individual was calculated. Similarly, mean feeding rate per 30 min was calculated. The aggressiveness of the control fish was subtracted from the aggressiveness of the test fish and this index was used in the final analysis. The same procedure was applied to the feeding rate.

Genetic analysis and statistics

We genotyped the offspring to ensure that the estimated genetic diversity values were accurate. After the behavioural trials, we clipped a small piece of fin from the test fish, and preserved it in 96% alcohol. These samples were analysed with 10 of the 11 microsatellite loci (except FGT1) described in Tiira et al. (2003) (Table 11.3). Twenty-two fish were successfully analysed for eight or more of the 10 loci. For these individuals we calculated the mean individual d^2, d^2_{scaled} (Coltman et al. 1998; Coulson et al. 1998), average heterozygosity (H) and internal relatedness (IR).

Spearman rank correlation analysis (r_S) was used to estimate the correlation between genetic diversity indices estimated from parental data and the observed genetic diversity values of offspring. The correlation between the observed genetic diversity and behavioural factors was done using Pearson product moment correlation analysis (r). Furthermore, we used generalized linear mixed models (GLMMs: McCullagh and Nelder 1989) to analyse the associations between observed and estimated genetic diversity and foraging. This was done using PROC MIXED of the SAS Statistical Package according to Littell et al. (1996). A GLMM was appropriate for this analyses as it allows the analysis of the data where the response variable (in this case the fitness-related behavioural trait) is determined both by random (family) and fixed effects (genetic variability estimator) (e.g. McCullagh and Nelder 1989; Merilä et al. 2001).

Table 11.3 *Microsatellite loci used in Experiment II and diversity indices. A is the total number of alleles observed at the locus and H is observed heterozygosity. Size range (bp) refers to the size range of the particular markers in this study. In addition, locus-specific mean d² and its variance V of the markers are given*

	Total[a]		This study[b]				
Microsatellite locus	A	H	A	H	Size range	Mean d^{2b}	Variance[b]
Ssa171	6	0.51	4	0.85	220–242	12.6	103.777
Ssa197	5	0.43	4	0.63	169–207	11.9	793.27
Ssa202	5	0.73	4	0.62	247–259	0.81	1.39
Ssa289	2	0.49	2	0.38	123–127	1.5	3.88
SSOSL417	6	0.60	2	0.48	184–190	4.1	21.04
SSOSL438	7	0.56	2	0.54	137–139	0.4	0.35
MST15	6	0.25	2	0.43	216–218	0.4	0.25
543AE	2	0.40	2	0.38	145–148	0.2	0.37
Omy27	4	0.58	3	0.41	139–151	3.7	75.37
Sfo8	5	0.34	2	0.48	205–217	16.3	388.23

[a] Figures based on analysis of >500 individuals (C. Primmer, unpubl.).
[b] Figures based on the juvenile fish used in this study.

To further investigate the mechanism behind the observed connection between genetic diversity and behaviour, we performed two extra analyses. It has been suggested that if heterozygosity is correlated between independent loci, then heterozygosity carries an inbreeding signal and the mechanism behind the observed result would be identity disequilibrium (i.e. general effect) (Balloux *et al.* 2004). This was studied by dividing the genotypes of an individual randomly into two groups and calculating mean d^2_{scaled} and heterozygosity for these two groups. The correlation between the genetic diversity indices of these two groups was then calculated. This was repeated 1000 times. To investigate whether the observed association between genetic diversity and foraging would originate from the effect of one single locus, we analysed the association between foraging index and mean d^2_{scaled} value calculated for each locius separately, using Pearson product moment correlation analysis (*r*). If the effect originates from one or few loci then these loci are potentially linked with fitness loci in the local chromosomal vicinity of the markers (local effect) (Hansson *et al.* 2004).

Results and discussion

The estimated values of offspring genetic diversity (mean $d^2_{scaled\text{-}EST}$ and mean d^2_{EST} values) correlated positively with observed genetic diversity (d^2_{scaled} and d^2 values) of individual fish ($r \geq 0.666$, $P \leq 0.004$) assuring

Table 11.4 *Associations between the observed genetic variation and competitive ability (foraging index, aggression index) among 20 salmon fry*

	Foraging index[a]	Aggression index
Mean d^2	0.408	0.032
Mean d^2_{scaled}	0.603**	0.132
H	0.526*	0.111

[a] Statistical significance of correlation coefficients * < 0.05, ** < 0.01.

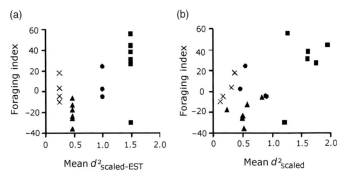

Figure 11.3. Relationship between individual foraging index and genetic diversity, as estimated by (a) average family mean $d^2_{scaled\text{-}EST}$, based on parental genotype data; and (b) individual mean d^2_{scaled}, based on offspring genotype data. Different symbols indicate offspring from each family. Within family correlation coefficients based on mean d^2_{scaled} are as follows: × = 0.955; black triangle = 0.432, black circle = −0.661, black square = 0.443. (From Primmer *et al.* 2003.)

that the group division of experimental fish based on the estimated values was accurate (Primmer *et al.* 2003).

Foraging index was found to be positively associated with mean d^2_{scaled} (Fig. 11.3b) and heterozygosity (Table 11.4) (Primmer *et al.* 2003). This indicates that fish with higher genetic diversity foraged with a higher rate in the presence of a competitor as compared to fish with less genetic variation. Aggressiveness, however, was not associated with any of the studied variables (Table 11.4), although aggression index and foraging index were positively correlated ($r = 0.425$, $N = 26$, $p = 0.030$). Reanalysing the data using internal relatedness (IR) found a similar trend, but no significant association with either foraging ($r = −0.389$, $N = 20$, NS) or aggressiveness ($r = −0.100$, $N = 20$, NS).

We had offspring from four families in our experiment, and thus family effect (maternal or paternal) could influence our results. These are effects

Table 11.5 *Relationship between estimated (based on parental genetic diversity) and observed offspring genetic diversity and foraging*

Measure of genetic diversity	df	Foraging[a] b (±SE)	t	p
Estimated mean d^2	16	3.7 (2.5)	1.5	0.15
Estimated mean d^2_{scaled}	16	28.4 (15.3)	1.9	0.08
Estimated H	16	57.5 (45.9)	1.3	0.23
Observed mean d^2	16	2.22 (1.57)	1.4	0.18
Observed mean d^2_{scaled}	16	**27.0 (11.5)**	**2.4**	**0.03**
Observed H	16	42.9 (32.0)	1.3	0.20

[a] Statistically significant ($P < 0.05$) associations, as estimated using generalized linear mixed model analysis, are highlighted in bold.
Source: Adapted from Primmer *et al.* (2003).

which influence the offspring's phenotype through the female's/male's phenotype or environment (Mousseau and Fox 1998; Einum and Fleming 1999). The relationship between foraging and genetic diversity was furthermore analysed using generalized linear mixed model analysis, where GLMM, with family as random factor, takes into account the non-independent data structure caused by replication within families. Also, in this analysis the foraging index was found to be significantly higher in offspring having higher genetic diversity (Table 11.5 and Figs. 11.3a, b).

The body lengths of the test fish and the opponent did not differ (paired *t*-test: t_{22} = 0.94, P = 0.357) and thus size difference cannot explain the differences in the behaviour. There was no association between the length of the test fish and foraging index (r = −0.078, N = 26, P = 0.704), nor between length and aggression index (r = −0.220, N = 26, P = 0.280). Similarly, no correlation was found among the genetic estimates and the size of the fish (mean d^2: r = −0.190, N = 22, P = 0.397; mean d^2_{scaled}: r = −0.233, N = 22, P = 0.297; heterozygosity: r = −0.202, N = 22, P = 0.368) (Primmer *et al.* 2003).

Individuals within a salmonid population tend to differ in their competitive ability. This study, together with Experiment I in this chapter, suggests that one of the factors generating variation in competitive ability is an individual's genetic diversity. We found that juvenile salmon with a low amount of genetic variation foraged less in the presence of a competitor. This study complements our earlier results in Experiment I, where genetic variation was found to be associated with individual aggressiveness. Several studies have thereafter reached similar results; high heterozygosity (or IR) has been found to be associated with higher competitive ability (Välimäki *et al.* 2007), aggression or dominance (Callardo and Neira 2005; Tiira *et al.* 2005). Competitive ability in salmonids consists of the ability to forage in the presence of a competitor and the ability to behave aggressively. Although aggressiveness was not associated with genetic diversity in this study, foraging ability and aggressiveness were positively correlated. The lack of a significant correlation between aggression and genetic variation in this study may have been due to small sample size. Furthermore, the overall aggression level in this study was lower compared with that found in Experiment I. Individuals with higher metabolic rates behave more aggressively (Metcalfe *et al.* 1995; Lahti *et al.* 2002) and thus the genetic variation may have a stronger association with metabolic rate than aggressiveness. This hypothesis needs further investigation.

As mentioned in the Introduction, the two most likely explanations for finding a positive association between genetic variation and fitness are local and/or general effects (Hansson and Westerberg 2002). According to the general effect hypothesis, there is a correlation between the homozygosity of different loci, also between unlinked loci (Charlesworth 1991), where heterozygosity in neutral loci reflects the variation in coding loci via correlation. If the observed result originates from general effects, then the heterozygosity of microsatellite markers should correlate thus signalling inbreeding in a population (Balloux *et al.* 2004). This is what we found in our study; heterozygosity values correlated positively among the markers ($r = 0.525$, $N = 20$, $p = 0.017$) in our study population. According to Balloux *et al.* (2004), high heterozygosity–heterozygosity correlation values suggest that the inbreeding coefficient and heterozygosity are strongly correlated. In other words, in this study inbreeding appears to be an important factor behind the observed correlation between fitness and genetic variation, where fish with low competitive ability are also more inbred.

Positive correlations between genetic diversity and fitness can also appear as a result of the co-segregation of neutral microsatellites with genes under selection (local effect) (Bierne *et al.* 1998; Hansson and Westerberg 2002, 2008), and as the local and general effect hypothesis are not mutually exclusive, we performed also a single locus analysis to investigate the local effect. Interestingly, the results of the single locus analysis revealed a significant effect for one locus, namely Ssa202 ($r = 0.576$, $P = 0.008$), thus giving support also for the linkage disequilibrium/local effect hypothesis as one of the factors causing the observed association between foraging index and genetic variability. In other words, microsatellite locus Ssa202 either has a physical linkage, or linkage through variety of demographic processes, with a gene, which is important for the fitness trait in question. We also analysed the association between mean d^2_{scaled} without locus SSa202 and foraging index, and found that after removing locus Ssa202, the significant association between mean d^2_{scaled} and foraging still existed (r = 0.528, p = 0.017). This indicates that although a local effect seems to influence on the association between foraging and genetic variation, the general effect and thus inbreeding is also responsible for the observed effect. Partial inbreeding, small population sizes, bottlenecks and population admixture (Lynch and Walsh 1998) are important factors creating LD and correlation in the genome; the same factors often appear in hatchery rearing (Allendorf and Phelps 1980; Ryman and Ståhl 1980; Altukhov et al. 2000) and thus the hatchery background of the study population may have an important role in the appearance of heterozygosity–fitness correlations. In support of this notion, we have later discovered socially dominant brown trout (*Salmo trutta*) to indeed have higher heterozygosity compared with subordinates; however this was evident only in trout populations having the longest hatchery background (Tiira et al. 2005).

Another point to consider is that the fitness effects of individual genetic diversity may not be evident in hatchery conditions, where the food is abundant and no predators exist, but appear in more stressful natural environments. Meagher et al. (2000) found that poor competitive ability of inbred mice had a negligible effect on fitness in the laboratory, but had a strong effect on reproductive success in a semi-natural environment. Furthermore, the effect of genetic variability on fitness related traits is often found in young individuals (Tiira et al. 2006), but is absent in older ones (David 1998). The life-history stage after which the effect of genetic variation evens out is probably very context- and species-dependent. The difference in competitive ability among juveniles differing in genetic diversity appeared in controlled laboratory conditions in this study. In wild salmonid populations, individual differences in competitive ability resulting from differences in genetic variation could be smaller than in laboratory conditions due to a higher selective mortality working via other traits affected by genetic variation already in the early life history phases in the wild. Alternatively, the effect of genetic variation might be mainly mediated through competition among juveniles eliminating the individuals with the lowest competitive ability. Thus, competition in the juvenile stage may be a strong selective force against individuals with a low amount of genetic variation. Behavioural syndromes (correlated behaviours: Sih et al. 2004) can, however, complicate the picture. Dominant and aggressive individuals often behave more boldly in risky situations, for example when confronted with a predator. We recently discovered that brown trout behaving reck-lessly while under threat from predation had a higher heterozygosity compared to individuals with good abilities for avoiding predators (Vilhunen et al. 2008). Thus, even though fish with high heterozygosity may have fitness benefits from better competitive ability, they may also suffer higher mortality as a result of their risk-prone behaviour.

EXPERIMENTS I AND II: DISCUSSION AND CONCLUSIONS

The two case studies presented above demonstrate positive associations between genetic diversity and fitness-related traits in Saimaa salmon. Salmon with low levels of genetic variation were less aggressive than individuals with higher levels of genetic diversity (Experiment I); in addi-tion, fry with a higher level of genetic variation foraged more in the presence of a competitor (Experiment II). Behavioural differences associated with the level of genetic diversity have been increasingly documented recently (Höglund et al. 2001; Tiira et al. 2003, 2005; van Oosterhout et al. 2003; Välimäki et al. 2007; Charpentier et al. 2008; Vilhunen et al. 2008).

The estimation of the offspring's genetic diversity using only parental genotypic information could have practical applications for captive breeding programmes. From a conservation perspective, this ability to assess the future success of offspring resulting from specific male–female combina-tions, prior to them actually mating, offers interesting possibilities for the conservation of captive species and populations (Primmer et al. 2003). The possibility to gain an insight into offspring fitness without having to analyse the offspring themselves could be particularly useful for species where family sizes can be extremely large such as many fishes and amphibians, which makes analysis of all offspring essentially impossible (Primmer et al. 2003; Lesbarréres et al. 2005). In addition to resulting in significant savings of genotyping resources, the ability to predict offspring genetic diversity prior to fertilization also enables experimental evaluation of predicted genetic diversity–fitness associations based on specific hypotheses (Lahti 2001; Tiira et al. 2003).

Studies based on meta-analyses, theoretical expectations and empirical data suggest a weak correlation between microsatellite measures of genetic variation and inbreeding coefficient (Balloux et al. 2004; Slate et al. 2004). However, if a correlation exists between multilocus heterozygosity and inbreeding, then the heterozygosity across markers should be correlated within individuals, and thus a random sample of markers can reflect inbreeding (Balloux *et al.* 2004). This hypothesis was tested in the second case study, and our results suggest that both inbreeding and linkage disequilibrium (local effect) are responsible for the significant heterozygosity–fitness association in Experiment II. The fact that inbreeding depression is one of the main reasons for the observed result is not surprising as Saimaa Lake salmon have a very low level of genetic variation (Vuorinen 1982; Primmer *et al.* 2000; Tonteri *et al.* 2005). In general, the level of genetic variation in captive salmonid populations has been a great cause of concern (Ryman and Ståhl 1980) and the general trend seems to be the decline in the amount of genetic diversity in hatcheries compared with the wild population (Allendorf and Phelps 1980; Ryman and Ståhl 1980; Fleming and Einum 1997; Crozier 1998). This is alarming as many of these captive populations are used to enhance the wild populations. Low numbers of founder individuals, as well as intentional and unintentional selection in breeding, appears to be the reason for the observed increase in homozygosity.

Several studies have found mean d^2 and/or heterozygosity to be associated with fitness traits (Coltman *et al.* 1998; Coulson *et al.* 1998, 1999; Hansson *et al.* 2004). However, two large surveys comparing mean d^2 (Hedrick *et al.* 2001; Slate and Pemberton 2002) and heterozygosity found mean d^2 to be less predictive in reflecting inbreeding/outbreeding in a population than heterozygosity. Support for heterozygosity outperforming mean d^2 came also from Tsitrone *et al.* (2001), who modelled genotype–fitness correlations taking into account different mutation rates and population scenarios. They argue that heterozygosity usually provides a better correlation with fitness compared to other indices (with the exception of the situation when the size of the subpopulation is large and markers have a very high mutation rate). In our studies we had the possibility to select offspring with large differences in the amount of genetic diversity, and therefore the likelihood that all the measures of genetic variation show the same signal increased. For example, in Experiment I, the LOW and HIGH groups differed in all genetic indices, namely mean d^2 $_{EST}$, d^2 $_{scaled -EST}$ and in H_{EST}. In Experiment II the foraging of the salmon fry in the presence of competitor was associated with both mean d^2 $_{scaled}$ and heterozygosity; however, in a more robust GLMM analysis only mean d^2 $_{scaled}$ showed significant effects with foraging.

We have presented two case studies, where low levels of genetic variation were found to be associated with low competitive ability in captive salmon. Low level of genetic variation is a cause of concern in the hatcheries, but what is the situation in the wild? Is inbreeding a threat in wild salmonid populations? The structure of a salmonid population can have several hierarchical levels. Under species and continents there is a regional level, which can be furthermore separated into river level. Also within rivers there can be reproductively separated populations. This structure is called a 'population system' (Altukhov et al. 2000), and it has a common historical background. Distinctive features of population systems in salmonids are the stability of their biological and genetic parameters through time, despite the great variability of the subpopulations (Altukhov et al. 2000). These features may lead to inbreeding, as the homing ability of migrating salmonids into their natal river is 98% according to the recalculation made by Altukhov et al. (2000) on the data of eight salmonid species. Natural straying, however, is considered to provide sufficient amount of outbreeding (Allendorf and Waples 1996). The lack of fixed allele frequencies throughout the world in wild salmon supports this view, as the fixation of allele frequencies (frequency = 1) is often a result of inbreeding (Allendorf and Waples 1996). Support for this hypothesis is also provided by the finding that very low levels of migration have been shown to be very effective in 'restoring' genetic variation (Ingvarsson 2001). Nevertheless, locally adapted small populations, with little gene flow, are very vulnerable to human actions like over-fishing and environmental degradation and the status of many salmonid populations can quickly develop into the state where inbreeding depression is a serious threat to wild salmonid populations.

ACKNOWLEDGEMENTS

We thank T. Aho and T. Mäkinen for assistance with microsatellite genotyping and the Finnish Game and Fisheries Research Institute for providing fish rearing facilities. J. Piironen and N. Peuhkuri provided useful comments on earlier versions of the manuscript. Special thanks to Sami Aikio and Bill Amos for the help in calculating correlations between heterozygosities. Our research has been funded by the Academy of Finland to CRP (project #201499 and Centre of Excellence in Evolutionary Genetics and Physiology) and to KT (project #80705), the Finnish ministry of Education (to KT) and the Finnish Game and Fisheries Research Institute (to CRP).

REFERENCES

Acevedo-Whitehouse K., Gulland F., Greig D. and Amos, W. (2003). Inbreeding: disease susceptibility in California sea lions. *Nature*, **422**, 35.

Allendorf, F. W. and Leary, R. F. (1988). Conservation and distribution of genetic variation in a polytypic species, the cutthroat trout. *Conservation Biology*, **2**, 170–184.

Allendorf, F. W. and Phelps, S. R. (1980). Loss of genetic variation in a hatchery stock of cutthroat trout. *Transactions of American Fisheries Society*, **109**, 537–543.

Allendorf, F. W. and Waples, R. S. (1996). Conservation and genetics of salmonid fishes. In *Conservation Genetics: Case Histories from Nature*, ed. J. C. Avise and J. L. Hamrick. New York: Chapman and Hall, pp. 238–280.

Altukhov, Y. P., Salmenkova, E. A. and Omelchenko, V. T. (2000). *Salmonid Fishes: Population Biology, Genetics and Management*, English translation ed. G. R. Carvalho and J. E. Thorpe. Oxford: Blackwell Science.

Amos, W., Worthington Wilmer, J., Fullard, K. *et al.* (2001). The influence of parental relatedness on reproductive success. *Proceedings of the Royal Society Series B*, **268**, 2021–2027.

Balloux, F., Amos, W. and Coulson, T. (2004). Does inbreeding estimate heterozygosity in real populations? *Molecular Ecology*, **13**, 3021–3031.

Beaumont, M. A. and Bruford, M. W. (1999). Microsatellites in conservation genetics. In *Microsatellites: Evolution and Applications*, ed. D. B. Goldstein and C. Schlötterer. Oxford: Oxford University Press, pp. 165–182.

Bernardo, J. (1996). Maternal effects in animal ecology. *American Zoologist*, **36**, 83–105.

Bierne, N., Launey, S., Naciri-Graven, Y. and Bonhomme, F. (1998). Early effects of inbreeding as revealed by microsatellite analysed on *Ostrea edulis* larvae. *Genetics*, **148**, 1893–1906.

Bohlin, T., Dellefors, C. and Faremo, U. (1993). Optimal time and size for smolt migration in wild sea trout (*Salmo trutta*). *Canadian Journal of Fisheries and Aquatic Sciences*, **50**, 224–232.

Brännäs, E. (1995). First access to territorial space and exposure to strong predation pressure: a conflict in early emerging Atlantic salmon (*Salmo salar*, L) fry. *Evolutionary Ecology*, **9**, 411–420.

Britten, H. B. (1996). Meta-analysis of the association between multilocus heterozygosity and fitness. *Evolution*, **50**, 2158–2164.

Callardo, J. R. and Neira, R. (2005). Environmental dependence of inbreeding depression in cultured Coho salmon (*Oncorhynchus kisutch*): aggressiveness, dominance and intraspecific competition. *Heredity*, **95**, 449–456.

Charlesworth, D. (1991). The apparent selection on neutral marker loci in partially inbreeding populations. *Genetical Research*, **57**, 159–175.

Charlesworth, D. and Charlesworth, B. (1987). Inbreeding depression and its evolutionary consequences. *Annual Reviews of Ecology and Systematics*, **18**, 237–268.

Charpentier, M. J. E., Prugnolle, F., Gimenez, O. and Widding, A. (2008). Genetic heterozgosity and sociality in primate species. *Behavioural Genetics*, **38**, 151–158.

Coltman, D. W. and Slate, J. (2003). Microsatellite measures of inbreeding: a meta-analysis. *Evolution*, **57**, 971–983.

Coltman, D. W., Bowen, W. D. and Wright, J. M. (1998). Birth weight and neonatal survival of harbour seal pups are positively correlated with genetic variation measured by microsatellites. *Proceedings of the Royal Society Series B*, **265**, 803–809.

Coulson, T. N., Pemberton, J. M., Albon, S. D. *et al.* (1998). Microsatellites reveal heterosis in red deer. *Proceedings of the Royal Society Series B*, **265**, 489–495.

Coulson, T., Albon, S., Slate, J. and Pemberton, J. (1999). Microsatellites loci reveal sex-dependent responses to inbreeding and outbreeding in red deer calves. *Evolution*, **53**, 1951–1960.

Crnokrak, P. and Roff, D. A. (1999). Inbreeding depression in the wild. *Heredity*, **83**, 260–270.

Crozier, W. W. (1998). Genetic implications of hatchery rearing in Atlantic salmon: effects of rearing environment on genetic composition. *Journal of Fish Biology*, **52**, 1014–1025.

Cutts, C. J., Brembs, B., Metcalfe, N. B. and Taylor, A. C. (1999). Prior residence, territory quality and life-history strategies in juvenile Atlantic salmon (*Salmo salar* L). *Journal of Fish Biology*, **55**, 784–794.

Danzmann, R. G., Ferguson, M. M. and Allendorf, F. W. (1987). Heterozygosity and oxygen-consumption rate as predictors of growth and developmental rate in rainbow trout. *Physiological Zoology*, **60**, 211–220.

Danzmann, R. G., Ferguson, M. M. and Allendorf, F. W. (1989). Genetic variability and components of fitness in hatchery strains of rainbow trout. *Journal of Fish Biology*, **35**, 313–319.

David, P. (1998). Heterozygosity–fitness correlations: new perspectives on old problem. *Heredity*, **80**, 531–537.

David, P., Delay, B., Berthou, P. and Jarne, P. (1995). Alternative models for allozyme-associated heterosis in the marine bivalve *Spisula ovalis*. *Genetics*, **139**, 1719–1726.

Einum, S. and Fleming, I. A. (1999). Maternal effects of egg size in brown trout (*Salmo trutta*): norms of reaction to environmental quality. *Proceedings of the Royal Society Series B*, **266**, 2095–2100.

Eklund, A. (1996). The effects of inbreeding on aggression in wild male house mice (*Mus domesticus*). *Behaviour*, **133**, 883–901.

Ellegren, H. (2000). Microsatellite mutations in the germline: implications for evolutionary inference. *Trends in Genetics*, **16**, 551–558.

Elliott, J. M. (1986). Spatial distribution and behavioural movements of migratory trout, *Salmo trutta*, in a Lake District stream. *Journal of Animal Ecology*, **55**, 907–922.

Elliott, J. M. (1994). *Quantitative Ecology and the Brown Trout*. Oxford: Oxford University Press.

Estoup, A. and Cornuet, J.-M. (1999). Microsatellite evolution: inferences from population data. In *Microsatellites: Evolution and Applications*, ed. D. B. Goldstein and C. Schlötterer. Oxford: Oxford University Press, pp. 49–65.

Falconer, D. S. and Mackay, T. F. C. (1996). *Introduction to Quantitative Genetics*. Harlow, UK: Longman.

Fleming, I. A. and Einum, S. (1997). Experimental tests of genetic divergence of farmed from wild Atlantic salmon due to domestication. *ICES Journal of Marine Sciences*, **54**, 1051–1063.

Frankham, R. (1995). Conservation genetics. *Annual Review of Genetics*, **29**, 305–327.

Griffiths, S. W. and Armstrong, J. D. (2001). The benefits of genetic diversity outweigh those of kin association in a territorial animal. *Proceedings of the Royal Society Series B*, **268**, 1293–1296.

Griffiths, S. W. and Armstrong, J. D. (2002). Kin-biased territory overlap and food sharing among Atlantic salmon juveniles. *Journal of Animal Ecology*, **71**, 480–486.

Haig, S. M. and Avise, J. C. (1996). Avian conservation genetics. In *Conservation Genetics: Case Histories from Nature*, ed. J. C. Avise and J. L. Hamrick. New York: Chapman and Hall, pp. 160–189.

Hansson, B. and Westerberg, L. (2002). On the correlation between heterozygosity and fitness in natural populations. *Molecular Ecology*, **11**, 2467–2474.

Hansson, B. and Westerberg, L. (2008). Heterozygosity–fitness correlation within inbreeding classes: local or genome-wide effects? *Conservation Genetics*, **9**, 73–83.

Hansson, B., Bensch, S., Hasselquist, D. and Åkesson, M. (2001). Microsatellite diversity predicts recruitment of sibling great reed warblers. *Proceedings of the Royal Society Series B*, **268**, 1287–1291.

Hansson, B., Westerdahl, H., Hasselquist, D., Åkesson, M. and Bensch, S. (2004). Does linkage disequilibrium generate heterozygosity–fitness correlations in great reed warblers? *Evolution*, **58**, 870–879.

Hedrick, P., Fredrickson, R. and Ellegren, H. (2001). Evaluation of $^-d^2$, a microsatellite measure of inbreeding and outbreeding, in wolves with a known pedigree. *Evolution*, **55**, 1256–1260.

Hoffman, J. I., Forcada, J., Trathan, P. N. and Amos, W. (2007). Female fur seals show active choice for males that are heterozygous and unrelated. *Nature*, **445**, 912–914.

Höglund, J., Piertney, S. B., Alatalo, R. V., Lindell, J., Lundberg, A., Rintamäki, P. T. (2002). Inbreeding depression and male fitness in black grouse. *Proceedings of the Royal Society of London Series B*, **269**, 711–715.

Hsu, Y. and Wolf, L. L. (1999). The winner and loser effect: integrating multiple experiences. *Animal Behaviour*, **57**, 903–910.

Ingvarsson, P. K. (2001). Restoration of genetic variation lost: the genetic rescue hypothesis. *Trends in Ecology and Evolution*, **16**, 62–63.

Johnsson, J. and Åkeman, A. (1998). Watch and learn: preview of the fighting ability of opponents alters contest behaviour in rainbow trout. *Animal Behaviour*, **56**, 771–776.

Karhu, A., Hurme, P., Karjalainen, M. *et al.* (1996). Do molecular markers reflect patterns of differentiation in adaptive traits in conifers. *Theoretical and Applied Genetics*, **96**, 215–221.

Keenleyside, M. H. and Yamamoto F. T. (1962). Territorial behaviour of juvenile Atlantic salmon (*Salmo salar* L.). *Behaviour*, **19**, 139–169.

Knaepkens, G., Knapen, D., Bervoets, L., Hänfling, B., Verheyen, E. and Eens, M. (2002). Genetic diversity and condition factor: a significant relationship in Flemish but not in German populations of the European bullhead (*Gottus gobio* L.). *Heredity*, **89**, 280–287.

Kruuk, L. E. B., Sheldon, B. C. and Merilä, J. (2002). Severe inbreeding depression in collared flycatchers (*Ficedula albicollis*). *Proceedings of the Royal Society Series B*, **269**, 1581–1589.

Kuusisto, E. 1999. Basin and balance. In *Saimaa: A Living Lake*, ed. E. Kuusisto. Helsinki: Tammi, pp. 21–40.

Lacy, R., C., Petric, A. and Warneke, M. (1993). Inbreeding and outbreeding in captive populations of wild animal species. In *Natural History of Inbreeding and Outbreeding*, ed. N.W. Thornhill. Chicago: University of Chicago Press, pp. 352–374.

Lahti, K. (2001). Integrated analysis of aggression in salmonids. Ph.D. thesis, University of Helsinki, Finland. Introduction available at http://ethesis. helsinki.fi/julkaisut/mat/ekolo/vk/lahti/integrat.pdf

Lahti, K., Huuskonen, H., Laurila, A. and Piironen, J. (2002). Metabolic rate and aggressiveness among brown trout populations. *Functional Ecology*, **16**, 167–174.

Latter, B. D. H. and Robertson, A. (1962). The effects of inbreeding and artificial selection on reproductive fitness. *Genetical Research*, **3**, 110–138.

Latter, B. D. H., and Sved, J. A. (1994). A re-evaluation of data from competitive tests shows high levels of heterosis in *Drosophila melanogaster*. *Genetics*, **137**, 509–511.

Lesbarréres, D., Primmer, C. R., Laurila, A. and Merilä, J. (2005). Environmental and population dependency of genetic variability–fitness correlations in *Rana temporaria*. *Molecular Ecology*, **14**, 311–323.

Liskauskas, A. P. and Ferguson, M. M. (1991). Genetic variation and fitness: a test in naturalized population of brook trout (*Salvelinus fontinalis*). *Canadian Journal of Fisheries and Aquatic Sciences*, **48**, 2152–2162.

Littell, R. C., Milliken, G. A., Stroup, W. W. and Wolfinger, R. D. (1996). SAS System for mixed models. Cary, NC: SAS Institute.

Lynch, M. and Walsh, B. (1998). *Genetics and Analysis of Quantitative Traits*. Sunderland: Sinauer Associates.

McCullagh, P. and Nelder, J. A. (1989). *Generalized Linear Models*, 2nd edn. London: Chapman and Hall.

Meagher, S., Penn, D. J. and Potts, W. K. (2000). Male–male competition magnifies inbreeding depression in wild house mice. *Proceedings of the National Academy of Sciences USA*, **97**, 3324–3329.

Merilä, J. and Crnokrak, P. (2001). Comparison of genetic differentiation at marker loci and quantitative traits. *Journal of Evolutionary Biology*, **14**, 892–903.

Merilä, J., Kruuk, L. E. and Sheldon, B. C. (2001). Cryptic evolution in a wild bird population. *Nature*, **412**, 76–79.

Metcalfe, N. B. (1991). Competitive ability influences seaward migration age in Atlantic salmon. *Canadian Journal of Zoology*, **69**, 815–817.

Metcalfe, N. B. (1998). The interaction between behaviour and physiology in determining the life history patterns in Atlantic salmon (*Salmo salar*). *Canadian Journal of Fisheries and Aquatic Sciences*, **55**, 93–103.

Metcalfe, N. B., Huntingford, F. A., Graham, W. D. and Thorpe, J. E. (1989). Early social status and the development of life-history strategies in Atlantic salmon. *Proceedings of the Royal Society Series B*, **236**, 7–19.

Metcalfe, N. B., Taylor, A. C. and Thorpe, J. E. (1995). Metabolic rate, social status and life-history strategies in Atlantic salmon. *Animal Behaviour*, **49**, 431–436.

Mitton, J. B. (1993). Theory and data pertinent to the relationship between heterozygosity and fitness. In *The Natural History of Inbreeding and Outbreeding*, ed. N.W. Thornhill. Chicago: University of Chicago Press, pp. 17–41.

Mitton, J. B. and Grant, M. C. (1984). Associations among protein heterozygosity, growth rate, and developmental homeostasis. *Annual Reviews of Ecology and Systematics*, **15**, 479–499.

Mousseau, T. A. and Fox, C. W. (1998). The adaptive significance of maternal effects. *Trends in Ecology and Evolution*, **13**, 403–407.

Nevison, C. M., Barnard, C. J., Beynon, R. J. and Hurst, J. L. (2000). The consequences of inbreeding for recognizing competitors. *Proceedings of the Royal Society Series B*, **267**, 687–694.

Nicieza, A. G. and Metcalfe, N. B. (1999). Cost of rapid growth: risk of aggression is high for fast-growing salmon. *Functional Ecology*, **13**, 793–800.

Olsén, K. H., Grahn, M., Lohm, J. and Langefors, Å. (1998). MHC and kin discrimination in juvenile Artic charr, *Salvelinus alpinus* (L.). *Animal Behaviour*, **56**, 319–327.

Porter, R. H., Matochik, J. A. and Makin, J. W. (1983). Evidence for phenotype matching in spiny mice (*Acomys cahirinus*). *Animal Behaviour*, **31**, 978–984.

Primmer, C. R., Koskinen, M. T. and Piironen, J. (2000). The one that did not get away: individual assignment using microsatellite data detects a case of fishing competition fraud. *Proceedings of the Royal Society Series B*, **267**, 1699–1704.

Primmer, C. R., Landry, P.-A., Ranta, E. *et al.* (2003). Prediction of offspring fitness based on parental genetic diversity in endangered salmonid populations. *Journal of Fish Biology*, **63**, 909–927.

Pursiainen, M., Makkonen, J. and Piironen, J. (1998). Maintenance and exploitation of landlocked salmon, *Salmo salar* m. *Sebago*, in the Vuoksi Watercourse. In *Stocking and Introduction of Fish* ed. I. G. Cowx. London: Blackwell Science (Fishing News Books), pp. 46–58.

Queller, D. C. and Goodnight, K. F. (1989). Estimation of genetic relatedness using allozyme data. *Evolution*, **43**, 258–275.

Ralls, K. and Ballou, J. (1983). Extinction: lessons from zoos. In *Genetics and Conservation: A Reference for Managing Wild Animal and Plant Populations*, ed. C. M. Schonewald-Cox, S. M. Chambers, B. MacBryde and L. Thomas. Menlo Park: Benjamin Cummings, pp. 164–184.

Reed, D. H. and Frankham, R. (2001). How correlated are molecular and quantitative measures of genetic variation? A meta-analysis. *Evolution*, **55**, 1095–1103.

Reid, J. M., Arcese, P. and Keller, L. V. (2003). Inbreeding depresses immune response in song sparrows (*Melospiza melodia*): direct and inter-generational effects. *Proceedings of the Royal Society Series B*, **270**, 2151–2157.

Ritland, K. (1996). Estimators of pairwise relatedness and individual inbreeding coefficients. *Genetical Research*, **67**, 175–185.

Røskaft, E., Järvi, T., Bakken, M., Bech, C. and Reinertsen, R. E. (1986). The relationship between social status and resting metabolic rate in great tits (*Parus major*) and pied flycatchers (*Ficedula hypoleuca*). *Animal Behaviour*, **34**, 838–842.

Ryman, N. and Ståhl, G. (1980). Genetic changes in hatchery stocks of brown trout (*Salmo trutta*). *Canadian Journal of Fisheries and Aquatic Sciences*, **37**, 82–87.

Schuett, G. W. (1997). Body size and agonistic experience affect dominance and mating success in male copperheads. *Animal Behaviour*, **54**, 213–224.

Shields, W. M. (1993). The natural and unnatural history of inbreeding and outbreeding. In *The Natural History of Inbreeding and Outbreeding*, ed. N. W. Thornhill. Chicago: University of Chicago Press, pp. 143–169.

Sih, A., Bell, A. and Johnson, C. (2004). Behavioural syndromes: an ecological and evolutionary overview. *Trends in Ecology and Evolution*, **19**, 372–378.

Slate, J. and Pemberton, J. M. (2002). Comparing molecular measures for detecting inbreeding depression. *Journal of Evolutionary Biology*, **15**, 20–31.

Slate, J., Kruuk, L. E. B., Marshall, T. C., Pemberton, J. M. and Clutton-Brock, T. H. (2000). Inbreeding depression influences lifetime breeding success in a wild population of red deer (*Cervus elaphus*). *Proceedings of the Royal Society Series B*, **267**, 1657–1662.

Slate, J., David, P., Dodss, K. G., et al. (2004). Understanding the relationship between the inbreeding coefficient and multilocus heterozygosity: theoretical expectations and empirical data. Heredity, **93**, 255–265.

Slater, P. J. B. (1994). Kinship and altruism. In Behaviour and Evolution, ed. P. J. B. Slater and T. R. Halliday. Cambridge: Cambridge University Press, pp. 193–222.

Symons, P. E. K. (1968). Increase in aggression and in social hierarchy among juvenile Atlantic salmon deprived of food. Journal of the Fisheries Research Board of Canada, **25**, 2387–2401.

Thorpe, J. E. (1989). Developmental variation in salmonid populations. Journal of Fish Biology, **35**, 295–303.

Tiira, K., Laurila, A., Peuhkuri, N., et al. (2003). Aggressiveness is associated with genetic diversity in landlocked salmon (Salmo salar). Molecular Ecology, **12**, 2399–2407.

Tiira, K., Laurila, A., Enberg, K., et al. (2005). Do dominants have higher heterozygosity? Social status and genetic variation in brown trout, Salmo trutta. Behavioral Ecology and Sociobiology, **59**, 657–665.

Tiira, K., Piironen, J. and Primmer, C. R. (2006). Evidence for reduced genetic variation in severely deformed juvenile salmonids. Canadian Journal of Fisheries and Aquatic Sciences, **63**, 2700–2707.

Tonteri, A., Titov, S., Veselov, A. et al. (2005). Phylogeography of anadromous and non-anadromous Atlantic salmon (Salmo salar) from northern Europe. Annales Zoologici Fennici, **42**, 1–22.

Tsitrone, A., Rousset, F. and David, P. (2001). Heterosis, marker mutational processes and population inbreeding history. Genetics, **159**, 1845–1859.

van Oosterhout, C., Trigg, R. E., Carvalho, G. R. et al. (2003). Inbreeding depression and genetic load of sexually selected traits: how the guppy lost its spots. Journal of Evolutionary Biology, **16**, 273–281.

Välimäki, K., Hinten, G. and Hanski, I. (2007). Inbreeding and competitive ability in common shrew (Sorex araneus). Behavioral Ecology and Sociobiology, DOI 10.1007/s00265-006-0332-8.

Vilhunen, S., Tiira, K., Laurila, A. and Hirvonen, H. (2008). The bold and the variable: fish with high heterozygosity act recklessly in the vicinity of predators. Ethology, doi: 10.1111/j.1439-0310.2007.01449.x.

Vuorinen, J. (1982). Little genetic variation in the Finnish Lake salmon, Salmo salar sebago (Girard). Hereditas, **97**, 189–192.

Wang, S., Hard, J. and Utter, F. (2002). Salmonid inbreeding: a review. Reviews in Fish Biology and Fisheries, **11**, 301–319.

12

Genetics and conservation on islands: the Galápagos giant tortoise as a case study

CLAUDIO CIOFI, ADALGISA CACCONE, LUCIANO
B. BEHEREGARAY, MICHEL C. MILINKOVITCH,
MICHAEL RUSSELLO AND JEFFREY R. POWELL

INTRODUCTION

The study of intraspecific genetic variation has demonstrated a vast potential to reconstruct phylogeographic patterns, infer historical demographic
processes and define levels of gene flow of conservation relevance
(Avise 2004). Evolutionary and demographic studies, along with evidence
of current genetic and ecological diversity can, in fact, describe levels of
population distinctiveness and direct management initiatives of importance
to the retention of intraspecific genetic variability and the long-term fitness
of endangered species (Fraser and Bernatchez 2001).

Population divergence and taxonomy

Molecular genetics is a particularly valuable tool for the study of island
systems where different selective pressures and dispersal ability of endemic
species can hamper clear patterns of morphological and ecological diversification for populations of taxonomic importance. In the Galápagos giant
tortoise *Geochelone nigra* (or *G. elephantopus*: see Zug 1997), the taxonomy
first proposed by Van Denburgh (1914) has been somewhat controversial.
Taxon designation was originally based on two main tortoise morphologies
and their variants: a large, dome morphotype with rounded carapace and
short limbs, and a smaller saddlebacked form with a highly elevated ante-
rior part of the carapace, longer neck and limbs, and thinner shell. Five
saddlebacked subspecies were described, on the islands of Española (*hoodensis*), San Cristóbal (*chatamensis*), Pinzón (*ephippium*), Fernandina (*phan-
tastica*) and Pinta (*abingdoni*). Domed tortoises were instead reported from
Santa Cruz (*porteri*), Rábida (*wallacei*) and in Isabela on Volcan Darwin
(*microphyes*), Volcan Alcedo (*vandenburghi*), Sierra Negra (*guntheri*) and
Cerro Azul (*vicina*).

Population Genetics for Animal Conservation, eds. G. Bertorelle, M. W. Bruford, H. C. Hauffe, A. Rizzoli
and C. Vernesi. Published by Cambridge University Press. © Cambridge University Press 2009.

Tortoises from Santiago (*darwini*) are of intermediate morphology. Similarly, heterogeneous morphotypes, assigned to the *becki* subspecies, were described on Volcan Wolf, in northern Isabela. For the majority of island populations recent genetic analysis validated the proposed taxonomy, while for others new patterns were recovered which were inconsistent with previous morphologically based nomenclature (Caccone *et al.* 2002; Beheregaray *et al.* 2003a; Russello *et al.* 2005; Ciofi *et al.* 2006).

Restoration genetics

Genetic analysis can also provide key data for breeding programmes to enhance genetic variation and viability through optimized matings (e.g. Blouin 2003; Russello and Amato 2004). Similarly, genetic assignment of individuals with unknown history is important to restore captive bred animals to their population of origin (Allendorf and Luikart 2006).

Extant populations of Galápagos giant tortoises are survivors of reiterated exploitation that removed about 200 000 animals from 11 islands in the Galápagos (Towsend 1925; MacFarland *et al.* 1974). By the early 1900s, *Geochelone* was extinct on the islands of Floreana, Fernandina, Santa Fé and Rábida (Fig. 12.1). The population of Pinta had its last tortoise transferred, in 1972, to the breeding facilities of the Charles Darwin Research Station (CDRS), in Santa Cruz. On Española, 14 tortoises survived and were also translocated to CDRS. All the other island populations were heavily depleted.

Among the different restoration initiatives conducted since the early 1970s, Española has been one of the most successful efforts (Merlen 1999). More than 1200 tortoises were relocated from captive breeding of 15 founders and second generation offspring were first documented on the island in 1994. Although reproduction of repatriated individuals is a first sign of successful reestablishment in the original habitat, maintenance of genetic diversity and population viability strongly depends on the gene pool of founders. A recent investigation looked at the genetic contribution of each captive breeder to the reintroduced population, and in particular to the effect of preferential breeding on effective population size, a major determinant of long-term viability of reintroduced population (Milinkovitch *et al.* 2004). A parallel study was performed to test for genetic assignment to population of origin for most captive tortoises currently held at CDRS (Burns *et al.* 2003). Although differences in carapace shape do exist, it was difficult to unambiguously assign captive individuals based on morphological traits alone. Genetic tests were therefore essential for individual identification and provided important data for reintroduction plans, where disruption of population genetic distinctiveness is minimized.

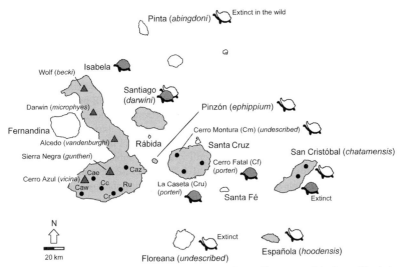

Figure 12.1. Distribution of giant tortoises in the Galápagos archipelago. Shaded islands indicate current presence of tortoise populations. Taxon designation is reported in italics. Triangles represents volcanoes and bullets are sampling locations on San Cristóbal, Santa Cruz and southern Isabela (see Table 12.1 for full names of sampling sites of southern Isabela). Domed and saddlebacked tortoise morphologies are indicated by shaded and white caricatures, respectively. Pinta is represented by a single captive male.

Aim of this chapter

In this chapter we review how population genetic approaches based on mitochondrial DNA (mtDNA) and nuclear DNA microsatellite analyses have aided in reconstructing the evolutionary history of Galápagos giant tortoise populations, assessing demographic patterns and identifying units with substantial demographic and genetic independence to assist taxonomic designation and direct conservation efforts. Sequence data from museum specimens of two extinct but biogeographically important taxa were included in the analysis to better address the origin of extant lineages. This body of work integrates information on population structure, parentage analysis and individual assignment tests on captive individuals to guide maintenance of viable effective population size for breeding programmes and assist management plans involving repatriation of tortoises to populations of origin.

SAMPLE COLLECTION AND GENETIC METHODS

Blood samples for genetic analysis were collected from 1128 wild Galápagos tortoises, 134 samples of reintroduced offspring from Española and 74 samples of captive individuals kept at CDRS. Bone samples were also obtained from two museum specimens representing the extinct population of Floreana and the western, domed population of San Cristóbal. Methods used for extracting DNA from blood and bone tissues are described in Ciofi *et al.* (2002) and Russello *et al.* (2005), respectively.

Phylogenetic relationships were recovered by sequence analysis of six mtDNA regions in the 12S (430 bp) and 16S ribosomal RNA (553 bp) genes, the cytocrome *b* (416 bp) gene, the control region (934 bp), the ND5 (1790 bp) and the ND6 (520 bp) genes as described in Caccone *et al.* (2002). Phylogeography, demographic history and assessment of geographic origin of captive individuals were resolved based on 705 bp of the mtDNA control region using the single-stranded conformation polymorphism technique and direct sequencing as described in Beheregaray *et al.* (2003a, 2004). We used 17 microsatellite loci for population genetic analysis and assignment of captive individuals to population of origin (see Ciofi *et al.* 2002, 2006). Fifteen loci were used for parentage analysis of tortoises from Española (Milinkovitch *et al.* 2004). Isolation of species-specific microsatellite loci was performed as described in Milinkovitch *et al.* (2004). Measures of mtDNA and microsatellite genetic diversity are reported in Table 12.1.

Choice of molecular markers was dictated by the relatively young history of tortoise colonization of the Galápagos (Caccone *et al.* 2002) and the necessity of using fast evolving genome sequences, particularly the mtDNA control region and the highly variable microsatellites, to explain microevolutionary processes and fine-scale patterns of genetic structure for recently diverging populations (e.g. Luikart and England 1999; Hedrick 2001). Nuclear DNA introns and ribosomal DNA internal transcribed spacer regions did not provide sufficient levels of variation or resolution to be of any use (Caccone *et al.* 2004). Moreover, asexual inheritance of mtDNA and a lack of recombination result in a relatively smaller effective population size of mitochondrial genes and allow a better understanding of the actual strength of genetic drift and long-term patterns of population dynamics (Avise 2004; but see Ballard and Whitlock 2004).

GENETIC DIVERGENCE AND PHYLOGENY OF ISLAND POPULATIONS

Biological isolation, a consequence of the substantial distance of Galápagos islands from mainland South America, geophysical and climatic variation resulting from volcanic activity, and consequent differences in habitat

Table 12.1 *Genetic diversity of Galápagos giant tortoise populations. Analyses were preformed using* ARLEQUIN *(Schneider et al. 2000) and* GENEPOP 3.3 *(Raymond and Rousset 1995). For Española, only captive breeders were considered for population genetic analysis. Estimated population size includes reintroduced offspring. The extinct specimen from San Cristóbal, not included in this table, had a different haplotype from the one recorded in the extant population*

ISLAND and sampling site (abbreviation)	N	MtDNA control region			Microsatellites		
		Sample size	n	h	Sample size	A	H_E
ESPAÑOLA	400	16	1	–	15	3.4	0.55
SAN CRISTÓBAL	600	27	1	–	27	5.4	0.70
SANTA CRUZ	2500						
La Caseta (Cru)		65	12	0.80	65	15.8	0.71
Cerro Fatal (Cf)		71	2	0.08	71	5.0	0.55
Cerro Montura (Cm)		3	2	0.67	3	3.6	0.74
PINZÓN	200	53	8	0.76	53	10.0	0.69
SANTIAGO	600	48	8	0.82	50	12.2	0.82
ISABELA							
Volcan Wolf	1500	83	4	0.69	68	12.8	0.84
Volcan Darwin	800	29	2	0.50	24	6.7	0.74
Volcan Alcedo	4000	84	5	0.17	94	7.3	0.66
Sierra Negra	300						
La Cazuela (Caz)		57	3	0.22	56	6.9	0.65
Roca Union (Ru)		53	5	0.61	64	8.5	0.75
Cabo Rosa (Cr)		38	6	0.78	44	7.7	0.72
Cerro Azul	600						
Cinco Cerros (Cc)		43	17	0.89	95	9.5	0.72
East Cerro Azul (Cae)		91	5	0.54	188	9.6	0.71
West Cerro Azul (Caw)		77	9	0.83	184	9.1	0.68

N, estimated population size (Pritchard 1996); n, number of haplotypes; h, haplotypic diversity; A, mean number of alleles per locus; H_E, mean expected heterozygosity. See Fig. 12.1 for geographical location of sampling sites.

conditions found by colonizing tortoises (Colinvaux 1972), most probably facilitated the evolution of different morphologies and distinct genetic lineages across islands.

Genetic diversity

Substantial genetic divergences and population structure were in fact recorded among all island populations at both mtDNA genes and microsatellite loci by analyses of genetic divergence implemented in ARLEQUIN (Schneider *et al.* 2000) and GENETIX (Belkhir *et al.* 2000). In particular, analysis of molecular variance recovered significant levels of population structuring for tortoises of Española, San Cristóbal, Santa Cruz and Pinzón, with more than 90% of total genetic variation found among, rather than within, island populations. High and significant (P < 0.01) values of genetic divergence were estimated by mtDNA control region sequence (overall Φ_{ST} = 0.71) and microsatellite loci (θ = 0.35) analysis, with relatively lower values between Santiago and Volcan Wolf, northern Isabela (θ = 0.034; P < 0.05).

A model-based clustering analysis implemented in STRUCTURE (Pritchard *et al.* 2000) defined the posterior probability of different numbers of populations (K), or clusters, needed for interpreting the observed multilocus microsatellite genotypes. The number of populations with the combined highest posterior probability and lowest variance was K = 13 with $P(K|X)$ = 0.998. The proportional membership of the genotype of each tortoise to each cluster was then calculated and used to determine similarity or distinctiveness among populations. Most populations had more than 80% of the proportions of tortoise genotypes assigned to distinct clusters, indicating a high level of population structuring across islands. A high proportion of the tortoises' genomes from Santiago was assigned to a cluster shared with a third of the tortoises from Volcan Wolf, corroborating evidence of genetic similarity between northern Isabela and Santiago, in accord with the geographic proximity and the evidence of a dual colonization of Isabela (Beheregaray *et al.* 2004).

Phylogeny and taxon designation

Current taxonomy based on differences in carapace and carapace scute morphology among populations (Pritchard 1996) was considered along with patterns of mtDNA haplotype distinction and microsatellite allelic divergence to either corroborate or confute taxon designation.

Phylogenetic analyses of the combined data set were congruent for maximum parsimony, maximum likelihood and neighbour-joining implemented in PAUP* (Swofford 2003), as well as a Bayesian estimation of the posterior probability distribution of trees based on a Markov chain Monte Carlo approximation performed using MRBAYES (Huelsenbeck 2000). The Galápagos complex is divided into divergent clades each composed of haplotypes from a single island (Fig. 12.2). Notable exceptions are the Santa Cruz and Isabela island populations, comprised of polyphyletic groups that showed closer affinity with taxa found on other islands. Statistical parsimony analyses, performed using the TCS program (Clement *et al.* 2000), corroborated the pattern described by consensus phylogenetic trees, in particular the similarity between the northern

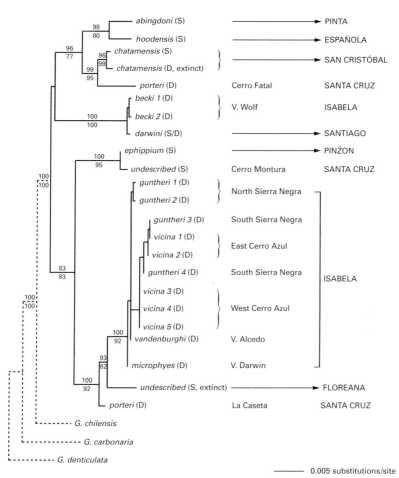

Figure 12.2. Best tree among Galápagos taxa obtained under a Bayesian approach to phylogeny inference. For each taxon, saddlebacked (S), domed (D) or mixed morphology (S/D) is indicated. Sampling site and island of origin are reported on the right. Arrows show island of origin of taxa for which no specific sampling site was identified. Different haplogroups of *becki*, *guntheri* and *vicina* are indicated with numbers. Numbers above and below branches are Bayesian posterior probabilities and maximum likelihood bootstrap values, respectively. Only bootstrap proportions greater than 50% are reported. For illustration purposes, branch lengths leading to outgroup taxa (dashed lines) are reported with no direct reference to the number of nucleotide substitutions per site (see Caccone *et al.* 1999 for details). (Adapted from Russello *et al.* 2005.)

population of Isabela (Volcan Wolf) and Santiago, between eastern Santa Cruz (Cerro Fatal) and San Cristóbal (which diverged for up to 36 mutational steps from haplotypes of other tortoise populations), and between northern Santa Cruz (Cerro Montura) and Pinzón.

Lineage sorting supported taxonomic distinction among island populations advocated by previous morphological and, in part, allozyme analysis (Marlow and Patton 1981; Fritts 1983; Pritchard 1996). Saddleback tortoises from Española and San Cristóbal represent ancestral lineages defining divergent clades and supported former designation of the taxon *hoodensis* and *chatamensis*. Monophyletic maternal lineages were also recovered in tortoises from the island of Pinzón, assigned to the saddlebacked morphotype *ephippium*, and from Santiago (*darwini*), characterized by intermediate morphologies.

The lone survivor of Pinta (genetically distinct for 20 to 21 mtDNA nucleotide substitutions) forms a clade with *hoodensis* on Española and two Española-like matrilines from the geographically closer population on Volcan Wolf, north Isabela (Caccone *et al.* 2002). Historical DNA analysis of museum specimens further confirmed the taxonomic distinctiveness of the Pinta lineage, recovering three novel mtDNA haplotypes and a distinct partition based on microsatellite genotype data and Bayesian clustering algorithms (Russello *et al.* 2007).

PATTERNS OF GENETIC DIFFERENTIATION WITHIN ISLANDS

Santa Cruz

Three genetically distinct ($\Phi_{ST} = 0.80$, $\theta = 0.22$, $P < 0.01$) non-monophyletic groups, originally assigned to the taxon *porteri*, were defined for the island of Santa Cruz (Fig. 12.2). La Caseta and Cerro Fatal comprise two distinct haplogroups in the mtDNA network based on statistical parsimony and are differentiated by 31 diagnostic characters across four mtDNA gene regions as well as 100 private microsatellite alleles. Also, an assignment test of individual tortoises based on a Bayesian method (Rannala and Mountain 1997) implemented in GENECLASS (Cornuet *et al.* 1999) placed 88% and 96% of tortoises from La Caseta and Cerro Fatal, respectively, to the populations from which they were collected.

A third haplogroup included three saddlebacked individuals from Cerro Montura and formed a well-supported sister group with the *ephippium* taxon of Pinzón, also with saddlebacked morphotypes and geographically very close to Cerro Montura. These two populations shared all mtDNA haplotypes and more than 80% of microsatellite alleles. Interestingly, the haplotype identified by statistical parsimony as the ancestral sequence of Pinzón was also detected in the tortoise sampled at Cerro Montura. According to our results, the *ephippium* lineage may have been founded by migrants from Santa Cruz, a scenario consistent with the sequel of island formation by which Santa Cruz predates the volcanic emergence of Pinzón (White *et al.* 1993).

Moreover, distinct demographic histories were recovered by analysis of mismatch distribution of haplotype differences (Rogers and Harpending 1992) using test statistics computed in ARLEQUIN. In particular, the haplogroup from La Caseta is characterized by a multimodal mismatch distribution indicative of a relatively old population with stable demography (Fig. 12.3). Tortoises from Pinzón, on the other hand, showed a mismatch curve that fitted with the expected unimodal pattern for an expanding population, probably a remnant of a pattern of demographic growth that would have followed colonization from Cerro Montura (Beheregaray *et al.* 2003a).

A pattern of strong haplotype similarity was recovered between tortoises of Cerro Fatal and San Cristóbal, suggesting an independent colonization event from San Cristóbal to eastern Santa Cruz (Beheregaray *et al.* 2003a; Russello *et al.* 2005). This seems intriguing because Cerro Fatal tortoises are morphologically domed, while tortoises of the *chatamensis* lineage are saddlebacked. However, San Cristóbal was once inhabited by a domed population, which was heavily collected by whalers and became extinct in the 1930s. In fact, phylogenetic analysis (Fig. 12.2) and relatively low genetic distances (Russello *et al.* 2005) indicated a close affinity between tortoise samples from Cerro Fatal and museum specimens from an extinct domed population from San Cristóbal. In this case, habitat segregation may have accounted for the divergence of tortoise populations in San Cristóbal and for the subsequent colonization event of Cerro Fatal (Pritchard 1996).

Isabela

In Isabela, phylogenetic analyses identified the population from Volcan Wolf, assigned to the taxon *becki*, as a distinct unit yet supporting the strong similarity with tortoises from Santiago. Although little resolution was found within the monophyletic group containing all the populations from central and southern Isabela, haplotype differentiation, the TCS network and definition of separate clusters by microsatellite analyses support distinction of tortoises from Volcan Darwin and Alcedo, assigned to *microphyes* and

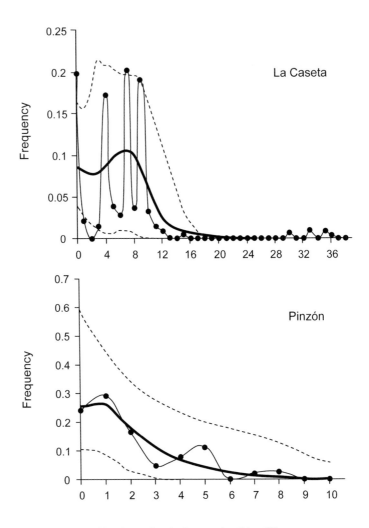

Number of pairwise nucleotide differences

Figure 12.3. Mismatch distributions of mtDNA control region haplotypes of Galápagos giant tortoises from La Caseta and Pinzón. Bullets are the observed relative frequencies of nucleotide differences between pairs of individuals. Heavy lines show the distribution expected under a model of population expansion. Dashed lines are 2.5 and 97.5 percentile values of 1000 simulations. (Modified from Beheregaray *et al.* 2003a.)

vandenburghi, respectively (Ciofi *et al.* 2002; Beheregaray *et al.* 2004). For Alcedo tortoises, low genetic diversity has been associated with a prehistorical bottleneck due to an explosive volcanic eruption 100 000 years ago and subsequent population expansion by individuals carrying the same haplotype. This is an example of how population contraction can influence evolutionary diversification (Beheregaray *et al.* 2003b).

In southern Isabela, patterns of haplotype differentiation were less consistent and probably linked to the relatively young and almost contemporary age of volcanoes (Nordlie 1973) and consequent recent history of colonization and dispersal. Phylogeographic patterns in Isabela were estimated by testing for geographical association between haplotypes reported in the TCS network using nested clade analysis (NCA) implemented in GEoDIs 2.0 (Posada *et al.* 2000) as described by Beheregaray *et al.* (2004) and Ciofi *et al.* (2006). NCA was conducted for Isabela and Santiago only (Fig. 12.4, also Plate 4, colour plate section). Clades from other island populations showed insufficient geographic divergence or genetic variability. Inferences based on the nested clade design identified Sierra Negra as an ancestral colonization site, and defined subsequent episodes of range expansions and long-distance colonization to Volcan Alcedo, Darwin and Cerro Azul. Detailed analyses recovered two temporally distinct range expansions and long distance dispersal events from Sierra Negra to Cerro Azul and more recent migration episodes between Cerro Azul and southwestern Sierra Negra. For clade B, NCA inferred relatively recent events of gene flow and possible range expansion from Santiago to Volcan Wolf, supporting phylogeographic affinities between the island of Santiago and northern Isabela.

Significant levels of genetic differentiation were also recorded among sampling locations in southern Isabela, with Φ_{ST} values ranging from 0.11 to 0.85 for mtDNA control region and values of θ varying between 0.01 and 0.26 (all tests with $P < 0.01$). Analysis of molecular variance based on nuclear and mtDNA data also recovered significant genetic differentiation among all sampling sites in southern Isabela, but did not support a simple two-unit genetic distinction between Sierra Negra and Cerro Azul. Genetic variation within and among sampling sites was significantly higher ($P < 0.01$) than among groups of populations pooled according to volcanic area, indicating that genetic diversity among haplotypes was unlikely to depend on specific geographical assemblage of sampling sites. Genetic distinction was nevertheless most evident for tortoises of La Cazuela, a result supported by principal component analysis of multilocus genotypes performed using PCA-GEN (Goudet 1999). Evidence of migration was also recorded across sampling sites using the Bayesian multilocus genotypic method implemented in BAYESAss 1.3 (Wilson and Rannala 2003). A significant proportion of migrants (>75%) was in fact recorded from Cinco Cerros to Cabo Rosa (Table 12.2), suggesting recent source–sink movement patterns from Cerro Azul to south Sierra Negra.

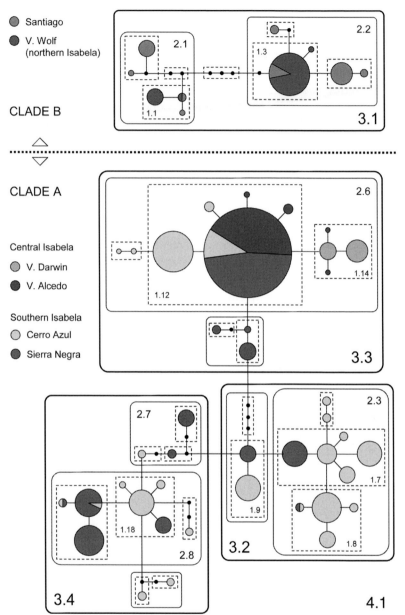

Figure 12.4. Nested clade design superimposed on the unrooted haplotype network estimated by statistical parsimony analysis for Santiago and Isabela. Size of circles is proportional to haplotype frequency in each population. Bullets are intermediate haplotype states that were not observed in the data set. Broken, solid and bold line boxes represent clades of increasing number of mutational steps between haplotypes. Numbers with decimals indicate the serial order of a clade. Only clades with significant geographic association are numbered. Clades including the islands of Española, San Cristóbal, Santa Cruz, Pinzón and Pinta exhibited no genetic variation or geographic differences and are not shown in this figure. Minimum number of mutation steps between boxed clades (data not shown, not supported by statistical parsimony) ranges between 13 and 31. (See also colour plate.)

Table 12.2 *Means of the posterior distributions for migration rates between Galápagos giant tortoise populations of southern Isabela based on microsatellite data*

	Gene flow from to:					
	Caz	Ru	Cr	Cc	Cae	Caw
Gene flow to from:						
La Cazuela (Caz)	0.994	0.001	0.001	0.001	0.001	0.001
Roca Union(Ru)	0.033	0.931	0.002	0.024	0.004	0.005
Cabo Rosa (Cr)	0.005	0.009	0.673	0.304	0.004	0.004
Cinco Cerros (Cc)	0.004	0.013	0.002	0.896	0.083	0.003
East Cerro Azul (Cae)	0.001	0.002	0.002	0.002	0.990	0.003
West Cerro Azul (Caw)	0.001	0.001	0.001	0.004	0.008	0.985

Values are the proportions of individuals derived from the source population each generation. Migration rates >0.10 are underlined.

A gradient of domed morphological variants occurs at different localities across Cerro Azul and Sierra Negra suggesting that body size and details of carapace morphologies represent adaptive responses to different habitat conditions, from dry to moist environments, which vary with altitude and slope exposure (Fritts 1983; Pritchard 1996). Larger animals with dorso-ventral compression, typical of the *guntheri* morphotype, are in fact better adapted to mesic conditions, whilst dry environments would harbour smaller and more domed *vicina* morphologies. Tortoises from east and west Cerro Azul are distributed along an altitudinal gradient with different habitat types. Nevertheless, previous studies showed that there was no significant difference among tortoises from different altitudes and ecotypes across either eastern or western Cerro Azul (Ciofi *et al.* 2002). Similarly, tortoises sampled at low altitudes across Sierra Negra and Cerro Azul exhibited significant population differentiation despite the fact that all sampling locations were characterized by similar habitat type. Overall patterns of genetic diversity across southern Isabela appear therefore not to mirror either geographical or ecological distribution of named subspecies. Similar results have been recovered for other Galápagos reptiles such as marine iguanas, for which morphological and coloration characters were thought to be the result of phenotypic plasticity (Rassmann *et al.* 1997).

PARENTAGE ANALYSIS AND GENETIC ASSIGNMENT

The Española captive breeding system
Parentage analysis of reintroduced offspring of captive breeders from Española was conducted using PAPA (Duchesne and Bernatchez 2002)

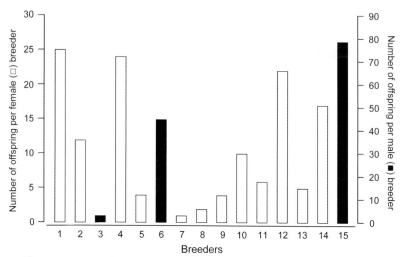

Figure 12.5. Reproductive success of female (white bars) and male (black bars) captive breeders from Española. (Data from Milinkovitch *et al.* 2004.)

and the probability of correct allocations was assessed by calculating likelihoods of parental pairs for a set of simulated offspring genotypes. Out of 134 offspring reintroduced on Española, 132 were assigned to captive Española breeders (Milinkovitch *et al.* 2004). Contribution of breeders was significantly different and skewed towards two males with more than 95% of offspring assigned, and towards four females with more than 15 offspring each (Fig. 12.5). The unequal contribution of the breeding population resulted in a reduced effective population size (N_e) of 5.7 computed from the observed variance in reproductive success and biased sex ratios (Milinkovitch *et al.* 2004).

The biased contribution of the breeding founders to the offspring gene pool results in a long-term reduction of N_e because this parameter is determined by the harmonic mean of N_e across generations, so that low values of N_e in the founders will affect the N_e of the repatriated individuals and their offspring. For repatriated tortoises of Española, even if the N_e of offspring is maintained at, for instance, 1000 individuals for 10 generations, the overall long-term N_e would be equal to just 59 tortoises. Increasing N_e in the founders by equalizing reproductive success would therefore result in a significant higher long-term genetic variability of reintroduced offspring.

Furthermore, analysis of an additional 473 captive-bred tortoises released on Española (Milinkovitch *et al.* 2007) revealed an individual with nuclear microsatellite alleles not found in any of the 15 breeders. Statistical analyses incorporating genotypes of 304 field-sampled individuals from populations of the major islands indicated that this contaminant individual was most probably a hybrid between an Española female tortoise and a male from the island of Pinzón, most probably the result of human transport to Española (Milinkovitch et al. 2007). Removal of the hybrid individual as well as its father and possible (half-)siblings would prevent further contamination within this taxon of particular conservation significance.

Genetic assignment of tortoises of unknown origin

Genetic identification of tortoises of unknown origin kept at the CDRS facilities was first conducted by maximum parsimony and neighbour-joining analysis on mtDNA control region sequences as described in Burns et al. (2003). An assignment test was then performed on multilocus genotypes as implemented in GENECLASS. All individuals were assigned with greater than 95% confidence based on statistical parsimony and with high bootstrap support (>85%, data not shown) to one of the known wild haplotypes (Table 12.3). Assignment of tortoises to the island population where their multilocus microsatellite genotype was most likely to occur matched haplotype designation for all but one individual (no. 38 in Table 12.3) for the parental enclosure and three individuals (no. 3, no. 25 and no. 63) from the progeny enclosure. Discrepancies between mtDNA and nuclear tests could be attributed to matings with males whose alleles were not sampled or to errors in microsatellite assignment due to either a limited number of individuals genotyped as reference populations or an insufficient number of loci analysed. Moreover, female tortoises of some species can store viable sperm for months or years (Pearse and Avise 2001), so that some contributing males could also not be part of the breeder enclosure. Four mtDNA haplotypes were identified among the six females kept in the parental enclosure and eight haplotypes among the 36 offspring from the progeny enclosure. Unfortunately, only two offspring haplotypes, represented in 23 individuals, matched those in the female breeders. Moreover, between two and seven microsatellite alleles per offspring were not represented in the parental multilocus genotypes, suggesting that not all of the original breeders were still alive at the time of sampling. This pattern may be accounted for by recorded deaths of at least four individuals in the captive-breeding population between 1984 and the first sampling period in 1998.

The majority of captive tortoises were assigned to Isabela on the basis of their mtDNA haplotypes. In particular, about 20% of the founders and 25% of offspring had their haplotypes assigned to southern Isabela, where

Table 12.3 *Island and population of origin of captive Galápagos giant tortoises from the Charles Darwin Research Station based on mtDNA control region sequences and microsatellite data*

		mtDNA control region			Microsatellites				
Animal number	Carapace morphology	Island	Location	Distance	Island	Location	L_1	Location	L_2
Parental									
5	D	Isabela	V. Alcedo Sierra Negra Cerro Azul	0 0 0	Isabela	Sierra Negra	22.77		
36	D/S	Isabela	V. Wolf	0	Isabela	V. Wolf	19.93		
37	D/S	San Cristóbal		0	San Cristóbal		22.61		
38	D	Isabela	V. Wolf	0	Isabela	V. Alcedo	19.44	V. Darwin	22.51
39	D/S	Isabela	V. Wolf	0	Isabela	V. Wolf	20.35		
40	S	Isabela	V. Wolf	0	Isabela	V. Wolf	23.36		
41	D	Santa Cruz	La Caseta	1	Santa Cruz	La Caseta	21.96		
42	S	Isabela	V. Wolf	0	Isabela	V. Wolf	22.35		
43	S	Isabela	V. Wolf	0	Isabela	V. Wolf	23.35		
44	S	Isabela	V. Wolf	0	Isabela	V. Wolf	26.77		
45	S	Isabela	V. Wolf	0	Isabela	V. Wolf	24.94		
46	D/S	Isabela	V. Wolf	0	Isabela	Cerro Azul	15.34	V. Wolf	19.11
47	S	Isabela	V. Wolf Sierra Negra	0 0	Isabela	V. Wolf	19.63		
48	D	Isabela	V. Wolf	0	Isabela	V. Wolf	18.05		
49	D/S	Isabela	V. Wolf	0	Isabela	V. Wolf	19.12		
50	D	Isabela	Sierra Negra	0	Isabela	Sierra Negra	15.07		
51	D	Isabela	Cerro Azul	0	Isabela	Cerro Azul	16.85		
53	D	Isabela	Sierra Negra	0	Isabela	Sierra Negra	19.66		

55	D	Santa Cruz	La Caseta	2	Santa Cruz	La Caseta	19.96		
56	D	Santa Cruz	La Caseta	3	Santa Cruz	La Caseta	22.59		
57	D/S	Isabela	V. Wolf	0	Isabela	V. Wolf	24.22		
58	D	Santa Cruz	La Caseta	0	Santa Cruz	La Caseta	18.34		
59	D	Isabela	V. Darwin	0	Isabela	Sierra Negra	10.32	V. Darwin	12.21
Progeny									
1	D	Isabela	Sierra Negra	0	Isabela	Sierra Negra	25.17	Sierra Negra	21.40
2	D	Isabela	Sierra Negra	0	Isabela	V. Alcedo	21.12	Santa Cruz	25.77
3	D/S	Isabela	V. Wolf	0	San Cristóbal	V. Wolf	19.97		
4	D/S	Isabela	V. Wolf	0	Isabela	V. Wolf	18.56		
6	D/S	Isabela	V. Wolf	0	Isabela	Cerro Azul	22.76	V. Wolf	23.26
7	D/S	Isabela	V. Wolf	0	San Cristóbal	V. Wolf	22.33	V. Wolf	23.76
8	D/S	Isabela	V. Wolf	0	Isabela	Cerro Azul	22.53		
9	D	Isabela	Sierra Negra / Cerro Azul	0	Isabela	V. Wolf	14.95		
10	D/S	Isabela	V. Wolf	0	Isabela	V. Wolf	21.67		
11	D/S	Isabela	V. Wolf	0	Isabela	V. Wolf	24.25		
12	D/S	Isabela	V. Wolf	0	Isabela	V. Wolf	28.29		
13	D/S	Isabela	V. Wolf	0	San Cristóbal	V. Wolf	24.84	V. Wolf (Isabela)	28.49
14	D/S	Isabela	V. Wolf	0	Isabela	V. Wolf	26.92		
15	D	Isabela	V. Alcedo	1	Isabela	V. Wolf	23.36	V. Alcedo	25.41
16	D	Isabela	Sierra Negra / Sierra Negra / Cerro Azul	0	Isabela	Sierra Negra	16.90		
17	D/S	Isabela	V. Wolf	0	Isabela	V. Wolf	20.89		
18	D/S	Isabela	V. Wolf	0	Santa Cruz	La Caseta	25.57	Isabela (V. Wolf)	30.01
19	D	Isabela	Sierra Negra	0	Isabela	V. Wolf	23.73	Sierra Negra	24.60
21	D/S	Isabela	V. Wolf	0	Isabela	V. Wolf	26.63		
22	D	Santa Cruz	La Caseta	1	Santa Cruz	La Caseta	17.28		
23	D	Isabela	Sierra Negra	0	Isabela	Sierra Negra	19.16		

Table 12.3 (cont.)

Animal number	Carapace morphology	mtDNA control region			Microsatellites				
		Island	Location	Distance	Island	Location	L_1	Location	L_2
24	D/S	Isabela	Cerro Azul / V. Wolf	0	Isabela	V. Wolf	25.78		
25	D/S	Isabela	V. Wolf	0	San Cristóbal	V. Wolf	24.34	Santa Cruz	26.31
26	S	Isabela	V. Wolf	0	Isabela	V. Wolf	27.57		
27	D	Santa Cruz	La Caseta	0	Santa Cruz	La Caseta	17.43		
28	D/S	Isabela	V. Wolf	0	San Cristóbal	V. Wolf	23.45	Isabela (V. Wolf)	26.71
29	D/S	Isabela	V. Wolf	0	Isabela	V. Wolf	28.70		
30	D	Isabela	Sierra Negra / Cerro Azul	0	Isabela	Cerro Azul	18.74		
31	D	Isabela	Sierra Negra / Cerro Azul	0	Isabela	Cerro Azul	16.27	Sierra Negra	17.42
32	D/S	Isabela	V. Wolf	0	Isabela	V. Wolf	21.97		
33	D	Santa Cruz	La Caseta	1	Santa Cruz	La Caseta	19.48		
34	D/S	Isabela	V. Wolf	0	Isabela	V. Wolf	28.30		
35	D/S	Isabela	V. Wolf	0	Isabela	V. Wolf	27.03		
60	D	Isabela	V. Wolf	0	Isabela	Cerro Azul	18.27	V. Wolf	20.86
61	D	Santa Cruz	La Caseta	1	Santa Cruz	La Caseta	21.49		
63	D	Isabela	V. Wolf	0	Isabela	V. Alcedo	27.33	Santa Cruz	27.36

Notes: Individuals are sorted into parental and progeny enclosures. Carapace morphology is based on visual inspection and measurements (D, domed; S, saddlebacked; D/S, intermediate morphology). Distance between each unknown mtDNA haplotype and the closest wild haplotype is reported in number of base pairs. Likelihood values based on microsatellite multilocus genotypes are given between each individual and the two closest natural populations (L_1, L_2). Tortoises with a likelihood <1% were not assigned to the sampled locality.

Source: Modified from Burns *et al.* (2003).

tortoises endured significant reduction in population size due to human-related threats (Pritchard 1996). Also of conservation importance was the assignment of one of the founders (no. 37) to San Cristóbal, an island with a small tortoise population with low nuclear DNA variation and no apparent mtDNA diversity.

CONSERVATION AND MANAGEMENT IMPLICATIONS

Island populations represent excellent models to study the relative effects of vicariance, colonization and dispersal and eventually for the identification of evolutionary and biologically divergent units of conservation importance (Grant 1997; Emerson 2002; Funk and Fa 2006). In the Galápagos, most of the original taxonomy based on morphological distinctions and geographic considerations for the islands of Española, San Cristóbal, Pinzón, Santiago, and for the sole representative of the island of Pinta was corroborated by our molecular genetic data. This is certainly important for management purposes that should aim at preserving adaptive diversity across the current geographic range of giant tortoises. For other island populations, however, a multidisciplinary approach provided better insights into conservation units that were not resolved by morphological characters alone. This is the case of Santa Cruz, where the taxonomic assignment of tortoise populations to a single lineage has been contentious, especially after a few, undescribed saddlebacked tortoises were reported from the xeric habitats of Cerro Montura. Two distinct units for conservation were identified in eastern (Cerro Fatal) and southern (La Caseta) Santa Cruz on the basis of different long-term evolutionary history and demographic features. Tortoises from Cerro Montura showed divergent morphological, molecular genetic and ecological characteristics, which warrant separate management effort for the few tortoises of northern Santa Cruz. On the other hand, considering that Cerro Montura and Pinzón share identical haplotypes, lineage preservation should be secured by protection of the Pinzón population. This is nevertheless a finding that may lead to a formal taxonomic revision of the single described *porteri* lineage on Santa Cruz. Other results with implications for conservation include the detection of contrasting levels of genetic diversity among populations and in particular, the very low levels of genetic variability found on Cerro Fatal and Española. In contrast, the high genetic variation and demographic stability detected in the abundant population from La Caseta suggests potential for natural persistence and diversification, and should be warranted protection.

Difficulties in the delineation of distinct units of conservation on islands is even more challenging if diversification occurred rapidly due to the relatively recent geological history, such as for the younger island of Isabela. For Isabela, molecular tools were pivotal for the recognition of taxonomy based on previous morphological investigation for central and northern volcanoes. In southern Isabela, in particular, molecular data gave perhaps a clearer picture than ecological and phenotypical traits for population management. Tortoises from different locations of Cerro Azul and Sierra Negra, in fact, warrant protection efforts given their degree of genetic diversity and evolutionary history regardless of supposed morphological separation (Ferrier 2002; Moritz 2002). In this case, evolutionary processes should be preserved by attaining a natural network of genetic connections between populations if different management units have to be recognized (Crandall *et al.* 2000; Sherwin *et al.* 2000; Frankham *et al.* 2002; Moritz 2002). A decrease in gene flow exacerbates genetic drift, a process suggested by the relatively low genetic variability found at La Cazuela, and edge effects linked to human-related threats and stochastic events. Restoration should be attempted more often for populations, such as La Cazuela, that have become isolated as a result of both vicariance and/or recent anthropogenic activities (e.g. Powell and Gibbs 1995; Naumann and Geist 2000; Kaiser 2001).

In any case, a clear differentiation between volcanoes or altitudinal ranges based on morphological grounds is difficult to support for tortoises from southern Isabela. Each population has a distinct genetic structure and, although some extreme circumstances may necessitate mixing of individuals from genetically different stocks (Hedrick 1995), genetic assignment tests should be a requirement prior to reintroduction. The assignment test conducted on most captive tortoises currently kept at CDRS provides a good example for demographic augmentation of southern Isabela without disruption of current genetic differentiation (Merlen 1999). Similarly, identification of captive individuals from San Cristóbal constitutes a valuable source to increase the genetic diversity of tortoises on that island without compromising their genetic integrity.

Finally, description of reproductive success among breeders, as in the example of Española, is central to adjust breeding regimes to maximize genetic variation, long-term effective population size and, eventually, fitness in the repatriated population (Reed 2005). Increased reproductive success of individuals that have contributed little to the gene pool of the reintroduced population can in fact be supported since successful breeding in captivity is not necessarily correlated with fitness or reproductive capacity in the wild. In this respect, it is worth stressing that even a small number of individuals can represent an important gene pool for demographic augmentation despite the possible undesirable effect of inbreeding (e.g. Saccheri *et al.* 1998). The natural recovery of the Alcedo population following an intensive prehistorical bottleneck provides a pertinent example for the continuation and implementation of conservation programmes based on a small number of breeders.

Despite these evident problems from a solely genetic standpoint, it must be emphasized that the Española captive-breeding programme has been, arguably, one of the most successful repatriation programmes ever attempted. More than 2000 offspring of the 15 captive breeders have been repatriated to Española and they are reproducing on their genetic-native island. The longer-term success of the Española reintroduction programme remains to be determined and whether the lack of genetic variation will eventually come into play remains to be seen. Given the generation time (20–30 years), it will be some time before the health of this repatriated population after, say, ten generations can be assessed.

In conclusion, our research on the Galápagos giant tortoise provides an example of the importance of integrating molecular population genetic analyses and ecological approaches in conservation biology. This is in line with population management planning which, in recent years, has shifted towards policies guided by multidisciplinary approaches where ecological and molecular genetic studies are combined in an effort to identify intervention priorities for conservation of threatened wildlife (Paetkau 1999; Roman *et al.* 1999; Crandall *et al.* 2000).

ACKNOWLEDGEMENTS

We would like to thank the Parque Nacional de Galápagos and Charles Darwin Research Station (CDRS) for logistical support during fieldwork and help in obtaining CITES permit for sample collection. We extend our gratitude to those who provided invaluable assistance and help in the field during sample collection, in particular W. Tapia and C. Marquez. A special thanks to J. Gibbs, H. Snell, T. Fritts and L. Cayot who provided useful and stimulating comments, and either facilitated our study or were directly involved in the research. We also thank the many park rangers and CDRS staff who helped us through the years and to the various US and Ecuadorian students who participated in field or laboratory work. We could not have done some of our work without the help of zoo and museum curators who made possible the sampling of valuable museum specimens or captive animals. This work was funded by the Yale Institute for Biospheric Studies with the Donnelley Environmental Fellowship and other financial support, National Science Foundation, National Geographic Society, The Paul and

Bay Foundation, the Communauté Française de Belgique, the National Fund for Scientific Research Belgium, the Van Buuren, Defay, BNB and Internationale Brachet Stiftung Funds.

REFERENCES

Allendorf, F. W. and Luikart, G. (2006). *Conservation and the Genetics of Populations.* Oxford: Blackwell.

Avise, J. C. (2004). *Molecular Markers, Natural History and Evolution.* Sunderland: Sinauer Associates.

Ballard, J. W. O. and Whitlock, M. C. (2004). The incomplete natural history of mitochondria. *Molecular Ecology*, **13**, 729–744.

Beheregaray, L. B., Ciofi, C., Caccone, A., Gibbs, J. P. and Powell, J. R. (2003a). Genetic divergence, phylogeography and conservation units of giant tortoises from Santa Cruz and Pinzón, Galápagos islands. *Conservation Genetics*, **4**, 31–46.

Beheregaray, L. B., Ciofi, C., Geist, D. *et al.* (2003b). Genes record a prehistoric volcano eruption in the Galápagos. *Science*, **302**, 75.

Beheregaray, L. B., Gibbs, J. P., Havill, N. *et al.* (2004). Giant tortoises are not so slow: rapid diversification and biogeographic consensus in the Galápagos. *Proceedings of the National Academy of Sciences USA*, **101**, 6514–6519.

Belkhir, K., Borsa, P., Chikhi, L., Goudet, J. and Bonhomme, F. (2000). GENETIX 4.01, Windows™ software for population genetics. Laboratoire Génome, Populations, Interactions, CNRS UPR 9060, Université de Montpellier II, Montpellier, France.

Blouin, M. S. (2003). DNA-based methods for pedigree reconstruction and kinship analysis in natural populations. *Trends in Ecology and Evolution*, **18**, 503–511.

Burns, C. E., Ciofi, C., Beheregaray, L. B. *et al.* (2003). The origin of captive Galápagos tortoises based on DNA analysis: implications for the management of natural populations. *Animal Conservation*, **6**, 329–337.

Caccone, A., Gibbs, J. P., Ketmaier, V., Suatoni, E. and Powell, J. R. (1999). Origin and evolutionary relationships of giant Galápagos tortoises. *Proceedings of the National Academy of Sciences USA*, **96**, 13 223–13 228.

Caccone, A., Gentile, G., Gibbs, J. P. *et al.* (2002). Phylogeography and history of giant Galápagos tortoises. *Evolution*, **56**, 2052–2066.

Caccone, A., Gentile, G., Burns, C. E. *et al.* (2004). Extreme difference in rate of mitochondrial and nuclear DNA evolution in a large ectotherm, Galápagos tortoises. *Molecular Phylogenetics and Evolution*, **31**, 794–798.

Ciofi, C., Milinkovitch, M. C., Gibbs, J. P., Caccone, A. and Powell, J. R. (2002). Microsatellite analysis of genetic divergence among populations of giant Galápagos tortoises. *Molecular Ecology*, **11**, 2265–2283.

Ciofi, C., Wilson, G. A., Beheregaray, L. B. *et al.* (2006). Phylogeographic history and gene flow among giant Galápagos tortoises on southern Isabela island. *Genetics*, **172**, 1–18.

Clement, M., Posada, D. and Crandall, K. A. (2000). TCS: a computer program to estimate gene genealogies. *Molecular Ecology*, **9**, 1657–1659.

Colinvaux, P. A. (1972). Climate and the Galapagos Islands. *Nature*, **240**, 17–20.

Cornuet, J.-M., Piry, S., Luikart, G., Estoup, A. and Solignac, M. (1999). New methods employing multilocus genotypes to select or exclude populations as origins of individuals. *Genetics*, 153, 1989–2000.

Crandall, K. A., Bininda-Emonds, O. R. P., Mace, G. M. and Wayne, R. K. (2000). Considering evolutionary processes in conservation biology. *Trends in Ecology and Evolution*, 15, 290–295.

Duchesne, P. and Bernatchez, L. (2002). An analytical investigation of the dynamics of inbreeding in multi-generation supportive breeding. *Conservation Genetics* 3, 47–60.

Emerson, B. C. (2002). Evolution on oceanic islands: molecular phylogenetic approaches to understanding pattern and process. *Molecular Ecology*, 11, 951–966.

Frankham, R., Ballou, J. D. and Briscoe, D. A. (2002). *Introduction to Conservation Genetics*. Cambridge: Cambridge University Press.

Fraser, D. J. and Bernatchez, L. (2001). Adaptive evolutionary conservation: towards a unified concept for defining conservation units. *Molecular Ecology*, 10, 2741–2752.

Ferrier, S. (2002). Mapping spatial pattern in biodiversity for regional conservation planning: where to from here? *Systematic Biology*, 51, 331–363.

Fritts, T. H. (1983). Morphometrics of Galápagos tortoises: evolutionary implications. In *Patterns of Evolution in Galápagos Organisms*, ed. R. I. Bowman, M. Berson and A. E. Leviton. San Francisco: Pacific Division of the American Associations for the Advancement of Science, pp. 107–122.

Funk, S. M. and Fa, J. E. (2006). Phylogeography of the endemic St Lucia whiptail lizard *Cnemidophorus vanzoi*: conservation genetics at the species boundary. *Conservation Genetics*, 7, 651–663.

Goudet, J. (1999). PCA-GEN for Windows. Lausanne, Switzerland: University of Lausanne.

Grant, P. R. (1997). *Evolution on Islands*. Oxford: Oxford University Press.

Hedrick, P. W. (1995). Gene flow and genetic restoration: the Florida panther as a case study. *Conservation Biology*, 9, 995–1007.

Hedrick, P. W. (2001). Conservation genetics: where are we now? *Trends in Ecology and Evolution*, 16, 629–636.

Huelsenbeck, J. P. (2000). MrBayes: Bayesian inference of phylogeny. Department of Biology, University of Rochester, distributed by the author.

Kaiser, J. (2001). Galápagos takes aim at alien invaders. *Nature*, 293, 590–592.

Luikart, G. and England, P. R. (1999). Statistical analysis of microsatellite DNA data. *Trends in Ecology and Evolution*, 14, 253–56.

MacFarland, C. G., Villa, J. and Toro, B. (1974). The Galapagos giant tortoises *Geochelone elephantopus*. I. Status of the surviving populations. *Biological Conservation*, 6, 118–133.

Marlow, R. W. and Patton, J. L. (1981). Biochemical relationships of the Galápagos giant tortoises (*Geochelone elephantopus*). *Journal of Zoology*, 195, 413–422.

Merlen, G. (1999). *Restoring the Tortoise Dynasty*. Quito, Ecuador: The Charles Darwin Foundation for the Galápagos Islands.

Milinkovitch, M. C., Monteyne, D., Gibbs, J. P. et al. (2004). Genetic analysis of a successful repatriation programme: giant Galápagos tortoises. *Proceedings of the Royal Society Series B*, 271, 341–345.

Milinkovitch, M. C., Monteyne, D., Russello M. *et al.* (2007). Giant Galápagos tortoises: molecular genetic analysis reveals contamination in a repatriation program of an endangered taxon. *BMC Ecology*, **7**, 2.

Moritz, C. (2002). Strategies to protect biological diversity and the evolutionary processes that sustain it. *Systematic Biology*, **51**, 238–254.

Naumann, T. and Geist, D. (2000). Physical volcanology and structural development of Cerro Azul volcano, Isabela island, Galápagos: implications for the development of Galápagos-type shield volcanoes. *Bulletin of Volcanology*, **61**, 497–514.

Nordlie, B. E. (1973). Geology and structure of the western Galápagos volcanoes and a model for their origin. *Geological Society of America Bulletin*, **84**, 2931–2956.

Paetkau, D. (1999). Using genetics to identify intraspecific conservation units: a critique of current methods. *Conservation Biology*, **13**, 1507–1509.

Pearse, D. E. and Avise, J. C. (2001). Turtle mating systems: behavior, sperm storage, and genetic paternity. *Journal of Heredity*, **92**, 206–211.

Posada, D., Crandall, K. A. and Templeton, A. R. (2000). GEODIS: a program for the cladistic nested analysis of the geographical distribution of genetic haplotypes. *Molecular Ecology*, **9**, 487–488.

Powell, J. R. and Gibbs, J. P. (1995). A report from Galápagos. *Trends in Ecology and Evolution*, **10**, 351–354.

Pritchard, J. K., Stephens, M. and Donnelly, P. (2000). Inference of population structure using multilocus genotype data. *Genetics*, **155**, 945–959.

Pritchard, P. C. H. (1996). *The Galápagos Tortoises: Nomenclatural and Survival Status*. Lunenburg, Mass.: Chelonian Research Foundation.

Rannala, B. and Mountain, J. L. (1997). Detecting immigration by using multilocus genotypes. *Proceedings of the National Academy of Sciences USA*, **94**, 9197–9221.

Rassmann, K., Trillmich, F. and Tautz, D. (1997). Hybridization between the Galápagos land and marine iguana (*Conolophus subcristatus* and *Amblyrhynchus cristatus*) on Plaza Sur. *Journal of Zoology*, **242**, 729–739.

Raymond, M. and Rousset, F. (1995). GENEPOP, v. 1.2: population genetics software for exact tests and ecumenicism. *Journal of Heredity*, **86**, 248–249.

Reed, D. H. (2005). Relationship between population size and fitness. *Conservation Biology*, **19**, 563–568.

Rogers, A. R. and Harpending, H. (1992). Population growth makes waves in the distribution of pairwise genetic differences. *Molecular Biology and Evolution*, **9**, 552–569.

Roman, J., Santhuff, S. D., Moler, P. E. and Bowen, B. W. (1999). Population structure and cryptic evolutionary units in the alligator snapping turtle. *Conservation Biology*, **13**, 135–142.

Russello, M. and Amato, G. (2004). Ex situ population management in the absence of pedigree information. *Molecular Ecology*, **13**, 2829–2840.

Russello, M., Beheregeray, L., Gibbs, J. P. *et al.* (2007). Lonesome George is not alone among Galápagos tortoises. *Current Biology*, **17**, R317–R318.

Russello, M., Glaberman, S., Gibbs, J. P., Marquez, C., Powell, J. R. and Caccone, A. (2005). A novel taxon of Giant tortoises in conservation peril. *Biology Letters*, **1**.

Saccheri, I., Kuussaari, M., Kankare, M. *et al.* (1998). Inbreeding and extinction in a butterfly metapopulation. *Nature*, **392**, 491–494.

Schneider, S., Roessli, D. and Excoffier, L. (2000). ARLEQUIN, v. 2.000: a software for population genetic data analysis. Genetics and Biometry Laboratory, University of Geneva, Switzerland.

Sherwin, W. B., Timms, P., Wilcken, J. and Houlden, B. (2000). Analysis and conservation implications of koala genetics. *Conservation Biology*, **14**, 639–649.

Swofford, D. L. (2003). PAUP*: phylogenetic analysis using parsimony (*and other methods), v. 4.0b10. Sunderland: Sinauer Associates.

Towsend, C. H. (1925). The Galápagos tortoises in their relation to the whaling industry: a study of old logbooks. *Zoological Journal of the Linnean Society*, **4**, 55–135.

Van Denburgh, J. (1914). The gigantic land tortoises of the Galápagos archipelago. *Proceedings of the California Academy of Sciences*, **4**, 203–374.

White, W. M., McBirney, A. R. and Duncan, R. A. (1993). Petrology and geochemistry of the Galápagos islands: portrait of a pathological mantle plume. *Journal of Geophysical Research*, **98**, 19 533–19 663.

Wilson, G. A. and Rannala, B. (2003). Bayesian inference of recent migration rates using multilocus genotypes. *Genetics*, **163**, 1177–1191.

Zug, G. R. (1997). Galápagos tortoise nomenclature: still unresolved. *Chelonian Conservation Biology*, **2**, 618–619.

13

Evolution of population genetic structure in marine mammal species

A. RUS HOELZEL

INTRODUCTION

Marine mammals are a taxonomically diverse group of species with evolutionary roots right back to the earliest mammalian radiations. The smallest species is a mustelid, the sea otter (*Enhydra lutris*), and the largest the blue whale (*Balaenoptera musculus*). The only things marine mammals have in common are the facts that they are all mammals (and therefore dependent on breathing air and constrained by the necessities of live birth and maternal care), and they are all dependent on an aquatic, typically marine environment. These two common attributes have meant that they are constrained in similar ways, though the different groups have met these challenges in different ways. The mustelid, the carnivore (polar bear, *Ursus maritimus*) and the pinnipeds (seals, sea lions and walrus) all meet thermoregulatory challenges with dense pelage. Most of these also still give birth on land, and are to varying extents amphibious. The cetaceans (whales, dolphins and porpoises) and sirenians (manatees and dugongs) are fully aquatic, and have little or no pelage. Instead they have adjusted to the high thermal conductivity of water and the generally cold temperatures by developing thick layers of subcutaneous fat, and in many cases, by becoming large (which provides a high volume to surface area ratio and conserves heat). All of these species, with the exception of the polar bear, have adapted to more efficient locomotion in water by acquiring a relatively fusiform shape – most extensively developed in the delphinid cetaceans (the dolphins).

In this chapter my focus will be on those features among the marine mammals that help to explain common patterns of population structure, or differences in these patterns among taxa. One feature shared by many is their high trophic position in the ecosystems they occupy. One exception is the sirenians, which are herbivorous. However, most marine mammals are predators, though their trophic position can vary dramatically (from baleen

Population Genetics for Animal Conservation, eds. G. Bertorelle, M. W. Bruford, H. C. Hauffe, A. Rizzoli and C. Vernesi. Published by Cambridge University Press. © Cambridge University Press 2009.

whales feeding on krill to killer whales feeding on other marine mammals). Another typical feature is large size. There are exceptions (such as the sea otter and some pinnipeds and cetaceans), but most marine mammals require large concentrations of prey to sustain them (see Boyd 2002). Therefore the global distribution and regional abundance of prey is especially important in determining the distribution and movement patterns of marine mammals, most dramatically exemplified in the long migrations of some baleen whales between nutrient-rich polar waters (to feeding grounds) and the tropics (for breeding). Another important factor is sociality. Many marine mammal species (most pinnipeds and cetaceans) are at least seasonally gregarious, and many species are social, especially among the delphinid cetaceans (which comprise about 35 species).

In this review I make no attempt to be inclusive, but instead include topics and examples that I feel are especially useful towards explaining patterns of population structure. I will focus on the two most species-rich marine mammal groups, the cetaceans and the pinnipeds, and further focus on those species for which relatively abundant data are available. Given the logistical difficulties with obtaining samples to conduct species-specific studies, the interests of conservation and management could be significantly advanced for these species by identifying some general patterns underlying the structuring of populations. In this way species of particular concern could be identified as a priority for study.

KIN ASSOCIATIONS

A tendency to associate with kin in either social groups or aggregations can reduce diversity within regions and increase differentiation among populations. Kin association is based on philopatry for one or both sexes, and may be reinforced by factors such as kin selection, reproductive skew, or environmental constraints (see Krebs and Davies 1997). The impact on population structure is, of course, strongest when both sexes tend to be philopatric. The most extreme example among cetacean species is the killer whale (*Orcinus orca*). Social groups in this species are defined along matrilines, with known long-term associations between adult whales lasting more than 30 years (see Ford *et al.* 2003). This cohesive group structure is probably related to resource exploitation at some level, either through cooperative prey location (e.g. Hoelzel 1993) or cooperative prey capture (Pitman *et al.* 2001; Ford *et al.* 2005). However, as has been shown for other social species (see Beauchamp and Fernandez-Juricic 2005), group size may be larger than required for prey capture. Individual killer whales hunting pup sea lions

(*Otaria flavescens*) in Argentina conducted most or all of the hunting in social groups of two to five whales, and then provisioned the other members of the group (Hoelzel 1991a). Alloparental care may be an additional factor in maintaining these groups (Haenel 1986), possibly supported by kin selection. However, kin selection is most clearly supported when there is preferential care given to kin. In killer whale social groups many of the interactions are likely to be with close kin, given the high degree of philopatry for both sexes. Among marine mammals, a study demonstrating preferential alloparental care towards kin has been reported for the Antarctic fur seal (*Arctocephalus gazella*: Gemmell 2003), though the interpretation of those data were later challenged (Hoffman and Amos 2005).

The consequence of killer whale social structure at the population level is fixed mtDNA haplotypes within regional populations (Hoelzel *et al.* 1998a, 2002a). Worldwide mtDNA diversity in this species is low (Fig. 13.1), and this together with a pattern of diversity among the extant control region haplotypes led Hoelzel *et al.* (2002a) to conclude that the species had been through an historical population bottleneck. For this reason, the magnitude of the differences among haplotypes in modern populations may reflect differences among remnant lineages that survived the bottleneck, rather than time in divergence. This is especially relevant to a fixed difference between foraging specialist populations of this species in the eastern North Pacific (Hoelzel *et al.* 1998a). Local populations of social groups appear to specialize exclusively on either marine mammal or fish prey, and this is especially well documented in the eastern North Pacific (e.g. Ford *et al.* 1998; Herman *et al.* 2005).

Identification of prey choice is based on stomach contents from stranded animals (Ford *et al.* 1998), fatty acid and stable isotope analyses (Herman *et al.* 2005), the identification of the remains of prey (e.g. Hoelzel 1993; Ford *et al.* 1998) and from visual identification during prey capture events (e.g. Fig. 13.2, also Plate 5, colour plate section). While the temporal and regional diets of populations in different parts of the world are mostly unknown, in the eastern North Pacific, foraging specializations seem to be strong and consistent (Ford *et al.* 1998; Herman *et al.* 2005). The comparatively large mtDNA genetic distance between these two ecotypes in this region (Hoelzel and Dover 1991; Hoelzel *et al.* 1998a) (Fig. 13.1) suggested the possibility of sympatric speciation between foraging specialists. However, nuclear genetic markers indicate a relatively even pattern of differentiation and continuing gene flow between and among populations of ecotypes (with isolation correlated to geographic distance within ecotypes), suggesting that the higher degree of mtDNA differentiation is

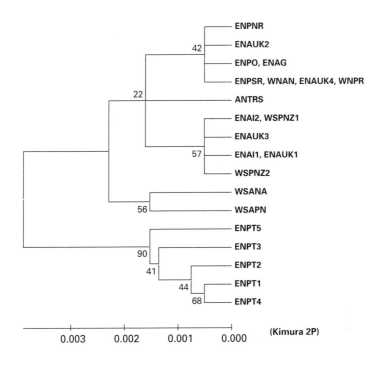

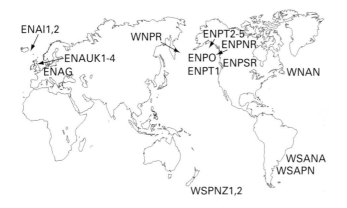

Figure 13.1. (a) Linearized neighbour-joining tree for the killer whale based on Kimura 2-parameter genetic distances and mtDNA control region sequences (after Hoelzel *et al.* 2002a). Bootstrap support shown at nodes. (b) Sample locations.

Figure 13.2. Killer whale preying on Pacific halibut (*Hippoglossus stenolepis*) off southeast Alaska. (Photo by A. R. Hoelzel.) (See also colour plate.)

instead a reflection of chance fixation of mtDNA haplotypes in local matri-lineal populations (Hoelzel *et al.* 2002a, 2007). The implication is that both resource specialization and geographic isolation can reduce the rate of male-mediated gene flow among matrifocal kin groups in this species.

The fish-eating killer whales studied in the eastern North Pacific, which show the highest level of philopatry for both sexes, could be a rare exception among marine mammals, however it is suspected that the long-finned pilot whale (*Globicephala melas*) shows a similar type of natal group philopatry (Ottensmeyer and Whitehead 2003). In a study of 322 individually identi-fied pilot whales off Nova Scotia, a best-fit model of the standardized lag association rate identified long-term associations lasting years in groups of 11–12 whales on average (Ottensmeyer and Whitehead 2003). Another species that shows evidence for matrifocal social structure is the sperm whale. Observational studies of recognized individuals suggest long-term relationships within matrilineal groups (Christal and Whitehead 2001), though Richard *et al.* (1996) showed that associating pairs were not neces-sarily close kin. However, at the population level, comparative analyses of mtDNA and nuclear markers suggested that females tend to be philopatric in regional social groups (Lyrholm *et al.* 1999). Sperm whales are deep divers, pursuing cephalopod prey (e.g. Evans and Hindell 2004). One theory about the evolution of this social structure is that deep diving has led to the communal care of young, which in turn resulted in long-term female bonds (Whitehead 1996). However, another deep diving species (the northern bottlenose whale, *Hyperoodon ampullatus*) has a social structure distinct from the sperm whale, with females and immature whales forming a loose network of associations and no apparent long-term bonds (Gowans *et al.* 2001).

Few dolphin species have been studied in sufficient detail to determine specific kin association patterns over time, though delphinids in general are known to include many highly social species (see Conner 2002). The bottlenose dolphin (*Tursiops* spp.) is an exception. Several long-term studies, especially in Florida (e.g. Reynolds *et al.* 2000), Australia (e.g. Conner *et al.* 2000a) and New Zealand (Lusseau *et al.* 2003) have documented association patterns over time in this genus. In general the social pattern is described as 'fission fusion', with few long-term associations, though there are exceptions. In Doubtful Sound, New Zealand bottlenose dolphins showed a high proportion of long-term associations, especially among males (Lusseau *et al.* 2003). Stable male associations have been shown in other populations as well, with some evidence that associating males tended to be close kin in the Bahamas (Parsons *et al.* 2003). It has been proposed that these male 'alliances' are involved in gaining access to females, perhaps in cooperation (Conner *et al.* 1992, 2000b). For striped dolphins (*Stenella coeruleoalba*) close kin associations have instead been shown for adult females within fission fusion social groups (Gaspari *et al.* 2007).

Many pinniped species (including all otariids, odobenids and some phocids) breed in colonies. These typically occur in the same locations at the same time of year (pinnipeds have delayed implantation of the embryo to permit synchronized parturition). Within the colony there is a hierarchical structure associated with clusters along the beach, and breeding units within the clusters (harems). Both males and females show site fidelity and philopatry to varying degrees (see review in Stevick *et al.* 2002). Fabiani *et al.* (2006) investigated kinship patterns among southern elephant seals (*Mirounga leonina*) in a breeding colony in the Falkland Islands (Malvenas). There was little evidence for fine-scale patterns of kin associations (e.g. among females within harems), though there was for some harems, and higher resolution analyses may show greater structure. However, there was greater kinship among females in general, consistent with females showing greater philopatry than males in this species.

HABITAT DEPENDENCE

Many marine mammals have a tremendous capacity for long-range movement, including seasonal migrations between breeding and feeding
grounds (for some pinnipeds and mysticete cetaceans) that can cover
thousands of miles (see review in Stevick *et al.* 2002). However, population
genetic structure at a comparatively fine geographic scale is common (see
reviews in Hoelzel 1998; Hoelzel *et al.* 2002b). In some cases genetic
differentiation is correlated to geographic distance, as is commonly seen
for terrestrial species. For example, such a pattern is proposed for
Franciscana (*Pontoporia blainvillei*) along the coast of Brazil based on
mtDNA control region sequence data (Lazaro *et al.* 2004). Within Shark
Bay in western Australia isolation by distance was also indicated for bottlenose dolphins based on ten microsatellite DNA loci and sequence data
from the mtDNA control region (Krutzen *et al.* 2004). This typically breaks
down at a larger geographic scale (e.g. for the bottlenose dolphin: Natoli
et al. 2004), however isolation by distance was reported for the harbour seal
(*Phoca vitulina*) for a geographic range spanning the North Pacific
(Westlake and O'Corry-Crow 2002). A total of 778 seals from 161 locations
distributed from northern Japan to southeast Alaska were sequenced for the
5′ segment of the mtDNA control region. For the most part differentiation
was least for proximate putative populations, with the exception of a region
along the Commander-Aleutian Island chain. There differentiation was
greater, and correlated to a proposed subspecies boundary (and possibly
the contact zone for expanding refugial populations after the last glacial
maximum).

However, it is more common for marine mammal populations to show
a discontinuous relationship between geographic and genetic distance. For
example, the southern elephant seal has been studied for both microsatellite DNA and mtDNA control region variation in the southern oceans.
Hoelzel *et al.* (2001) found both genetic (mtDNA sequence data) and
morphometric differentiation between the one extant mainland breeding
colony (on Peninsula Valdez, Argentina) and the oceanic island breeding
colony on South Georgia. Diversity at the mainland breeding colony was
relatively low, and no haplotypes were shared between the two colonies
(suggesting no recent female-mediated gene flow). A later study incorporating additional populations, but based solely on molecular markers, showed
an uneven pattern of differentiation among breeding colonies, with
comparatively high distances between the mainland and all oceanic
colonies, but also between the island colony on MacQuarie island and all

other colonies (Fabiani *et al.* 2003) (Fig. 13.3). Both mtDNA and micro-satellite DNA data showed a similar pattern with respect to the isolation of the mainland colony. Elephant seals breeding on the mainland initiate their foraging excursions over the continental shelf, and often feed either over the shelf, at its edge, or just beyond it (see Campagna *et al.* 1995, 1998, 1999), while seals from the island colonies forage over deep water, with excursions sometimes extending down to the Antarctic mainland (McConnell and Fedak 1996). Differences in foraging range could result in reduced gene flow if they affect dispersal behaviour, though the precise mechanism is not known for this species.

Differentiation between coastal and pelagic populations is common for delphinid cetaceans as well. For example, spotted dolphins (*Stenella attenuata* and *S. frontalis*) both in the Pacific and in the Atlantic are distributed in coastal and pelagic populations, and show morphotypic differentiation (Perrin *et al.* 1987; Perrin 2002a). The morphotypes differ with respect to tooth and jaw structure and pigmentation (Douglas *et al.* 1984; Perrin *et al.*

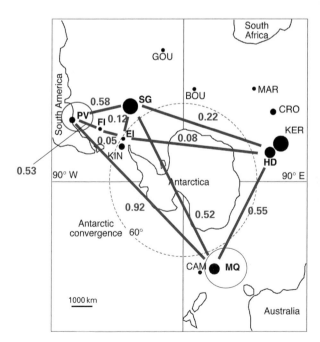

Figure 13.3. Southern elephant seal breeding colonies in the southern oceans and ϕ_{ST} values for mtDNA control region sequence data for pairwise comparisons. (After Fabiani *et al.* 2003.)

1987; Perrin 2002a). The common dolphin (*Delphinus* spp.) also shows morphotypic differentiation, with coastal populations having longer beaks and distinct colour patterns (Perrin 2002b). Heyning and Perrin (1994) proposed that the two common dolphin forms in the eastern tropical Pacific are sufficiently different to justify separate specific classification, and proposed *D. capensis* for the long-beaked form. A genetic study based on mtDNA sequences suggested significant differentiation between these forms off southern California, but the sample size was small (12 short-beaked and 11 long-beaked forms: Rosel *et al.* 1994). A recent study investigating mtDNA control region variation and microsatellite DNA diversity for putative long and short-beaked *Delphinus* populations worldwide found no correlation between morphotype and genetic differentiation (Natoli *et al.* 2006). The long-beaked animals from Rosel *et al.* (1994) formed a monophyletic lineage well separated from the rest of the samples, but long- and short-beaked forms from elsewhere in the world were fully polyphyletic. Natoli *et al.* (2006) suggested that regional coastal populations adapted to the habitat with similar morphology several times independently (convergent evolution).

The best-known example of coastal and pelagic morphotypes is for bottlenose dolphins (*Tursiops* spp.). In the western North Atlantic the coastal *Tursiops truncatus* form is relatively smaller and differs in cranial morphology from the pelagic form (Mead and Potter 1995; Hoelzel *et al.* 1998b). Parasite load also differs, consistent with the parasite species found in their respective habitats (Mead and Potter 1995). Coastal schools are typically found within 8 km of the coast, while the pelagic schools are usually found 34+ km from shore (though their ranges sometime overlap: Torres *et al.* 2003). Foraging differences were evident from both stomach contents and stable isotope analyses (Mead and Potter 1995; Cortese 2000), with the pelagic form taking more cephalopod prey, and the coastal form more fish. Hoelzel *et al.* (1998a) investigated the genetic structure of these populations, and found highly significant differentiation between the coastal and pelagic forms both for mtDNA control region sequences ($\phi_{ST} = 0.604$) and microsatellite DNA loci ($R_{ST} = 0.373$). A similar pattern has been found for the coastal and pelagic populations found off southern California (Lowther *et al.* 2005). Those populations also show morphometric differentiation, but in that case it is the pelagic form that is smaller (Walker 1981). In both studies the coastal form showed lower levels of diversity, suggesting a possible historical founder event. A preliminary study of putative coastal and pelagic forms of *T. truncatus* off the South African coast showed no differentiation, but the sample size was small

(Hoelzel *et al.* 1998b). However, a coastal population of morphologically distinct bottlenose dolphins (the 'aduncus' form, Ross 1977) inhabiting the Natal coast of South Africa was highly differentiated from both pelagic and coastal 'truncatus' forms (Natoli *et al.* 2004). The distinction was sufficient to justify classification as a new species, *T. aduncus*, though equally divergent from a morphologically similar form found in Asian waters, also designated '*T. aduncus*' (Wang *et al.* 1999; Natoli *et al.* 2004). This coastal form has a relatively long beak, as for the coastal *Delphinus* forms, which may reflect convergent evolution in similar habitat (see Natoli *et al.* 2006 and above).

Dolphin populations that are found in coastal habitat often show fine-scale population structure. In a study comparing bottlenose dolphin samples collected across the contiguous range from Scotland to the Black Sea, a likelihood assignment method (STRUCTURE: Pritchard *et al.* 2000) clustered samples to apparent habitat regions (Natoli *et al.* 2005) (Fig. 13.4, also Plate 2, colour plate section). When possible (power is relatively low), this type of clustering method (based on genotypes and equilibrium models, and not a priori assignment to putative populations) is especially useful for marine mammal species, since boundaries to gene flow are often cryptic in the marine environment. The identified boundaries for bottlenose dolphins between the North Atlantic and Mediterranean, and between the eastern and western basins of the Mediterranean are separated by open water (rather than boundaries defined by land mass), but oceanic conditions distinguish these regions. In each case a number of other species also show genetic differentiation across a similar geographic range (see Natoli *et al.* 2005). The diet of the dolphins in the study by Natoli *et al.* (2005) was not known, but other studies have shown clear differences in feeding behaviour for this species in different coastal habitats. For example, Gannon and Waples (2004) describe prey choice differences between populations in open coastal versus estuarine habitat in Florida. If local populations are dependent on different prey resources, and if that dependence affects dispersal behaviour, this could serve as a mechanism for reducing gene flow. One way foraging specializations could affect dispersal behaviour would be through the social facilitation of foraging. Remaining in natal social groups (or within a broader affiliation of social groups) to learn foraging strategies appropriate to a given habitat could be beneficial provided that some individuals still dispersed (to avoid inbreeding depression). This would suggest a frequency dependent strategy. It does not imply a bias in benefit for dispersal in one sex over the other, and consistent with this Natoli *et al.* (2005) found no evidence of sex-biased dispersal. Sellas *et al.* (2005) describe a similar pattern of fine-scale population structure in the coastal habitats of the eastern Gulf of Mexico, and no indication of sex-biased dispersal for the bottlenose dolphin (however, see Moller and Beheregaray (2004) on *T. aduncus* dispersal behaviour in Australia).

An assumption of this hypothesis is that bottlenose dolphins benefit from social interactions during foraging. In support of this idea, Krutzen *et al.* (2004) describe the cultural transmission of a strategy associated with using sponges to help probe for fish in the substratum. The strategy is apparently passed down from mother to offspring, and this interpretation is based on matching mtDNA haplotype among animals involved in this behaviour. In another study bottlenose dolphins within a social group showed a division of labour during foraging such that some acted as 'drivers' forcing fish into other dolphins who were serving as 'barriers' (Gazda *et al.* 2005). The idea that social behaviour is an important aspect of learning strategy in this species is further supported by a study with captive dolphins where their capacity to learn was shown to be affected by social interactions (Delfour and Marten 2005).

It is unclear how generally the behaviour documented for the bottlenose dolphin may apply to other dolphin species, as most have not been studied in as much detail. It seems likely that the complexities of coastal habitat are important. For example, considerably less evidence for population structure was seen for the common dolphin, which is in most cases more pelagically distributed (Natoli *et al.* 2006; Fig. 13.4). The strongest evidence for

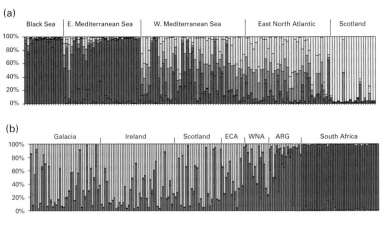

Figure 13.4. Assignment probabilities for individual genotypes to putative populations (in different colours) for (a) bottlenose dolphin and (b) common dolphin samples (after Natoli *et al.* 2005, 2006). Sampling locations are indicated above histograms. ECA, eastern central Atlantic; WNA, western North Atlantic; ARG, Argentina. (See also colour plate.)

structure detected was between a coastal population in South Africa in comparison with the predominantly pelagic populations sampled elsewhere in the world. Other coastal dolphin species show population genetic structure over a geographic range comparable to that seen for the bottlenose dolphin, though not all (see review in Hoelzel *et al.* 2002b). For example, Burmeister's porpoise (*Phocoena spinipinnis*) is differentiated between Chilean and Peruvian stocks (Rosa *et al.* 2005), and the sympatric Peruvian population of dusky dolphins (*Lagenorhynchus obscurus*) is also differentiated from other studied populations (Cassens *et al.* 2005). Rosa *et al.* (2005) suggest that this may be related to the relatively unstable Peruvian upwelling system (due to recurrent El Niño events), perhaps accelerated by genetic drift if El Niño cycles have caused demographic fluctuations.

Perhaps the other best-studied case next to the bottlenose dolphin is that of the killer whale, as described in some detail above (see Hoelzel 2002). Sympatric killer whale populations in the North Pacific that differ in foraging specializations show evidence of reduced gene flow comparable to that seen for populations in allopatry (Hoelzel *et al.* 2002a). These specializations involve learned strategies for finding prey with critical temporal and spatial components (see Hoelzel 1991a, 1993; Baird and Whitehead 2000). Such differences will result in distinct patterns of area use and searching behaviour, including differences in search strategies. For example, fish-eating killer whales in waters off Washington State were distributed over a smaller spatial range when actively feeding, suggesting coordinated searching and convergence on prey when located (Hoelzel 1993). In contrast, marine-mammal-eating populations sometimes return predictably to a known concentrated resource, such as southern sea lion colonies in Argentina when the pups are being weaned (Hoelzel 1991a). The main distinctions are probably in the scale over which prey patches are distributed, and the concentration of prey within a patch. There are also differences in the manner and extent to which whales within the pod may coordinate effort related to prey capture. A subset of the pod is typically involved in the capture of marine mammals (see Hoelzel 1991a, and above), though coordinated effort is sometimes necessary for larger prey (e.g. Pitman *et al.* 2001). The magnitude of the genetic differentiation, even in sympatry, suggests that foraging specialization is directly responsible for reduced gene flow in this species. As for the bottlenose dolphin, this could be due to social facilitation, and to benefit related to continuing the strategy learned in the natal pod (and so dispersing primarily to pods or populations that share a similar strategy).

There is evidence for investment in the training of young (Hoelzel 1991a; Guinet and Bouvier 1995), but more data are needed to establish its importance. Foraging specializations have also been suggested to influence population genetic structure in another highly social species, the gray wolf (*Canis lupus*: Carmichael *et al.* 2001).

MIGRATORY SPECIES AND PREDICTABLE HABITATS

Another factor likely to shape population structure is the divergent habitat requirements for breeding and foraging, in some species wholly or mostly separated into different periods of the annual cycle. This involves migration between breeding and feeding grounds, as seen for many avian species. Among cetaceans, the best understood migrations involve just four species: the humpback whale (*Megaptera novaeangliae*), the grey whale (*Eschrichtius robustus*), the North Atlantic right whale (*Eubalaena glacialis*) and the southern right whale (*Eubalaena australis*). It is presumed that the North Pacific right whale (*Eubalaena japonica*) behaves similarly, but only the summer feeding aggregation is currently known: a coastal breeding congregation has never been found for this species (Shelden *et al.* 2005). In each of the other four species well-documented breeding and/or feeding grounds are known in coastal waters. The distance between foraging and breeding sites is typically thousands of kilometres (e.g. up to 9000 km for the eastern Pacific for grey whales: Swartz 1986).

Fidelity to natal breeding grounds has resulted in population genetic structure that varies in spatial range with season, potentially seen over a very small geographic scale. For example, in the eastern North Pacific grey whales breed in lagoons along the coast of Baja California, and then migrate north along the coast to summer feeding grounds off British Colombia and Alaska. There are three main breeding areas (Scammon's Lagoon, Laguna San Ignacio and Bahia Magdelena-Almejas), separated by approximately 200 km between each site along the Pacific coast (Rice and Wolman 1971). Goerlitz *et al.* (2003) used mtDNA sequence data to compare grey whale cows from San Ignacio's Lagoon with females sampled on migration (or on feeding grounds), and found evidence for substructuring ($F_{ST} = 0.064$, $P < 0.01$) suggesting some level of breeding site fidelity. A larger sample size comparing breeding sites would be useful to more clearly define this pattern.

Fidelity between breeding and feeding grounds can also be very high, for example up to 90% for humpback whales migrating between the Caribbean and Gulf of Maine in the North Atlantic (Clapham *et al.* 1993). In the North

Pacific Baker *et al.* (1993, 1998) showed that this type of fidelity led to the differentiation of stocks based on evidence from both mtDNA and nuclear genetic markers. In that case one stock migrated between breeding grounds off Hawaii and feeding grounds off Alaska, while another migrated between Mexico and California (Fig. 13.5). This type of migratory profile maintains a geographic separation between stocks. However, in some cases (e.g. especially in the North Atlantic for humpback whales) a single breeding population distributes among multiple feeding populations (see review in Stevick *et al.* 2002; Fig. 13.5). In other cases multiple breeding populations converge on single feeding grounds (see Hoelzel 1998; Fig. 13.5).

The convergence of breeding stocks into mixed assemblages on feeding grounds is of special concern for the development of effective conservation and management strategies (see Hoelzel 1991b, 1998). This is because exploitation is common on the feeding grounds, and could lead to the uncontrolled (unrecognized) depletion of unique stocks. An important case in point is the mixing of minke whale (*Balaenoptera acutorostrata*) stocks in the Okhotsk Sea (Fig. 13.5). Studies based on allozyme markers (Wada 1991) and mtDNA (Goto and Pastene 1996) indicated mixing of the genetically depauperate Sea of Japan stock with a differentiated population from the east of Japan on feeding grounds in the Okhotsk Sea. Low genetic diversity in the Sea of Japan stock suggests both small effective population size and isolation from significant levels of migration. This mysticete

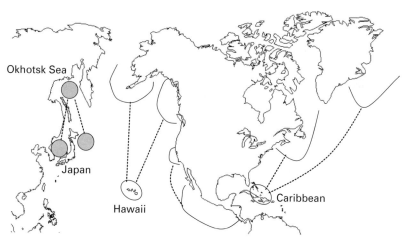

Figure 13.5. Minke whale (grey) and humpback whale (open) feeding and breeding grounds and the migration paths between them. See text for further explanation.

species is one of several where data on migration patterns are few, but for which there is evidence that only some populations are migratory (e.g. Dorsey *et al.* 1990). The fin whale (*Balaenoptera physalus*) is another species for which migratory routes are probably pelagic and poorly known, but for which there are genetic data in support of the mixing of genetically differentiated breeding stocks on feeding grounds (Danielsdottir *et al.* 1991, 1992; Bérubé *et al.* 1998).

For pinnipeds, feeding excursions are either primarily seasonal following a period of fasting during the breeding season, or continuous throughout the year. In the former case feeding excursions can be very long range. For example, the southern elephant seal breeds on one mainland and a number of island colony sites circumpolar to the Antarctic, and these populations are genetically differentiated from each other (see Fig. 13.3). However, satellite tag studies have shown foraging excursions that extensively overlap with other breeding colony sites, and with animals from different colonies, extending for up to 3000 km from the breeding sites (see review in Stevick *et al.* 2002). Breeding site resource is probably limiting (see below), and this may explain the apparent fidelity to site, in spite of the clear potential for dispersal to other sites, given the overlapping foraging ranges.

SEX-SPECIFIC DISPERSAL

Greenwood (1980) documented sex-biased dispersal behaviour for males and females in mammalian and avian species. While not exclusive, on balance male mammals dispersed while female birds dispersed. All of the examples were for terrestrial species, and the bias was explained on the basis of critical resources (the 'resource competition' hypothesis). Greenwood (1980) suggested that for birds male strategy would be selected to defend territories to attract females. In avian species both sexes can and typically do care for their young throughout the developmental stages. However, female mammals invest more heavily in parental care than males, with parturition coming after an extended period of development, and postpartum care based on the suckling of dependent young. Therefore, according to this theory females should be philopatric to secure suitable resource for raising their young. Males disperse because they can, and because one sex should disperse to avoid inbreeding depression.

Perrin and Mazalov (2000) further proposed that in polygynous species (which includes most mammals) male-biased dispersal is expected when local mate competition exceeds resource competition, but only when females are limited by intrinsic factors (the rate at which they can process resources, rather than the resources themselves). Numerous studies have confirmed the trend for male-biased dispersal in mammals (see reviews in Pusey 1987; Perrin and Mazalov 2000). This also broadly matches our expectations for some marine mammals based on mark–recapture studies. For example, sperm whale (*Physeter catadon*) males are known to travel great distances visiting female groups, while the females are re-sighted over a comparatively small geographic range (see Christal and Whitehead 1997).

In some cases genetic analyses can also clearly indicate sex-biased dispersal. For example, both killer whale (Hoelzel *et al.* 1998a, 2002a) and some southern elephant seal populations show fixed differences at mtDNA markers, indicating female philopatry. In the case of the killer whale, one haplotype is shared by all individuals and unique to local populations. In that case any gene flow must be male-mediated. One record of male-mediated gene flow in the southern elephant seal indicated a dispersal event of approximately 8000 km and that male's successful fathering of more than 10 offspring at the breeding site (Fabiani *et al.* 2003). The male, found breeding on Falkland Island, matched a mtDNA haplotype otherwise specific to MacQuarie Island, and microsatellite DNA assignment data excluded the Falklands as his natal colony.

Various molecular methods have been used to assess sex-biased dispersal, but there are important considerations about their efficacy and power in some cases (see Prugnolle and de Meeus 2002). By far the most common method is the comparison of markers with different modes of inheritance (especially the comparison of biparentally inherited nuclear markers with the matrilineal mtDNA markers). Some examples of these comparisons for marine mammal species are given in Table 13.1. Problems with interpretation are especially associated with differences in mutation rate and the effective population size represented by the different marker types. The latter is in turn affected by reproductive behaviour. Under random mating the expectation is that the effective population size for nuclear markers would be four times larger than for the mtDNA marker in gonochoric species (e.g. Seielstad *et al.* 1998). However, in natural populations the effective size of the population reflected by variation in diploid markers will depend on factors such as breeding behaviour (e.g. polygamy vs. monogamy), variation in family size, and sex-specific dispersal levels (Chesser and Baker 1996). For example, the southern elephant seal shows strong polygyny and very low variation in female reproductive success rate. In such cases the effective population size of bi-parentally inherited genes can be smaller than that for the maternally inherited

Table 13.1 *Comparing* F_{ST} *based on mtDNA control region sequence and microsatellite DNA loci*

| | Population structure | | | |
| | mtDNA | Microsatellite | | |
Species (populations)	F_{ST} (ϕ_{ST})	F_{ST} (R_{ST})	Loci	Reference
Orcinus orca (8)	0.910	0.135	16	Hoelzel *et al.* (2002a, unpubl.)
Tursiops truncatus (5)	0.114	0.104	9	Natoli *et al.* (2005)
Delphinus delphis (7)	0.156	0.030	9	Natoli *et al.* (2006)
Stenella frontalis (3)	(0.215)	(0.096)	5	Adams and Rosel (2006)
Phocoena phocoena (6)	0.047	0.007	12	Tolly *et al.* (2001); Andersen *et al.* (2001)
Balaenoptera acuterostrata (4)	(0.009)	0.008	16	Andersen *et al.* (2003)
Mirounga leonina (5)	0.341	0.020	9	Fabiani *et al.* (2003)
Phoca vitulina (24, 12)	0.248	0.187	7	Goodman (1998); Stanley *et al.* (1996)

genes (Chesser and Baker 1996). Smaller effective population size means greater susceptibility to differentiation by drift. Table 13.1 shows that for the southern elephant seal and a number of other species, measures of genetic differentiation (reflected in the magnitude of F_{ST}) are far greater for mtDNA than for microsatellite (bi-parental) markers. As mentioned above, differences in mutation rate can also impact interpretation, but for these comparisons the mutation rate is likely to be broadly similar (for microsatellite loci and the mtDNA control region, see Hoelzel *et al.* 1991; Dallas 1992).

Therefore the large differences indicated in Table 13.1 for some species do seem to imply greater male-mediated dispersal. Since polygyny would tend to increase the level of differentiation apparent in the nuclear markers, the large difference (order of magnitude in some cases) is seen in spite of this tendency. However, there are some exceptions. For example, the bottlenose dolphin and the harbour seal (*Phoca vitulina*) show more equivalent values for the two marker types. Migration is a powerful force towards the homogenization of diversity among populations (e.g. Hudson 1998), and therefore dispersal behaviour is a critical factor in the evolution of population structure. In marine mammals the key question probably comes down to the behaviour of males. Females typically show a strong tendency towards philopatry, and in some cases so do males. When both do, this will accelerate the differentiation of populations.

This brings us back to the question of why philopatry should be selected for in the first place. The resource competition hypothesis suggests that females should be philopatric to ensure access to sufficient resources to raise and protect their young. However, in marine mammal species there is often a decoupling between feeding and breeding activities. For example, in the southern elephant seal, where there is good evidence for female philopatry and male dispersal, foraging takes place primarily outside of the breeding season, thousands of kilometres from the breeding sites. Pups are weaned after they are suckled, at which time the female returns to the sea to forage (having fasted during the breeding season). In this species it seems likely that competition among males for access to females is a more important factor driving male-biased dispersal. In social species both males and females may benefit from the cooperative exploitation of a local resource, and therefore both benefit from philopatry. This appears to be true for both bottlenose dolphins (at least in the Mediterranean and eastern North Atlantic) and killer whales. Some killer whale populations living in the eastern North Pacific appear to take this to an extreme. Although dispersal is mediated by males, this is probably through temporary associations with other pods. Males have been recorded remaining in association with their natal pod for life (e.g. Ford *et al.* 2003).

CONCLUSIONS

Marine mammals represent a diversity of taxa poorly represented by the few species I have focused on here. However, I feel there are some important themes highlighted in these examples. Many of these species are either social or gregarious, sometimes seasonally, sometimes throughout the year. Those associations typically reflect the common exploitation of a resource, either suitable habitat for breeding and parturition, or prey (or sometimes both). In the marine environment, these two types of resources may be separated by a large distance, especially for the larger species that depend on high-density concentrations of prey (typically found in polar waters). In this case suitable habitats are likely to be limited in number, and navigation between sites must be learned, which provides some pressure for philopatry. Breeding site philopatry may be most important for females seeking predictably safe habitat suitable for giving birth. In this case the males could disperse among these sites to enhance reproductive success. The relative rate of male and female dispersal, together with the size and demographic history of breeding populations, will determine the degree of isolation. The pattern of seasonal movement will determine the spatial and temporal structure of populations, including the potential for mixed assemblages of breeding stocks on seasonal feeding grounds.

Some species associate in more permanent social groups, probably to facilitate both breeding and foraging success. In these associations (most common among dolphin species) groups learn temporal and spatial foraging strategies, sometimes facilitated by group searching or prey capture. The cultural transmission of this information makes philopatry advantageous for both sexes, and could lead to the fine-scale structuring of populations, especially when habitat is complex and varies over a relatively small geographic range. This appears to be most common in coastal habitat. Taken together these observations (considered in the context of other influences on population structure, such as demographic processes including population bottlenecks and periods of expansion/contraction; see Hoelzel *et al.* 2002b) suggest some transferable strategies for conservation that could be tested with further study involving some of the many marine mammal species about which little is known.

REFERENCES

Adams, L. D. and Rosel, P. E. (2006). Population differentiation of the Atlantic spotted dolphin (*Stenella frontalis*) in the western North Atlantic, including the Gulf of Mexico. *Marine Biology*, **148**, 671–681.

Andersen, L. W. Russante, D. E., Walton, M. *et al.* (2001). Conservation genetics of harbour porpoises, *Phocoena phocoena*, in eastern and central North Atlantic. *Conservation Genetics*, **2**, 309–324.

Andersen, L. W., Born, E. W., Dietz, R. *et al.* (2003). Genetic population structure of minke whales *Balaenoptera acutorostrata* from Greenland, the North East Atlantic and the North Sea probably reflects different ecological regions. *Marine Ecology Progress Series*, **247**, 263–280.

Baird, R. W. and Whitehead, H. (2000). Social organization of mammal-eating killer whales: group stability and dispersal patterns. *Canadian Journal of Zoology*, **78**, 2096–2105.

Baker, C. S., Perry, A., Bannister, J. L. *et al.* (1993). Abundant mitochondrial DNA variation and worldwide population-structure in humpback whales. *Proceedings of the National Academy of Sciences USA*, **90**, 8239–8243.

Baker, C. S., Medrano-Gonzalez, L., Calambokidis, J. *et al.* (1998). Population structure of nuclear and mitochondrial DNA variation among humpback whales in the North Pacific. *Molecular Ecology*, **7**, 695–707.

Beauchamp, G. and Fernandez-Juricic, E. (2005). The group-size paradox: effects of learning and patch departure rules. *Behavioral Ecology*, **16**, 352–357.

Bérubé, M., Aguilar, A., Dendanto, D. *et al.* (1998). Population genetic structure of North Atlantic, Mediterranean Sea and Sea of Cortez fin whales, *Balaenoptera physalus* (Linnaeus 1758): analysis of mitochondrial and nuclear loci. *Molecular Ecology*, **7**, 585–599.

Boyd, I. L. (2002). Energetics: consequences for fitness. In *Marine Mammal Biology: An Evolutionary Approach*, ed. A. R. Hoelzel. Oxford: Blackwell Science, pp. 247–277.

Campagna, C., Leboeuf, B. J., Blackwell, S. B., Crocker, D. E. and Quintana, F. (1995). Diving behavior and foraging location of female southern elephant seals from Patagonia. *Journal of Zoology*, **236**, 55–71.

Campagna, C., Quintana, F., Le Boeuf, B. J., Blackwell, S. Y. and Crocker, D. (1998). Diving behaviour and foraging ecology of female southern elephant seals from Patagonia. *Aquatic Mammals*, **4**, 1–11.

Campagna, C., Fedak, M. A. and McConnell, B. J. (1999). Post-breeding distribution and diving behaviour of adult male southern elephant seals from Patagonia. *Journal of Mammalogy*, **80**, 1341–1352.

Carmichael, L. E., Nagy, J. A., Larter, N. C. and Strobeck, C. (2001). Prey specialization may influence patterns of gene flow in wolves of the Canadian Northwest. *Molecular Ecology*, **10**, 2787–2798.

Cassens, I., Van Waerebeek, K., Best, P. B. *et al.* (2005). Evidence for male dispersal along the coasts but no migration in pelagic waters in dusky dolphins (*Lagenorhynchus obscurus*). *Molecular Ecology*, **14**, 107–121.

Chesser, R. K. and Baker, R. J. (1996). Effective sizes and dynamics of uniparentally and diparentally inherited genes. *Genetics*, **144**, 1225–1235.

Christal, J. and Whitehead, H. (1997). Aggregations of mature male sperm whales on the Galapagos Islands breeding ground. *Marine Mammal Science*, **13**, 59–69.

Christal, J. and Whitehead, H. (2001). Social affiliations within sperm whale (*Physeter macrocephalus*) groups. *Ethology*, **107**, 323–340.

Clapham, P. J., Baraff, L. S., Carson, C. A. *et al.* (1993). Seasonal occurrence and annual return of humpback whales in the southern Gulf of Maine. *Canadian Journal of Zoology*, **71**, 440–443.

Conner, R. C. (2002). Ecology of group living and social behaviour. In *Marine Mammal Biology: An Evolutionary Approach*, ed. A. R. Hoelzel. Oxford: Blackwell Science, pp. 353–370.

Conner, R. C., Smolker, R. A. and Richards, A. F. (1992). Dolphin alliances and coalitions. In *Coalitions and Alliances in Humans and Other Mammals*, ed. A. H. Harcourt and F. B. M. de Waal. Oxford: Oxford University Press, pp. 415–443.

Conner, R. C., Read, A. J. and Wrangham, R. W. (2000a). Male reproductive strategies and social bonds. In *Cetacean Societies: Field Studies of Dolphins and Whales*, ed. J. Mann, R. C. Conner, P. L. Tyack and H. Whitehead. Chicago: University of Chicago Press, pp. 247–269.

Conner, R. C., Wells, R. S., Mann, J. and Read, A. J. (2000b). The bottlenose dolphin: social relationships in a fission fusion society. In *Cetacean Societies: Field Studies of Dolphins and Whales*, ed. J. Mann, R. C. Conner, P. L. Tyack and H. Whitehead. Chicago: University of Chicago Press, pp. 91–126.

Cortese, N. A. (2000). Delineation of bottlenose dolphin populations in the western Atlantic Ocean using stable isotopes. M.S. thesis, University of Virginia.

Dallas, J. F. (1992). Estimation of microsatellite mutation rates in recombinant inbred strains of mouse. *Mammalian Genome*, **3**, 452–456.

Danielsdottir, A. K., Duke, E. J., Joyce, P. and Arnason, A. (1991). Preliminary studies on the genetic variation at enzyme loci in fin whales and sei whales from the North Atlantic. *International Whaling Commission Special Issue*, **13**, 115–124.

Danielsdottir, A. K., Duke, E. J. and Arnason, A. (1992). Mitochondrial DNA analysis of North Atlantic fin whales and comparison with four species of whales: sei, minke, pilot and sperm whales. *Reports of the International Whaling Commission* SC/F91/F17.

Delfour, F. and Marten, K. (2005). Inter-modal learning task in bottlenosed dolphins (*Tursiops truncatus*): a preliminary study showed that social factors might influence learning strategies. *Acta Ethologica*, **8**, 57–64.

Dorsey, E. M., Stern, S. J., Hoelzel, A. R. and Jacobsen, J. (1990). Recognition of individual minke whales from the west coast of North America. *Reports of the International Whaling Commission Special Issue*, **12**, 357–368.

Douglas, M. E., Schnell, G. D. and Hough, D. J. (1984). Differentiation between inshore and offshore spotted dolphins in the eastern tropical Pacific Ocean. *Journal of Mammalogy*, **65**, 375–387.

Evans, K. and Hindell, M. A. (2004). The diet of sperm whales (*Physeter macrocephalus*) in southern Australian waters. *ICES Journal of Marine Science*, **61**, 1313–1329.

Fabiani, A., Hoelzel, A. R., Galimberti, F. and Muelbert, M. M. C. (2003). Long-range paternal gene flow in the southern elephant seal. *Science*, **299**, 676.

Fabiani, A., Galimberti, F., Sanvito, S. and Hoelzel, A. R. (2006). Relatedness and site fidelity at the southern elephant seal (*Mirounga Leonina*) breeding colony in the Falkland Islands. *Animal Behaviour*, **72**, 617–626.

Ford, J. K. B., Ellis, G. M., Barrett-Lennard, L. G. *et al.* (1998). Dietary specialization in two sympatric populations of killer whales (*Orcinus orca*) in coastal British Columbia and adjacent waters. *Canadian Journal of Zoology*, **76**, 1456–1471.

Ford, J. K. B., Ellis, G. M. and Balcomb, K. C. (2003). *Killer Whales*, 2nd edn. Seattle: University of Washington Press.

Ford, J. K. B., Ellis, G. M., Matkin, D. R. *et al.* (2005). Killer whale attacks on minke whales: prey capture and antipredator tactics. *Marine Mammal Science*, **21**, 603–618.

Gannon, D. P. and Waples, D. M. (2004). Diets of coastal bottlenose dolphins from the US mid-Atlantic coast differ by habitat. *Marine Mammal Science*, **20**, 527–545.

Gaspari, S., Azzellino, A., Airoldi, S. and Hoelzel, A. R. (2007). Social kin associations and genetic structuring of striped dolphin populations (*Stenella coeruleoalba*) in the Mediterranean Sea. *Molecular Ecology*, **16**, 2922–2933.

Gazda, S. K., Conner, R. C., Edgar, R. K. and Cox, F. (2005). A division of labour with role specialization in group-hunting bottlenose dolphins (*Tursiops truncatus*) off Cedar Key, Florida. *Proceedings of the Royal Society Series B*, **272**, 135–140.

Gemmell, N. J. (2003). Kin selection may influence fostering behaviour in Antarctic fur seals (*Arctoc ephalus gazella*). *Proceedings of the Royal Society Series B*, **270**, 2033–2037.

Goerlitz, D. S., Urban, J., Rojas-Bracho, L., Belson, M. and Schaeff, C. M. (2003). Mitochondrial DNA variation among Eastern North Pacific gray whales (*Eschrichtius robustus*) on winter breeding grounds in Baja California. *Canadian Journal of Zoology*, **81**, 1965–1972.

Goodman, S. J. (1998). Patterns of extensive genetic differentiation and variation among European harbour seals (*Phoca vitulina* vitulina) revealed using microsatellite DNA polymorphisms. *Molecular Biology and Evolution*, **15**, 104–118.

Goto, M. and Pastene, L. A. (1996). Population genetic structure in the North Pacific minke whale examined by two independent RFLP analyses of mitochondrial DNA. *Report of the International Whaling Commission* SC/48/NP5.

Gowans, S., Whitehead, H. and Hooker, S. K. (2001). Social organization in northern bottlenose whales, *Hyperoodon ampullatus*: not driven by deep-water foraging? *Animal Behaviour*, **62**, 369–377.

Greenwood, P. J. (1980). Mating systems, philopatry and dispersal in birds and mammals. *Animal Behaviour*, **28**, 1140–1162.

Guinet, C. and Bouvier, J. (1995). Development of intentional stranding hunting techniques in killer whale (*Orcinus orca*) calves at Crozet archipelago. *Canadian Journal of Zoology*, **73**, 27–33.

Haenel, N. J. (1986). General notes on the behavioral ontogeny of Puget Sound killer whales and the occurrence of allomaternal behavior. In *Behavioral Biology of Killer Whales*, ed. B. C. Kirkevold and J. S. Lockard. New York: Alan R. Liss, pp. 285–302.

Herman, D. P., Burrows, D. G., Wade, P. R. *et al.* (2005). Feeding ecology of eastern North Pacific killer whales *Orcinus orca* from fatty acid, stable isotope, and organochlorine analyses of blubber biopsies. *Marine Ecology–Progress Series*, **302**, 275–291.

Heyning, J. E. and Perrin, W. F. (1994). *Evidence for Two Species of Common Dolphins (Genus Delphinus) from the Eastern North Pacific*. Los Angeles: Natural History Museum of Los Angeles County.

Hoelzel, A. R. (1991a). Killer whale predation on marine mammals at Punta Norte, Argentina; foraging strategy, provisioning and food sharing. *Behavioral Ecology and Sociobiology*, **29**, 197–204.

Hoelzel, A. R. (1991b). Whaling in the dark. *Nature*, **352**, 481.

Hoelzel, A. R. (1993). Foraging behaviour and social group dynamics in Puget Sound killer whales. *Animal Behaviour*, **45**, 581–591.

Hoelzel, A. R. (1998). Genetic structure of cetacean populations in sympatry, parapatry and mixed assemblages; implications for conservation policy. *Journal of Heredity*, **89**, 451–458.

Hoelzel, A. R. (2002). Resource specialization and the evolution of population genetic structure in delphinid species. In *Cell and Molecular Biology of Marine Mammals*, ed. C. J. Pfeiffer. New York: Krieger Publishing, pp. 12–20.

Hoelzel, A. R. and Dover, G. A. (1991). Genetic differentiation between sympatric killer whale populations. *Heredity*, **66**, 191–196.

Hoelzel, A. R., Hancock, J. M. and Dover, G. A. (1991). Evolution of the cetacean mitochondrial D-loop region. *Molecular Biology and Evolution*, **8**, 475–493.

Hoelzel, A. R., Dahlheim. M. and Stern, S. J. (1998a). Low genetic variation among killer whales (*Orcinus orca*) in the eastern North Pacific, and genetic differentiation between foraging specialists. *Journal of Heredity*, **89**, 121–128.

Hoelzel, A. R., Potter, C. W. and Best, P. (1998b). Genetic differentiation between parapatric 'nearshore' and 'offshore' populations of the bottlenose dolphin. *Proceedings of the Royal Society Series B*, **265**, 1–7.

Hoelzel, A. R., Campagna, C. and Arnbom, T. (2001). Genetic and morphometric differentiation between island and mainland southern elephant seal populations. *Proceedings of the Royal Society Series B*, **268**, 325–332.

Hoelzel, A. R., Natoli, A., Dahlheim, M. *et al.* (2002a). Low world-wide genetic diversity in the killer whale (*Orcinus orca*): implications for demographic history. *Proceedings of the Royal Society Series B*, **269**, 1467–1475.

Hoelzel, A. R., Goldsworthy, S. D. and Fleischer, R. C. (2002b). Population genetic structure. In *Marine Mammal Biology: An Evolutionary Approach*, ed. A. R. Hoelzel. Oxford: Blackwell Publishing, pp. 325–352.

Hoelzel, A. R., Hey, J., Dahlheim, M. E. *et al.* (2007). Evolution of population structure in a highly social top predator, the killer whale. *Molecular Biology and Evolution*, **24**, 1407–1415.

Hoffman, J. I. and Amos, W. (2005). Does kin selection influence fostering behaviour in Antarctic fur seals (*Arctocephalus gazella*)? *Proceedings of the Royal Society Series B*, **272**, 2017–2022.

Hudson, R. R. (1998). Island models and the coalescent process. *Molecular Ecology*, **7**, 413–418.

Krebs, J. R. and Davies, N. B. (1997). *Behavioural Ecology: an Evolutionary Approach*. Oxford: Blackwell Publishing.

Krutzen, M., Sherwin, W. B., Berggren, P. and Gales, N. (2004). Population structure in an inshore cetacean revealed by microsatellite and mtDNA analysis: bottlenose dolphins (*Tursiops* spp.) in Shark Bay, Western Australia. *Marine Mammal Science*, **20**, 28–47.

Lazaro, M., Lessa, E. P. and Hamilton, H. (2004). Geographic genetic structure in the franciscana dolphin (*Pontoporia blainvillei*). *Marine Mammal Science*, **20**, 201–214.

Lowther, J. L., Archer, F. I. and Weller, D. W. (2005). A genetic analysis of coastal and offshore bottlenose dolphins, *Tursiops truncatus*, off the western United States. Abstracts from the 16th Biennial Conference on the Biology of Marine Mammals.

Lusseau, D., Schneider, K., Boisseau, O. J. *et al.* (2003). The bottlenose dolphin community of Doubtful Sound features a large proportion of long-lasting associations: can geographic isolation explain this unique trait? *Behavioral Ecology and Sociobiology*, **54**, 396–405.

Lyrholm, T., Leimar, O., Johanneson, B. and Gyllensten, U. (1999). Sex-biased dispersal in sperm whales: contrasting mitochondrial and nuclear genetic structure of global populations. *Proceedings of the Royal Society Series B*, **266**, 347–354.

McConnell, B. J. and Fedak, M. A. (1996). Movements of southern elephant seals. *Canadian Journal of Zoology*, **74**, 1485–1496.

Mead, J. G. and Potter, C. W. (1995). Recognizing two populations of the bottlenose dolphin (*Tursiops truncatus*) off the coast of North America: morphologic and ecological considerations. *Index of Biological Integrity Report*, **5**, 31–44.

Moller, L. M. and Beheregaray, L. B. (2004). Genetic evidence for sex-biased dispersal in resident bottlenose dolphins (*Tursiops aduncus*). *Molecular Ecology*, **13**, 1607–1612.

Natoli, A., Peddemors, V. and Hoelzel A. R. (2004). Population structure and speciation in the genus *Tursiops* based on microsatellite and mitochondrial DNA analyses. *Journal of Evolutionary Biology*, **17**, 363–375.

Natoli, A., Birkin, A., Aquilar, A., Lopez, A. and Hoelzel, A. R. (2005). Habitat structure and the dispersal of male and female bottlenose dolphins (*Tursiops truncatus*). *Proceedings of the Royal Society Series B*, **272**, 1217–1226.

Natoli, A., Cañadas, A., Peddemors, V. M. *et al.* (2006). Phylogeography and alpha taxonomy of the common dolphin (*Delphinus* sp.). *Journal of Evolutionary Biology*, **19**, 943–954.

Ottensmeyer, C. A. and Whitehead, H. (2003). Behavioural evidence for social units in long-finned pilot whales. *Canadian Journal of Zoology*, **81**, 1327–1338.

Parsons, K. M., Durban, J. W., Claridge, D. E. *et al.* (2003). Kinship as a basis for alliance formation between male bottlenose dolphins, *Tursiops truncatus*, in the Bahamas. *Animal Behaviour*, **66**, 185–194.

Perrin, N. and Mazalov, V. (2000). Local competition, inbreeding, and the evolution of sex-biased dispersal. *American Naturalist*, **155**, 116–127.

Perrin, W. F. (2002a). Atlantic spotted dolphin. In *Encyclopedia of Marine Mammals*, ed. W. F. Perrin, B. Würsig and J. G. M. Thewissen. San Diego: Academic Press, pp. 47–49.

Perrin, W. F. (2002b). Common dolphins. In *Encyclopedia of Marine Mammals*, ed. W. F. Perrin, B. Würsig and J. G. M. Thewissen. San Diego: Academic Press, pp. 245–248.

Perrin, W. F., Mitchell, E. D., Mead, J. G. *et al.* (1987). Revision of the spotted dolphins, *Stenella* spp. *Marine Mammal Science*, **3**, 99–170.

Pitman, R. L., Balance, L. T., Mesnick, S. I. and Chivers, S. J. (2001). Killer whale predation on sperm whales: observations and implications. *Marine Mammal Science*, **17**, 494–507.

Pritchard, J. K., Stephan, M. and Donnelly, P. (2000). Inference of population structure using multilocus genotype data. *Genetics*, **155**, 945–959.

Prugnolle, F. and de Meeus, T. (2002). Inferring sex-biased dispersal from population genetic tools: a review. *Heredity*, **88**, 161–165.

Pusey, A. E. (1987). Sex-biased dispersal and inbreeding avoidance in birds and mammals. *Trends in Ecology and Evolution*, **2**, 295–299.

Reynolds, J. E., Wells, R. S. and Eide, S. D. (2000). *The Bottlenose Dolphin: Biology and Conservation*. Gainesville: Florida University Press.

Rice, D. W. and Wolman, R. (1971). The life history and ecology of the gray whale (*Eschrichtius robustus*). *Special Publication of the American Society of Mammalogy*, **3**, 1–142.

Richard, K. R., Dillon, M. C., Whitehead, H. and Wright, J. M. (1996). Patterns of kinship in groups of free-living sperm whales (*Physeter macrocephalus*) revealed by multiple molecular genetic analyses. *Proceedings of the National Academy of Sciences USA*, **93**, 8792–8795.

Rosa, S., Milinkovitch, M. C., Van Waerebeek, K. *et al.* (2005). Population structure of nuclear and mitochondrial DNA variation among South American Burmeister's porpoises (*Phocoena spinipinnis*). *Conservation Genetics*, **6**, 431–443.

Rosel P. E., Dizon A. E. and Heyning J. E. (1994). Genetic analysis of sympatric morphotypes of common dolphins (genus *Delphinus*). *Marine Biology*, **119**, 159–167.

Ross, G. J. B. (1977). The taxonomy of bottlenose dolphin *Tursiops* species in South African waters with notes on their biology. *Annals of Cape Provincial Museum*, **11**, 135–194.

Seielstad, M. T., Minch, E. and Cavalli-Sforza, L. L. (1998). Genetic evidence for a higher female migration rate in humans. *Nature Genetics*, 20, 278–280.

Sellas, A. B., Wells, R. S. and Rosel, P. E. (2005). Mitochondrial and nuclear DNA analyses reveal fine scale geographic structure in bottlenose dolphins (*Tursiops truncatus*) in the Gulf of Mexico. *Conservation Genetics*, 6, 715–728.

Shelden, K. E. W., Moore, S. E., Waite, J. M., Wade, P. R. and Rugh, D. J. (2005). Historic and current habitat use by North Pacific right whales, *Eubalaena japonica*, in the Bering Sea and Gulf of Alaska. *Mammal Review*, 3, 129–155.

Stanley, H. F., Casey, S., Carnahan, J. M. *et al.* (1996). Worldwide patterns of mitochondrial DNA differentiation in the harbor seal (*Phoca vitulina*). *Molecular Biology and Evolution*, 13, 368–382.

Stevick, P. T., McConnell, B. J. and Hammond, P. S. (2002). Patterns of movement. In *Marine Mammal Biology: An Evolutionary Approach*, ed. A. R. Hoelzel. Oxford: Blackwell Publishing, pp. 185–216.

Swartz, S. L. (1986). Gray whale migratory, social and breeding behavior. *Reports to the International Whaling Commission*, 8, 207–229.

Tolly, K. A., Vikingsson, G. A. and Rosel, P. E. (2001). Mitochondrial DNA sequence variation and phylogeographic patterns in harbour porpoises (*Phocoena phocoena*) from the North Atlantic. *Conservation Genetics*, 2, 349–361.

Torres, L. G., Rosel, P. E., D'Agrosa, C. and Read, A. J. (2003). Improving management of overlapping bottlenose dolphin ecotypes through spatial analysis and genetics. *Marine Mammal Science*, 19, 502–514.

Wada, S. (1991). Genetic distinction between two minke whale stocks in the Othotsk Sea coast of Japan. *Report of the International Whaling Commission* SC/43/Mi32.

Walker, W. A. (1981). Geographic variation in morphology and biology of bottlenose dolphins (*Tursiops*) in the eastern North Pacific. *NOAA/NMFS Southwest Fisheries Science Center Administrative Report* No. LJ-81-3c.

Wang, J. Y., Chou, L. S. and White, B. N. (1999). Mitochondrial DNA analysis of sympatric morphotypes of bottlenose dolphins (genus *Tursiops*) in Chinese waters. *Molecular Ecology*, 8, 1603–1612.

Westlake, R. L. and O'Corry-Crowe, G. M. (2002). Macrogeographic structure and patterns of genetic diversity in harbor seals (*Phoca vitulina*) from Alaska to Japan. *Journal of Mammalogy*, 83, 1111–1126.

Whitehead, H. (1996). Babysitting, dive synchrony, and indications of alloparental care in sperm whales. *Behavioral Ecology and Sociobiology*, 38, 237–244.

Future directions in conservation genetics

Recent developments in molecular tools for conservation

CRISTIANO VERNESI AND MICHAEL W. BRUFORD

INTRODUCTION

The availability and application of molecular tools for biodiversity conservation has advanced considerably over the last 15 years, as has been documented in a series of books (e.g. Loeschcke *et al.* 1994; Avise and Hamrick 1996; Smith and Wayne 1996; Frankham *et al.* 2002, 2004), the journal *Conservation Genetics* (in production since 2000) and a series of reviews, all of which give the impression of a maturing discipline (e.g. Hedrick 2001; DeSalle and Amato 2004). Dramatic advances in data analysis have occurred over the last five to seven years and these are well-documented elsewhere in this volume. However, in other respects, it could be argued that the field has been dominated by the use of a few tried-and-tested marker types and has not been as quick as it could have been to adopt new laboratory methodologies. For instance, the quantum leaps in high throughput molecular protocols for detecting and analysing DNA polymorphisms, applicable for example to rapid community-level biodiversity assessment using DNA barcoding methods (such as developments in rapid sequencing and large-scale SNP genotyping), are yet to make a significant appearance in the conservation genetics literature. In addition, a framework for the routine translation of conservation genetic data into population management and specific actions in the field remains in its infancy. Conservation genetics has largely remained a field where routinely, a relatively small number of molecular markers are isolated and applied to a few populations of a single species, some of which may be threatened. The data from such studies are then published in reports and peer-reviewed scientific publications such as *Molecular Ecology* and *Conservation Genetics*, wherein management recommendations may be made. However, it is not clear whether these suggestions are ever incorporated into species on habitat

Population Genetics for Animal Conservation, eds. G. Bertorelle, M. W. Bruford, H. C. Hauffe, A. Rizzoli and C. Vernesi. Published by Cambridge University Press. © Cambridge University Press 2009.

action plans; in fact, many conservation geneticists complain that their results are, on the whole, completely ignored by management authorities.

A number of factors could potentially contribute to this hiatus: (1) expense: molecular studies remain out of reach for many small governmental and non-governmental organizations and such studies are the first to be cut when budgets are set; (2) occasionally, when multiple studies are carried out on the same species or populations they produce apparently non-congruent results (e.g. King *et al.* 2006; see also chapter by Bruford, this volume), which leave managers and policy-makers questioning the value of such studies; (3) we, in the scientific community, appear to be doing a singularly poor job of convincing our colleagues at the coal-face of applied conservation that genetics really matters; (4) there seems to be genuine inertia and hostility towards genetics within the conservation community and a lack of appreciation of its relevance, including at the policy level, and it also seems that (5) there is a general perception in the wider community of the limited nature of the genetic research carried out in the name of conservation. In short, our approaches have a credibility problem!

How could these problems be addressed? Of the above points, 1, 2 and 5 appear largely to be issues that can be dealt with by further development at both the practical and technical level, and if these problems can be solved then it seems plausible that 3 and even 4 will no longer remain an issue. In other words, the interactions between 'pure' and 'applied' science depend critically on the esteem and credibility with which the former is held by the latter. This chapter sets out to take a prospective look at conservation genetics tools over the coming ten years and attempts to address some of the above-mentioned problems by highlighting where conservation genetics can improve technically and practically in the near future and how this might eventually lead to the routine, sound and relevant application of molecular biology to managing the extinction crisis we all face.

Dissecting the underlying problems of the above points may identify the problems encountered when interacting with conservation practitioners, and may also lead towards potential solutions and the more routine use of genetics in conservation. Taking point 1 (expense): although molecular reagents, salaries and field costs are not decreasing (quite the contrary), it seems clear that by applying high throughput approaches and generic methodologies, an 'economy-of-scale' and rapid isolation of informative markers can significantly reduce costs. DNA sequencing is already becoming remarkably cheap, with several companies offering single pass sequencing of PCR amplicons for just a few dollars, but with new developments in parallel sequencing methods (e.g. Margulies *et al.* 2005), the prospects for a further order-of-magnitude decrease in cost look promising. Furthermore, the almost exponential increase in genome, EST and SNP data in an ever-expanding group of organisms (e.g. 3×10^6 SNPs recently characterized in chickens: Ellegren 2005) means that there is a rapidly increasing fund of data with which to derive genetic markers and a wide availability of technology solutions for isolating such markers in threatened species. It seems clear that we must attempt to isolate, database and promulgate information on informative molecular markers in as many species as possible, as rapidly as possible. Even if expectations on the generic nature of specific markers within the microsatellite (see any issue of *Molecular Ecology Notes*) or SNP (e.g. Scotti-Saintagne *et al.* 2004) families are somewhat limited, the methodologies to rapidly isolate, bulk process and analyse such data in all species are now available and provide an imperative. In some species groups with 'recalcitrant' genomes (e.g. Lepidoptera: Zhang 2004), it may take many months and even occasionally longer, to generate a battery of markers. This problem is even clearer as the benefit of analysing kilo-markers is becoming evident in the new era of population genomics (Luikart *et al.* 2003).

The use of more markers, common markers and improved standards of sampling and analysis may obviate the occasional problem of conflicting results arising from similar studies on the same populations (point 2). Such occurrences are frequently technical (e.g. Gagneux *et al.* 2001), but may also arise from the use of different molecular markers, and/or debatable interpretation (e.g. Debruyne 2005; Roca *et al.* 2005). Clearly, scientific discussion and disagreements about both the validity and interpretation of data among research groups are normal; however, the size of the task facing conservationists is enormous, and the credibility of conservation genetics as a discipline depends on some measure of confidence within the conservation community. The question is also legitimately raised as to why there is so much overlap among research groups in terms of taxa and populations studied, when resources in the field of conservation are so scarce. These problems can be solved by greater data availability, transferability and sharing, especially for the less well-studied organisms and those phylogenetically distinct taxa which require conservation genetic approaches just as much as, if not more than, the more commonly studied species. While there is an obvious mechanism for such data sharing for DNA sequences, a solution remains far less obvious for microsatellite studies.

Finally, addressing point 5 requires a step-change in the scope of conservation genetics studies to encompass adaptive variation, at the genome, transcriptome and proteome levels. The use of hundreds to thousands of markers in future DNA studies of threatened populations (the population-omic approach), perhaps involving SNPs and microarray technology, is no longer far-fetched, and the integrative use of 'adaptive' and 'neutral' regions of the genome may allow several kinds of variation to be assayed at once (e.g. Ruvinsky 2002; Aitken *et al.* 2004; Turner *et al.* 2005): these approaches are already coming on stream in bacteria (e.g. Dorrell *et al.* 2005) and may soon involve expression arrays (e.g. Vasemägi and Primmer 2005). These shortcomings in the current state of the art form the focus of what follows and we hope by doing so here to highlight potential future avenues for truly bringing conservation genetics into the new millennium.

WHERE WE ARE

In most conservation genetics studies, sequence polymorphisms have been analysed using variants of classical Sanger sequencing (Sanger *et al.* 1974), which has been applied for more than 30 years. Thanks to the development of this technique, genome-sequencing projects of many organisms, from humans to, more recently, dogs (Lindblad-Toh *et al.* 2005) have been successfully carried out. Going beyond the description of the technical improvements of the Sanger sequencing approach (see e.g. Gibson and Muse 2004), it is important to remember that sequencing costs are becoming increasingly more affordable for most laboratories, even eliminating the need to purchase automated sequencers. Also, fragment analysis, such as microsatellites and amplified fragment length polymorphism (AFLP) can be easily analysed employing the same equipment used for sequencing. Therefore, in the following sections we aim to describe the advantages and disadvantages of alternative techniques that are currently not commonly used in conservation genetics. We present the techniques in loose chronological order (see Table 14.1).

As will be seen, for most of these techniques the leading principle is not to characterize long stretches of sequence but, instead, to maximize the number of samples analysed per unit time, directly focusing on shorter fragments containing the polymorphism to be surveyed. While most of the described techniques are suitable for sequencing, some of them can be applied to the analysis of fragment length polymorphism such as microsatellites and AFLP.

Minisequencing

This method dates back to the early 1990s (Syvänen *et al.* 1990), relatively soon after the advent of PCR. The rationale of this assay is to allow a primer

Table 14.1 *Chronological list of the techniques described in the text along with reference and the type of analysis provided by each*

Year	Technique	Analyses provided	Reference
1990	Minisequencing	SNP	Syvänen *et al.* (1990)
1991	Real-time PCR	Quantitation of DNA; SNP; gene expression	Holland *et al.* (1991)
1993	MALDI-TOF	SNP	Wu *et al.* (1993)
1995	AFLP	Anonymous polymorphic fragments; cDNA expression	Vos *et al.* (1995)
1995	Microarray	Gene expression; SNP	Schena *et al.* (1995)
1996	Pyrosequencing	SNP; sequencing	Ronaghi *et al.* (1996)

to anneal to the sequence immediately 3′ of the polymorphic position and to extend this primer by just a single labelled nucleoside triphosphate which is complementary to the variable nucleotide using a DNA polymerase. This technique is particularly suitable for single nucleotide polymorphism (SNP) analysis. The difference between homozyotes and heterozygotes can be unambiguously determined and, furthermore, due to the extreme specificity of the primer extension reaction, several loci can be simultaneously analysed (multiplexed), reducing costs and time. Originally, several techniques for separating minisequencing reaction products were used (see Syvänen 1999), while today gel or capillary electrophoresis via automated sequencers can be adopted. This is also the reason why several commercial kits for minisequencing are now available. In some cases, with slight modification of these kits, many polymorphisms can be investigated even starting with low quantity and quality DNA template (Makridakis *et al.* 2001; Salas *et al.* 2005).

To increase the throughput of minisequencing, two modifications are usually used: increasing the number of loci to be multiplexed in a single reaction, and immobilizing oligonucleotides on a solid support complementary to the polymorphism to be surveyed by minisequencing. The latter can be effectively regarded as a DNA chip, allowing minisequencing to enter the field of microarray analysis (Shumaker *et al.* 1996). Microarray-based minisequencing approaches have recently reached high-throughput levels: 80 or 14 specimens can be simultaneously analysed for 200 or 600 SNPs, respectively, on a single microscope slide (Lovmar and Syvänen 2005). The same methods can be successfully adopted in cases (non-invasively extracted, ancient DNA, etc.) where it is necessary to use short PCR products (<200 bp), by analysing high copy number DNA fragments such as mtDNA (Divine and Allen 2005).

Minisequencing techniques are now routinely employed in diagnostics, due to the ease with which most of the procedures can be automated, and to the high throughput offered by the concomitant use of chip technology. Moreover, minisequencing has recently been used for studying gene expression profiles for typing imbalanced expression linked to different alleles at the same SNP locus (Liljedahl *et al.* 2004). Limitations to this approach include the need to choose carefully primers for multiplex analysis of several SNPs in a single reaction. Different primers can easily cross-react, forming homo- and heterodimers and hairpins. To improve the situation, the aid provided by bioinformatics seems particularly promising: for example, the adoption of specific software for primer design allowed the simultaneous analysis of 45 different SNPs in the same minisequencing reaction (Kaderali *et al.* 2003).

Real-time polymerase chain reaction (RT-PCR)

This technique provides a measure of quantitation that, for the first time, allows precise estimates of DNA concentration in all elements of the PCR reaction, an extremely useful innovation for non-invasive genetics. A quantitative measurement of the fragments amplified in each cycle is obtained by a detector, which records the light emitted from specific fluorochromes incorporated in the newly synthesized PCR products, cycle by cycle (Holland *et al.* 1991).

We can subdivide the way a RT-PCR assay works into two main elements: (1) non-sequence-specific detection and (2) sequence-specific detection. In the first case, specific DNA fluorophore dyes are intercalated into the double helix to produce an increase in fluorescence. The main requisites for these dyes are stability and absence of inhibition during the PCR reaction. SYBR green I and LCV green are among the most commonly used dyes. Being non-sequence-specific, this method of RT-PCR is rarely adopted for polymorphism detection with the exceptions of amplification refractory mutation system (ARMS) PCR and amplicon melting assays (see Newton *et al.* 1989 and Gupta *et al.* 2004 for detailed descriptions of these methods).

One of the most common approaches used today is the 'TaqMan assay' (Holland *et al.* 1991), also known as the exonuclease 5′ assay. This reaction involves two primers and two allele-specific probes. Each probe is complementary to one of the two alleles of a single SNP locus and is labelled with a different fluorophore. Each probe specifically binds to its target sequence during PCR. The exonuclease activity of the DNA polymerase cleaves at the 5′ end of the probe that dissociates from its quencher linked at the 3′ end, thus increasing the fluorescence. Depending on which fluorescence signal is detected, the allelic status can be inferred. If two fluorescence signals are generated, it means that the specimen analysed is heterozygous (see Louis *et al.* 2004 for an example).

SNP typing by RT-PCR has been widely adopted in the diagnostic and biomedical field (see Gibson 2006 for a recent review). This technique has been successfully compared to well-established techniques such as restriction fragment length polymorphism (RFLP) (Schroell-Metzeger *et al.* 2003), demonstrating that it really can decrease time and costs. Recently it has been shown that RT-PCR is also effective in the determination of the haplotypic phase of a particular SNP: this is a very promising application as the phase is unambiguously determined without needing to subclone, greatly reducing cost and time. Another attractive feature of RT methods in comparison to end-point PCR is that all post-PCR processes such as electrophoresis and PCR products purification are eliminated, therefore speeding up the experiment and making it more feasible even in the field (Belgrader *et al.* 1999; Watanabe *et al.* 2004).

However, the greatest strength of RT-PCR is in its extreme sensitivity. It has been shown that as few as 1% of mutated sequences can be detected in a background of wild-type sequence (Hodgson *et al.* 2002). In the field of conservation biology, RT-PCR offers very promising opportunities to detect, for example, particular pathogen species and/or strains in the species of interest (Amar *et al.* 2002; Campsall *et al.* 2004; Yeh *et al.* 2004) even at very low concentrations. This means that the presence of some diseases could be detected and monitored at their outbreak, before massive spread and propagation. Since many conservation genetics studies are based on non-invasively extracted or ancient DNA, this research has often been hampered by the number of starting template molecules and very low copy numbers, resulting in PCR failure and severe contamination (Cooper and Poinar 2000). In these situations, the use of RT-PCR is highly recommended, as already shown in several studies (Morin *et al.* 2001, 2005; Poinar *et al.* 2003; Wandeler *et al.* 2003; Pruvost and Geigl 2004; Pruvost *et al.* 2005; von Wurmb-Schwark *et al.* 2005).

Although the vast majority of conservation genetics studies are based on the screening of neutral markers, the post-genomic era, with the accumulation of information about the function of many genes (and thanks to the completion of several genome sequencing projects in different organisms) will signal a progressive shift to the investigation of markers related to selection and expression profiles. Here, RT-PCR can also be advantageous: by using this technique, it is now possible to quantitatively assess the expression profile of a particular gene by comparing its profile to that of a housekeeping gene (for a review see Bustin 2000). This approach is very promising, even if in some cases the right choice of the housekeeping gene still represents a limiting factor (Klein 2002).

Matrix-assisted laser desorption/ionization time-of-flight mass spectrometry (MALDI-TOF)

In MALDI-TOF, a matrix solution is added to a stretch of DNA to be analysed. This matrix–analyte mix is then spotted on to a target plate and allowed to crystallize. A laser hits the resulting crystal so that the analyte is ionized and introduced into a flight tube. This ionization process determines a collision between the matrix and DNA molecules, generating DNA ions. These ions are made to pass across an electric field, causing them to fly through from the tube to the detector. Lighter ions (i.e. smaller DNA fragments) travel faster than heavier ions (larger fragments). Thus the fragments are separated on the basis of their mass difference. The time of flight is measured and, with a calibration factor, converted into the mass: charge ratio. Theoretically, separation capacity can be as low as 1 Da.

This technique was originally devised for protein analysis; only later did it become available for DNA analysis (Wu *et al.* 1993; Vestal *et al.* 1995; Little *et al.* 1997). The major advantages of MALDI-TOF are the speed and extreme accuracy of the spectrometric measurements. Currently available machines allow the recording of a single spectrum in less than one second. This means that, being a multi-channel instrument, the same machine can be used for recording many data points in a single experiment, and several SNPs can be simultaneously analysed. This translates into the theoretical opportunity of typing 30 000 genotypes in a single day (Gut 2004), with very low price per sample in comparison to other methods (Bray *et al.* 2001).

Although the technique was originally applied to sequence polymorphisms, some trials for the screening of length polymorphisms such as microsatellites have been performed (Butler *et al.* 1998). This greatly reduces the chance to multiplex several loci in a single reaction. MALDI-TOF is seemingly a very high-throughput technique, but the cost of the machine and, paradoxically, its capacity for analysing so many samples (also required for decreasing the cost per unit), make it less attractive for most of the conservation genetics laboratories currently. Nonetheless, for single projects requiring large scale SNP genotyping, it could be advisable to link with laboratories that already own an instrument and who will be able to offer analyses at very low costs.

Amplification fragment length polymorphism (AFLP) PCR

AFLP-PCR can be regarded as a combination of two existing techniques, namely random amplified polymorphic DNA (RAPD) and RFLP analyses. While the latter was developed long before PCR, the former is a PCR-based method in which random oligonucleotide primers are used to generate several anonymous polymorphisms. These polymorphisms are dominant markers; therefore, the distinction between homo- and heterozygotes is not possible.

Originally greeted with enthusiasm, as a simple, generic technique to assess genomic variation, RAPD analysis soon began to reveal problems, mainly related to reproducibility. To overcome this, AFLP-PCR was introduced in the mid 1990s (Vos *et al.* 1995). Extracted genomic DNA is cut with a four- or six-base-recognizing restriction enzyme and synthetic oligonucleotide adaptors are ligated to the sticky ends of the fragments generated. A PCR reaction with arbitrary primers is then performed. Usually, to reduce the complexity of the profile, the following steps are adopted: a preamplification is done where only fragments with one of the four azotate bases are amplified. In the next selective amplification, only a small aliquot of the preamplification product is employed, using two primers that extend inside the fragments for two other bases, one of the primers being labelled with specific fluorochromes so that the approximately 100 fragments of PCR product can be easily analysed through an automated sequencer.

Several commercial kits are now available and many improvements are being devised to analyse AFLP profiles using standard automated sequencers (Papa *et al.* 2005). The most striking feature of AFLP analysis is that a very large number of polymorphisms can be generated rapidly and at low cost even for species lacking genomic information. Perhaps the major drawback is the genetic characteristics of AFLP markers: they are dominant thus harbouring less information than classical co-dominant markers such as STR. Furthermore, not all co-migrating bands have the same nucleotide sequence (Vekemans *et al.* 2002). Another point to be considered is that AFLP-PCR requires reasonably large quantities of well purified DNA, since it is more sensitive to inhibition in comparison to other PCR-based techniques (Bensch and Åkesson 2005). This limits the use of AFLP in conservation studies based on non-invasively extracted DNA.

AFLP-PCR has rapidly gained popularity, especially in plant studies, yet with comparatively few studies in animals (Bensch and Åkesson 2005). This is surprising since AFLP possesses features suitable for animal conservation genetics. While for paternity testing and parentage analyses there are obvious limitations of dominant markers such as AFLPs, for surveys of intrapopulation variability and genetic structure among different populations, AFLPs have proved as effective as standard markers (i.e. mtDNA and microsatellites): in most studies it has been shown the F_{ST} values inferred from AFLPs are of the same order of magnitude (Nybom 2004; Baus *et al.* 2005). More generally, it appears that to reach the same degree of informativeness provided by classical markers, one has to use from four- to ten-fold more AFLP markers (Mariette *et al.* 2002). This does not seem to represent a major problem since to obtain a high number of AFLP markers is relatively easy, fast and cheap. On the contrary, the possibility to retrieve numerous AFLP markers means that in some cases, when dealing with very poorly differentiated populations, AFLPs are highly effective (Wang *et al.* 2004; Campbell and Bernatchez 2004). The same holds for individual assignment, since it is crucial to work with many polymorphic markers, especially when the information about the presumed source populations is scarce (Dearborn *et al.* 2003; Irwin *et al.* 2005). This also applies to hybridization studies: in some instances AFLPs have proved even better than microsatellites (Bensch *et al.* 2002). Furthermore, with the adoption of appropriate statistics, AFLPs can be successfully used for phylogenetic inferences (Sullivan *et al.* 2004).

As AFLP-PCR can easily scan vast genomic regions, markers can be applied for finding new polymorphisms in non-model organisms. AFLPs have thus been applied for the detection of new SNPs (Bensch *et al.* 2002; Nicod and Largiadèr 2003), new sex-specific markers (Ezaz *et al.* 2004) and microsatellites (Albertini *et al.* 2003). Finally, AFLPs seem very useful for finding new polymorphisms in different species or strains of pathogens (van den Braak *et al.* 2004; Fearnley *et al.* 2005). Being useful for screening large genomic regions without any prior sequence information, AFLPs are very effective in finding markers linked to genomic regions under selection. This field has recently benefited from the concomitant improvement in the statistical tools (see 'Perspective' section in this chapter and other chapters; Wilding *et al.* 2001; Bensch *et al.* 2002; Campbell and Bernatchez 2004). They have also proved useful in identifying quantitative trait loci (Beaumont *et al.* 2005).

Flexibility, no need for extensive technique development, and low cost make AFLPs potentially effective for conservation genetics studies (Lucchini 2003). Despite this, few laboratories routinely apply this technique. AFLPs perhaps ought to represent the first choice when rapid and cost-effective preliminary information about genetic diversity and differentiation among natural populations is needed in unstudied organisms.

Considering that most laboratories now have access to an automated sequencer, the limited use of AFLPs in animal conservation genetics seems even more paradoxical.

Pyrosequencing

This non-electrophoretic based sequencing technique requires real-time monitoring of an enzymatic reaction during which the nucleotide sequence of the fragment of interest is determined by synthesis. The enzymatic cascade starts by the release of a pyrophosphate (PPi) as result of the incorporation of a nucleotide by the DNA polymerase. Subsequently, ATP sulphurylase converts the PPi into ATP, thus providing the necessary energy for luciferase to oxidize luciferin and generate light, which is then detected. The unincorporated nucleotides are then degraded by apyrase prior to addition of the next nucleotide, so that addition becomes iterative. As the added nucleotide is known, the sequence of the template can be determined (Ronaghi et al. 1996). The PCR product to be used in the enzymatic cascade needs to be well purified, since four different enzymes are involved. When this technique first became available, only short DNA fragments (up to 150 bp) could be analysed. This made pyrosequencing especially attractive for SNP typing, especially because the determination of homo- and heterozygotes is unambiguous (Ronaghi 2001).

Recently, pyrosequencing has been vastly improved by the coupling of emulsion-based methods to isolate and amplify DNA fragments in vitro with optimization of pyrosequencing protocols for picolitre-sized wells (Margulies et al. 2005). In its first configuration, the system, originally devised by 454 Life Sciences, was able to generate over 25 million bases in a single run. The recent improvement of the entire workflow, concomitant with 454's acquisition by the worldwide leader biotech company Roche, makes it possible to increase the number of bases generated in a single run to 100 million in just 7.5 hours with single-read accuracies greater than 99.5% over 200–300 bases. Despite the fact that the cost of acquiring the machine is still high, the efficiency, productivity and accuracy of the so-called '454 sequence technology' allows the completion of entire genome sequencing projects at costs and time even lower than those offered by standard Sanger-based sequencing techniques (Swaminathan et al. 2007). The flexibility of the 454 method goes far beyond only genome projects, including several applications in resequencing, transcriptome analysis, metagenomics and palaeogenomics. The use of the 454 sequencing technique in the last two fields is holding great promise as highlighted in many recent papers (e.g. Angly *et al.* 2006; Green *et al.* 2006; Leininger *et al.* 2006; Poinar *et al.* 2006).

Provided the availability of commercial services offering 454 sequencing at increasingly more affordable prices becomes reality, the 454 pyrosequencing method is likely to become popular in conservation genetics. The possibility of completing shotgun genome projects without the expensive and time-consuming step of cloning opens up the possibility to unveil the genome of several non-model organisms like those typically encountered in most conservation genetics studies.

Microarray technology

Microarray techniques (Schena *et al.* 1995) have become very popular in the study of gene expression profiles and many methods are now available for interrogating labelled mRNA samples with nucleic acid probes arrayed at high density. Here, we present the approach that can be most usefully employed in non-model organisms.

The source is represented by a large library of cDNA to be arrayed. Usually the experiment is started by obtaining end-sequences for several thousand clones and then a selected unique set of these expressed sequence tags (ESTs) is selected for amplification. The PCR products are robotically spotted at a density of about 20–30 clones per square millimetre on the surface of a glass slide or filter, usually in batches of about 100 slides. The cDNA microarray is hybridized to radioactively or fluorescently labelled cDNA obtained by reverse transcription of mRNA isolated from the tissues or cells of interest. A competitive hybridization of two samples labelled with two different dyes (usually Cy3 and Cy5) allows the estimation of the ratio of transcript abundance in the two RNA samples to be compared, independently for each spot on the microarray. Changes in gene expression are therefore inferred from changes in the signal intensity of each clone relative to the sample mean, employing a common reference sample as the standard against which experimental specimens are compared.

Among the first cases of microarray-based evolutionary studies were studies such as surveys on yeast wild isolates (Cavalieri *et al.* 2000; Townsend *et al.* 2003). A teleost fish microarray showed a surprising variability in the expression profiles not only between populations but even between single individuals (Oleksiak *et al.* 2002). In the same species, microarray experiments indicated that there are several tissue-specific differences in gene expression that were unique to some populations (Whitehead and Crawford 2005). Since a shift to the study of adaptation processes seems highly advisable in the conservation genetics field, microarrays can potentially be very useful. It is important to remember that changes in expression profiles occur well before any recordable change at the phenotypic level in many ecologically important situations such as the response to environmental stress caused by pollutants (*Xenopus* tadpoles: Jelaso *et al.* 2003), and this kind of result means that new frontiers for ecological monitoring are now open.

The genomic information of model organisms can potentially be transferred to phylogenetically related non-model species for which there is conservation interest, as highlighted by a study on Eastern tiger salamanders (*Ambystoma tigrinum tigrinum*) and Mexican axolotls (*Ambystoma mexicanum*; Putta *et al.* 2004). Furthermore, identifying genomic regions implicated in evolutionarily important processes such as speciation in model species by means of microarrays (Turner *et al.* 2005), can give relevant information about non-model organisms, too. In addition, other less complex applications such as species identification can benefit from the application of microarrays (Pfunder *et al.* 2004). Microarrays are clearly very promising but their high-throughput and costs still make it difficult for most of labs involved in conservation genetics studies to use this technique.

Maximizing DNA retrieval: whole-genome amplification (WGA) and metagenomics approaches

Whole-genome amplification

A common problem in molecular laboratories is that, after several analyses and years since the original extraction, an important DNA sample runs out, making it impossible to continue using it. In some cases this problem could be overcome by establishing cell lines, but more often this is hindered by the biological tissues available for original sampling (faeces, bone fragments, blood droplets, etc). For this reason, soon after PCR became available, techniques were devised to non-selectively amplify entire genomic DNA (gDNA) so that large amounts of DNA could be made. These PCR-based techniques include primer extension preamplification (PEP: Zhang *et al.* 1992), degenerate-oligonucleotide-primed PCR (DOP: Telenius *et al.* 1992), and improved-primer extension preamplification (I-PEP: Dietmaier *et al.* 1996). With these methods, fragments of about 100–1000 bp can be obtained, but without a large genome coverage. Furthermore, these approaches are strongly affected by the features of the template being amplified (e.g. GC content).

Recently, a new method has been introduced: multiple displacement amplification (MDA: Dean *et al.* 2002), which does not rely on PCR but on enzymatic replication by the DNA polymerase of the bacteriophage phi29.

This polymerase, using hexamer primers, isothermically replicates the template DNA at 30 °C by a 'hyperbranching' mechanism of strand displacement synthesis, with the polymerase laying down a new copy as it displaces previously made copies. This mechanism is very efficient: micrograms of DNA are synthesized starting from only 10–100 ng of template DNA. This occurs with an error rate as low as a single nucleotide over 10^6–10^7 incorporated (Esteban et al. 1993). The reaction only needs a simple oven or water-bath and it is scalable to any volume, thus allowing easy generation of milligrams of genomic DNA.

Many studies (for a review see Lasken and Egholm 2003) have demonstrated the usefulness of WGA techniques to obtain gDNA of the quantity and quality required for SNP typing, even starting from very low amount (<1 ng) of template DNA (Lovmar et al. 2003). MDA, now available in commercial kits, proved particularly effective to generate DNA to be analysed by direct sequencing, TaqMan assays, pyrosequencing and microsatellite analysis (Holbrook et al. 2005). For the latter, it has been observed that probably I-PEP is more efficient than MDA and that it is better to start with at least 5 ng of template DNA (Sun et al. 2005), although reliable results have been obtained with MDA on forensic material (e.g. bloodstains) as old as 1 year and with starting amount of around 5 pg, that is the equivalent to 1–2 diploid cells (Hanson and Ballantyne 2005). This also means that biological material containing scarce traces of DNA could be employed in WGA, suggesting the use of these techniques in non-invasive, degraded or ancient DNA. However, in these cases, the extreme sensitivity of the method means the amplification of potentially contaminating DNA is a possibility (Sorensen et al. 2004).

WGA represents a potentially very efficient tool for conservation genetics studies, where it is often hard to obtain just a single sample, even from sources like faeces and urine. Therefore, it is likely that WGA will become more popular in the analysis of protected and elusive species, as highlighted in a recent study on some non-human primate species (Rönn et al. 2006). However, currently it seems that WGA is not proving very efficient at amplifying many non-invasive samples (Gunn et al. 2007), where the genomes of more than one species (e.g. animal, plant and bacteria) will be present and where fragment sizes are short.

Metagenomics

The term 'metagenomics' was first introduced in 1998 (Handelsman et al. 1998) and it refers to a habitat-based analysis of mixed microbial populations at the DNA level. Now the meaning is more general, describing the approach by which the entire DNA from a particular source (usually environmental, such as soil or sea water) is extracted and cloned, and then the nucleotide sequence of many clones is determined. Finally, by comparing the sequences obtained with those deposited in public repositories, the organisms of the original sample are identified.

Originally employed almost exclusively for soil and sea microorganisms (for reviews see Tringe and Rubin 2005 and Steele and Streit 2005), and for the opportunity to discover new natural products (Daniel 2004), this approach has proved very effective for aDNA surveys, as demonstrated in two recent studies on the cave bear (Noonan et al. 2005) and mammoth (Poinar et al. 2006). In such cases, avoiding the initial use of PCR is thought to reduce the risk of contamination. On average, only a small fraction (about 3–5%) of the cloned DNA contains nucleotide sequences of the species of interest: this means that the sequencing efforts are considerable and they could represent a serious limiting factor, although the development of high-throughput DNA sequencing techniques such as 454 pyrosequencing could be helpful (Poinar et al. 2006). In conservation biology, the metagenomics approach could be applied in other studies; for instance, for characterizing the microbial content of the gut of many species (Cann et al. 2005; Turnbaugh et al. 2006), gaining important information on studying gastro-intestinal diseases that often have a serious impact on natural populations.

Technical issues such as the influence of the method used for DNA isolation and the proper handling of sequence data for comparison with reference databases have to be taken into account. However, the meta-genomics approach is without doubt very promising, especially since the focus of such studies will not just be on a single species but on the entire spectrum of organisms living in a particular ecosystem; retrieving information on the level of genetic diversity at the ecosystem level may help to devise better conservation and management policies.

CONCLUSIONS AND PERSPECTIVE

The techniques briefly reviewed here do not represent a complete description of what is currently available; for example, denaturing high-performance liquid chromatography (DHPLC) (for a review see Xiao and Oefner 2001), which is particularly suited for SNP detection, has been excluded. However, considering that in most conservation genetics projects the budget is limited, and that flexibility, rather than high-throughput capacity is required, techniques like the RT-PCR and AFLP-PCR appear to be the most appropriate. These two approaches are not only cheap and rapid, they are already suitable for the so-called 'post-genomic era' (i.e. the study of gene expression profiles and non-neutral markers) that will presumably be the next horizon in conservation genetics.

Moreover, since the main interest in conservation genetics is for non-model organisms, we are still engaged in the discovery of genomic resources. To this end the advent of fast and accurate techniques for large-scale genome sequencing like 454 pyrosequencing seems very promising.

The genome era has seen quantum leaps in data-handling and bioinformatics applications, and it is true to say that conservation genetics analysis has come a long way since the 1980s and early 1990s when most researchers were using allozyme or morphometrics-designed packages such as BIOSYS (Swofford and Selander 1981) and NTSYS (Rohlf 1992). It is clear from chapters elsewhere in this volume that an enormous amount of work has been carried out to produce powerful software to implement phylogenetic, population genetic and individual-based analyses which now utilize much information gleanable from DNA sequences, microsatellite allelic configurations, SNP haplotypes, etc. However, unlike the genome-scale analytical tools that have been developed, many analytical software packages in conservation genetics have absolute (software) or practical (computation time) limitations on the number of individuals or loci that can be assessed simultaneously. This is a problem that will only intensify with the ever-increasing number of loci and individuals that will be used in future studies. Some innovations are directly aimed at overcoming these problems, such as the so-called approximate Bayesian approaches (e.g. Beaumont and Rannala 2004) and software that can replicate and distribute files for simultaneous analysis on multiple processor systems (such as, for example, 'REPMAKER' for use with the program PAUP*). However, it is evident that if kilo-locus analysis is to become commonplace then a number of commonly used applications will need to be modified or replaced in the near future.

Raw and transformed data availability and sharing have become more commonplace in the era of automated DNA analysis and this needs to continue in the future. More and more, journals are insisting on online appendices to conservation genetics papers, but the extent to which these contains raw data varies widely. With the terabyte storage capacity of institutional servers, it should be the aim of all groups to maintain (minimally) examples and preferably entire data sets on-line for others to utilize and analyse once papers are published. This issue has already become critical for microarray studies, since analysis and interpretation of array data can be problematic and sometimes controversial (e.g. Ideker et al. 2000; Spruill *et al.* 2002). While this is potentially a Herculean task in some cases, it remains an important step in establishing the credibility of conservation genetics to the global community and will most importantly facilitate comparative studies where currently few are published. In short, we need to apply the same standards for DNA fragment-based studies as we do for sequences.

REFERENCES

Aitken, N., Smith, S., Schwarz, C. and Morin, P. A. (2004). Single nucleotide polymorphism (SNP) discovery in mammals: a targeted-gene approach. *Molecular Ecology*, **13**, 1423–1431.

Albertini, E., Porceddu, A., Marconi, G. *et al.* (2003). Microsatellite-AFLP for genetic mapping of complex polyploids. *Genome*, **46**, 824–832.

Amar, C. F., Dear, P. H. and McLauchlin, J. (2002). Detection and genotyping by real-time PCR/RFLP analyses of *Giardia duodenalis* from human faeces. *Journal of Medical Microbiology*, **52**, 681–683.

Angly, F. E., Felts, B., Breitbart, M. *et al.* (2006). The marine viromes of four oceanic regions. *PLoS Biology*, **4**, e368.

Avise, J. C. and Hamrick, J. L., eds. (1996). *Conservation Genetics: Case Histories from Nature*. New York: Chapman and Hall.

Baus, E., Darrock, D. J. and Bruford, M. W. (2005). Gene-flow patterns in Atlantic and Mediterranean populations of the Lusitanian sea star *Asterina gibbosa*. *Molecular Ecology*, **14**, 3373–3382.

Beaumont, M. A. and Rannala, B. (2004). The Bayesian revolution in genetics. *Nature Reviews Genetics*, **5**, 251–261.

Beaumont, C., Roussot, O., Feve, K. *et al.* (2005). A genome scan with AFLP markers to detect fearfulness-related QTLs in Japanese quail. *Animal Genetics*, **36**, 401–407.

Belgrader, P., Benett, W., Hadley, D. *et al.* (1999). PCR detection of bacteria in seven minutes. *Science*, **284**, 449–450.

Bensch, S. and Åkesson, M. (2005). Ten years of AFLP in ecology and evolution: why so few animals? *Molecular Ecology*, **14**, 2899–2914.

Bensch, S., Åkesson, M. and Irwin, D. E. (2002). The use of AFLP to find an informative SNP: genetic differences across a migratory divide in willow warblers. *Molecular Ecology*, **11**, 2359–2366.

Bray, M. S., Boerwinkle, E. and Doris, P. E. (2001). High-throughput multiplex SNP genotyping with MALDI-TOF mass spectrometry: practice, problems and promise. *Human Mutation*, **17**, 296–304.

Bustin, S. A. (2000). Absolute quantification of mRNA using real-time reverse transcription polymerase chain reaction. *Journal of Molecular Endocrinology*, **25**, 169–193.

Butler, J. M., Li, J., Shaker, T. A. and Becker, C. H. (1998). Reliable genotyping of short tandem repeat loci without an allelic ladder using time-of-flight mass spectrometry. *International Journal of Legal Medicine*, **112**, 45–49.

Campbell, D. and Bernatchez, L. (2004). Generic scan using AFLP markers as a means to assess the role of directional selection in the divergence of sympatric whitefish ecotypes. *Molecular Biology and Evolution*, **21**, 945–956.

Campsall, P. A., Au, N. H., Prendiville, J. S. *et al.* (2004). Detection and genotyping of varicella-zoster virus by *TaqMan* allelic discrimination real-time PCR. *Journal of Clinical Microbiology*, **42**, 1409–1413.

Cann, A. J., Fandrich, S. E. and Heaphy, S. (2005). Analysis of the virus population present in equine faeces indicates the presence of hundreds of uncharacterized virus genomes. *Virus Genes*, **30**, 151–156.

Cavalieri, D., Townsend, J. and Hartl, D. (2000). Manifold anomalies in gene expression in a vineyard isolate of *Saccharomyces cerevisiae* revealed by DNA microarray analysis. *Proceedings of the National Academy of Sciences USA*, **97**, 12 369–12 374.

Cooper, A. R. and Poinar, H. (2000). Ancient DNA: do it right or not at all. *Science*, **289**, 1139.

Daniel, R. (2004). The soil metagenomics: a rich resource for the discovery of novel natural products. *Current Opinion in Biotechnology*, **15**, 199–204.

Dean, F. B., Hosono, S., Fang, L. *et al.* (2002). Comprehensive human genome amplification using multiple displacement amplification. *Proceedings of the National Academy of Sciences USA*, **99**, 5261–5266.

Dearborn, D. C., Anders, A. D., Schreiber, E. A., Adams, R. M. M. and Mueller, U. G. (2003). Inter-island movements and population differentiation in a pelagic seabird. *Molecular Ecology*, **12**, 2835–2843.

Debruyne, R. (2005). A case study of apparent conflict between molecular phylogenies: the interrelationships of African elephants. *Cladistics*, **21**, 31–50.

DeSalle, R. and Amato G. (2004). The expansion of conservation genetics. *Nature Reviews Genetics*, **5**, 702–712.

Dietmaier, W., Hartmann, A., Wallinger, S. *et al.* (1996). Multiple mutation analyses in single tumor cells with improved whole genome amplification. *American Journal of Pathology*, **154**, 83–95.

Divine, A. M. and Allen, M. (2005). A DNA microarray system for forensic SNP analysis. *Forensic Science International*, **154**, 111–121.

Dorrell, N., Hinchliffe, S. J. and Wren, B. W. (2005). Comparative phylogenomics of pathogenic bacteria by microarray analysis. *Current Opinion in Microbiology*, **8**, 620–626.

Ellegren, H. (2005). The avian genome uncovered. *Trends in Ecology and Evolution*, **20**, 180–186.

Esteban, J.A., Salas, M. and Blanco, L. (1993). Fidelity of phi 29 DNA polymerase: comparison between protein-primed initiation and DNA polymerization. *Journal of Biological Chemistry*, **268**, 2719–2726.

Ezaz, M. T., Harvey, S. C., Boonphakdee, C. *et al.* (2004). Isolation and physical mapping of sex-linked AFLP markers in Nile tilapia (*Oreochromis niloticus* L.). *Marine Biotechnology*, **6**, 435–445.

Fearnley, C., On, S. L. W., Kokotovic, B. *et al.* (2005). Application of fluorescent amplified fragment length polymorphism for comparison of human and animal isolates of *Yersinia enterocolitica*. *Applied and Environmental Microbiology*, **71**, 4960–4965.

Frankham, R., Ballou, J. D. and Briscoe, D. A. (2002). *Introduction to Conservation Genetics*. Cambridge: Cambridge University Press.

Frankham, R., Ballou, J. D. and Briscoe, D. A. (2004). *A Primer of Conservation Genetics*. Cambridge: Cambridge University Press.

Gagneux, P., Woodruff, D. S. and Boesch, C. (2001). Retraction: furtive mating in female chimpanzees. *Nature*, **387**, 358–359.

Gibson, G. and Muse, S. V. (2004). *A Primer of Genome Science*. Sunderland: Sinauer Associates.

Gibson, N. J. (2006). The use of real-time PCR in DNA sequence variation analysis. *Clinica Chimica Acta*, **363**, 32–47.

Green, R. E., Krause, J., Ptak, S. E. *et al.* (2006). Analysis of one million base pairs of Neanderthal DNA. *Nature*, **444**, 330–336.

Gunn, M. R., Hartnup, K., Boutin, S., Slate, J. and Coltman, D. W. (2007). A test of the efficacy of whole-genome amplification on DNA obtained from low-yield samples. *Molecular Ecology Notes*, **7**, 393–399.

Gupta, M., Song, P., Yates, C. R. and Meibohm, B. (2004). Real-time PCR-based genotyping assay for CXCR2 polymorphism. *Clinica Chimica Acta*, **341**, 93–100.

Gut, I. G. (2004). DNA analysis by MALDI-TOF mass spectrometry. *Human Mutation*, **23**, 437–441.

Handelsman, J., Rondon, M. R., Brady, S. F., Clardy, J. and Goodman, R. M. (1998). Molecular biological access to the chemistry of unknown soil microbes: a new frontier for natural products. *Chemistry and Biology*, **5**, R245–R249.

Hanson, E. K. and Ballantyne, J. (2005). Whole genome amplification strategy for forensic analysis using single cell or few cell equivalents of genomic DNA. *Analytical Biochemistry*, **346**, 246–257.

Hedrick, P. W. (2001). Conservation genetics: where are we now? *Trends in Ecology and Evolution*, **16**, 629–636.

Hodgson, D. R., Foy, C. A., Partridge, M., Pateromichelakis, S. and Gibson, N. J. (2002). Development of a facile fluorescent assay for the detection of 80 mutations within the p53 gene. *Molecular Medicine*, **8**, 227–237.

Holbrook, J. F., Stabley, D. and Sol-Church, K. (2005). Exploring whole genome amplification as a DNA recovery tool for molecular genetic studies. *Journal of Biomolecular Techniques*, **16**, 125–133.

Holland, P. M., Abramson, R. D., Watson, R. and Gelfand, D. H. (1991). Detection of specific polymerase chain reaction product by utilizing the 5′–3′ exonuclease activity of *Thermus acquaticus* DNA polymerase. *Proceedings of the National Academy of Sciences USA*, **88**, 7276–7280.

Ideker, T., Thorsson, V., Siegel, A. F. and Hood, L. E. (2000). Testing for differentially expressed genes by maximum-likelihood analysis of microarray data. *Journal of Computational Biology*, **7**, 805–817.

Irwin, D. E., Bensch, S., Irwin, J. H. and Price T. D. (2005). Speciation by distance in a ring species. *Science*, **307**, 414–416.

Jelaso, A. M., Lehigh-Shirey, E., Means, J. and Ide, C. F. (2003). Gene expression patterns predict exposure to PCBs in developing *Xenopus laevis* tadpoles. *Environmental and Molecular Mutagenesis*, **42**, 1–10.

Kaderali, L., Desphande, A., Nolan, J. P. and White, P. S. (2003). Primer-design for multiplexed genotyping. *Nucleic Acids Research*, **31**, 1796–1802.

King, T. L., Switzer, J. F., Morrison, C. L. *et al.* (2006). Comprehensive genetic analyses reveal evolutionary distinction of a mouse (*Zapus hudsonius preblei*) proposed for delisting from the US Endangered Species Act. *Molecular Ecology*, **15**, 4331–4359.

Klein, D. (2002). Quantification using real-time PCR technology: applications and limitations. *Trends in Molecular Medicine*, **8**, 257–260.

Lasken, R. S. and Egholm, M. (2003). Whole genome amplification: abundant supplies of DNA from precious samples or clinical specimens. *Trends in Biotechnology*, **21**, 531–535.

Leininger, S., Urich, T., Schloter, M. *et al.* (2006). Archea predominate among ammonia-oxidizing prokaryotes in soils. *Nature*, **442**, 806–809.

Liljedahl, U., Fredriksson, M., Dahlgren, A. and Syvänen, A. C. (2004). Detecting imbalanced expression of SNP alleles by minisequencing on microarrays. *BMC Biotechnology*, **22**, 4–24.

Lindblad-Toh, K., Wade, C. M., Mikkelsen. T. S. *et al.* (2005). Genome sequence, comparative analysis and haplotype structure of the domestic dog. *Nature*, **438**, 803–819.

Little, D. P., Braun, A., Darnhofer-Demar, B. and Köster, H. (1997). Identification of apolipoprotein E polymorphisms using temperature cycled primer oligo base extension and mass spectrometry. *European Journal of Clinical Chemistry and Clinical Biochemistry*, **35**, 545–548.

Loeschcke, V., Tomiuk, J. and Jain, S. K. (1994). *Conservation Genetics*, Experientia Special Volume No. EXS 68. Basel: Birkhäuser-Verlag.

Louis, M., Dekairelle, A. F. and Gala, J. L. (2004). Rapid combined genotyping of factor V, prothrombin and methylenetetrahydrofolate reductase single nucleotide polymorphism using minor groove binding DNA oligonucleotides (MGB probes) and real-time polymerase chain reaction. *Clinical Chemistry and Laboratory Medicine*, **42**, 1364–1369.

Lovmar, L. and Syvänen, A. C. (2005). Genotyping single-nucleotide polymorphisms by minisequencing using tag arrays. *Methods in Molecular Medicine*, **114**, 79–92.

Lovmar, L., Fredriksson, M., Liljedah, U., Sigurdsson, S. and Syvänen, A. C. (2003). Quantitative evaluation by minisequencing and microarrays reveals accurate multiplexed SNP genotyping of whole genome amplified DNA. *Nucleic Acids Research*, **31**, e129.

Lucchini, V. (2003). AFLP: a useful approach for biodiversity, conservation and management. *Comptes Rendus Biologies*, **326**, S43–S48.

Luikart, G., England, P. R., Tallmon, D., Jordan, S. and Taberlet, P. (2003). The power and promise of population genomics: from genotyping to genome typing. *Nature Reviews Genetics*, **4**, 981–994.

Makridakis, N. M. and Reichardt, J. K. (2001). Multiplex automated primer extension analysis: simultaneous genotyping of several polymorphisms. *Biotechniques*, **31**, 1374–1380.

Margulies, M., Egholm, M., Altman, W. E. *et al.* (2005). Genome sequencing in microfabricated high-density picolitre reactors. *Nature*, **437**, 376–380.

Mariette, S., Le Corre, V., Austerlitz, F. and Kremer, A. (2002). Sampling within the genome for measuring within-population diversity: trade-offs between markers. *Molecular Ecology*, **11**, 1145–1156.

Morin, P. A., Chambers, K. E., Boesch, C. and Vigilant, L. (2001). Quantitative polymerase chain reaction analysis of DNA from noninvasive samples for accurate microsatellite genotyping of wild chimpanzees (*Pan troglodytes verus*). *Molecular Ecology*, **10**, 1835–1842.

Morin, P. A., Nestler, A., Rubio-Cisneros, N. T., Robertson, K. M. and Mesnick, S. L. (2005). Interfamilial characterization of a region of the ZFX and ZFY genes facilitates sex determination in cetaceans and other mammals. *Molecular Ecology*, **14**, 3275–3286.

Newton, C. R., Graham, A., Heptinstall, L. E. *et al.* (1989). Analysis of any point mutation in DNA: the amplification refractory mutation system (ARMS). *Nucleic Acids Research*, **17**, 2503–2516.

Nicod, J. C. and Largiadèr, C. R. (2003). SNPs by AFLP (SBA): a rapid SNP isolation strategy for non-model organisms. *Nucleic Acids Research*, **31**, e19.

Noonan, J. P., Hofreiter, M., Smith, D. *et al.* (2005). Genomic sequencing of Pleistocene cave bears. *Science*, **309**, 597–599.

Nybom, H. (2004). Comparison of different nuclear markers for estimating intraspecific genetic diversity in plants. *Molecular Ecology*, **13**, 1143–1155.

Oleksiak, M., Churchill, G. A. and Crawford, D. L. (2002). Variation in gene expression within and among natural populations. *Nature Genetics*, **32**, 261–266.

Papa, R., Troggio, M., Ajmone-Marsan, P. and Nonnis Marzano, F. (2005). An improved protocol for the production of AFLP markers in complex genomes by means of capillary electrophoresis. *Journal of Animal Breeding and Genetics*, **122**, 62–68.

Pfunder, M., Holzgang, O. and Frey, J. E. (2004). Development of microarray-based diagnostics of voles and shrews for use in biodiversity monitoring studies, and evaluation of mitochondrial cytochrome oxidase I vs. cytochrome b as genetic markers. *Molecular Ecology*, **13**, 1277–1286.

Poinar, H., Kuch, M., McDonald, G., Martin, P. and Pääbo, S. (2003). Nuclear gene sequences from a late Pleistocene sloth coprolite. *Current Biology*, **13**, 1150–1152.

Poinar, H. N., Schwarz, C., Qi, J. *et al.* (2006). Metagenomics to paleogenomics: large-scale sequencing of mammoth DNA. *Science*, **5759**, 392–394.

Pruvost, M. and Geigl, E. M. (2004). Real-time quantitative PCR to assess the authenticity of ancient DNA amplification. *Journal of Archaeological Science*, **31**, 1191–1197.

Pruvost, M., Grange, T. and Geigl, E. M. (2005). Minimizing DNA contamination by using UNG-coupled quantitative real-time PCR on degraded DNA samples: application to ancient DNA studies. *Biotechniques*, **38**, 569–575.

Putta, S., Smith, J. J., Walker, J. A. *et al.* (2004). From biomedicine to natural history research: EST resources for ambystomatid salamanders. *BMC Genomics*, **5**, 54.

Roca, A. L., Georgiadis, N. and O'Brien, S. J. (2005). Cytonuclear genomic dissociation in African elephant species. *Nature Genetics*, **37**, 96–100.

Rohlf, F. J. (1992). *NTSYS-PC: numerical taxonomy and multivariate analysis system*, v. 1.7. New York: Exeter Software.

Ronaghi, M. (2001). Pyrosequencing shed lights on DNA sequencing. *Genome Research*, **11**, 3–11.

Ronaghi, M., Karamohamed, S., Petterson, B., Uhlen, M. and Nyren, P. (1996). Real-time DNA sequencing using detection of pyrophosphate release. *Analytical Biochemistry*, **242**, 84–89.

Rönn, A. C., Andrés, O., Bruford, M. W. *et al.* (2006). Multiple displacement amplification for generating an unlimited source of DNA for genotyping in non-human primates. *International Journal of Primatology*, **27**, 1145–1169.

Ruvinsky, J. (2002). Metersticks and microarrays. *Trends in Ecology and Evolution*, **17**, 496–497.

Salas, A., Quintans, B. and Alvarez-Iglesias, V. (2005). SNaPshot typing of mitochondrial DNA coding region variants. *Methods in Molecular Biology*, **297**, 197–208.

Sanger, F. J., Donelson, J., Coulson, A., Kossel, H. and Fischer, D. (1974). Determination of a nucleotide sequence in bacteriophage f₁ DNA by primed synthesis with DNA polymerase. *Journal of Molecular Biology*, **90**, 315–333.

Schena, M., Shalon, D., Davis, R. W. *et al.* (1995). Quantitative monitoring of gene expression patterns with a cDNA microarray. *Science*, **270**, 467–470.

Schroell-Metzeger, B., Dicato, M., Bosseler, M. and Berchem, G. (2003). Comparison of standard PCR and the LightCycler technique to determine the trombophilic mutations: an efficiency and cost study. *Clinical Chemistry and Laboratory Medicine*, **41**, 482–485.

Scotti-Saintagne, C., Mariette, S., Porth, I. *et al.* (2004). Genome scanning for interspecific differentiation between two closely related oak species [*Quercus robur* L. and *Q petraea* (Matt.) Liebl.]. *Genetics*, **168**, 1615–1626.

Shumaker, J. M., Metspalu. A. and Caskey, C. T. (1996). Mutation detection by solid-phase primer extension. *Human Mutation*, **7**, 346–354.

Smith, T. B. and Wayne, R. K. (1996). *Molecular Genetic Approaches in Conservation*. Oxford: Oxford University Press.

Sorensen, K. J., Turteltaub, K., Vrannkovich, G., Williams, J. and Christian, A. T. (2004). Whole-genome amplification from residual cells left by incidental contact. *Analytical Biochemistry*, **324**, 312–314.

Spruill, S. E., Lu. J., Hardy, S. and Weir, B. (2002). Assessing sources of variability in microarray gene expression data. *Biotechniques*, **33**, 916.

Steele, H. L. and Streit, W. R. (2005). Metagenomics: advances in ecology and biotechnology. *FEMS Microbiology Letters*, **247**, 105–111.

Sullivan, J. P., Lavoue, S., Arnegard, M. E. and Hopkins C. D. (2004). AFLPs resolve phylogeny and reveal mitochondrial introgression within a species flock of African electric fish (Mormyroidea: Teleostei). *Evolution*, **58**, 825–841.

Sun, G., Kaushal, R., Pal, P. *et al.* (2005). Whole-genome amplification: relative efficiencies of the current methods. *Legal Medicine*, **7**, 279–286.

Swanithan, K., Varal, K. and Hudson, M. E. (2007). Global repeat discovery and estimation of genomic copy number in large, complex genome using a high-throughput 454 sequence survey. *BMC Genomics*, **8**, 132.

Swofford, D.L. and Selander, R.B. (1981). BIOSYS-1: a FORTRAN program for the comprehensive analysis of electrophoretic data in population genetics and systematics. *Journal of Heredity*, **72**, 281–283.

Syvänen, A. C. (1999). From gels to chips: 'minisequencing' primer extension for analysis of point mutations and single nucleotide polymorphisms. *Human Mutation*, **13**, 1–10.

Syvänen, A. C., Aalto-Setälä, K., Harju, L., Kontula, K. and Söderlund, H. (1990). A primer-guided nucleotide incorporation assay in the genotyping of apolipoprotein E. *Genomics*, **8**, 684–692.

Telenius, S., Carter, N. P., Bebb, C. E., Nordenskjold, M., Ponder, B. A., Turncliff, A. (1992). Degenerate oligonucleotide-primed PCR: general amplification of target DNA by a single degenerate primer. *Genomics*, **13**, 718–725.

Townsend, J. P., Cavalieri, D. and Hartl, D. L. (2003). Population genetic variation in genome-wide gene expression. *Molecular Biology and Evolution*, **20**, 955–963.

Tringe, S. G. and Rubin, E. M. (2005). Metagenomics: DNA sequencing of environmental samples. *Nature Reviews Genetics*, **6**, 805–813.

Turnbaugh, P. J., Ley, R. E., Mahowald, M. A. *et al.* (2006). An obesity-associated gut microbiome with increased capacity for energy harvest. *Nature*, **444**, 1027–1031.

Turner, T. L., Hahn, M. W. and Nuzhdin, S. V. (2005). Genomic islands of speciation in *Anopheles gambiae*. *PLoS Biology*, **3**, e285.

van den Braak, N., Simons, G., Gorkink, R. *et al.* (2004). A new high-throughput AFLP approach for identification of new genetic polymorphism in the genome of the clonal microorganism *Mycobacterium tuberculosis*. *Journal of Microbiological Methods*, **56**, 49–62.

Vasemägi, A. and Primmer, C. R. (2005). Challenges for identifying functionally important genetic variation: the promise of combining complementary research strategies. *Molecular Ecology*, **14**, 3623–3642.

Vekemans, X., Beauwens, T., Lemaire, M. and Roldán-Ruiz, I. (2002). Data from amplified length polymorphism (AFLP) markers show indication of size homoplasy and of a relationship between degree of homoplasy and fragment size. *Molecular Ecology*, **11**, 139–151.

Vestal, M. L., Juhasz, P. and Martin, S. A. (1995). Delayed extraction assisted laser desorption time-of-flight mass spectrometry. *Rapid Communications in Mass Spectrometry*, **9**, 1044–1050.

von Wurmb-Schwark, N., Ringleb, A., Gebuhr, M. and Simeoni, E. (2005). Genetic analysis of modern and historical burned human remains. *Anthropologischer Anzeiger*, **63**, 1–12.

Vos, P., Hogers, R., Bleeker, M. *et al.* (1995). AFLP: a new technique for DNA fingerprinting. *Nucleic Acids Research*, **23**, 4407–4414.

Wandeler, P., Smith, S., Morin, P. A., Pettifor, R. A. and Funk, S. M. (2003). Patterns of nuclear DNA degeneration over time: a case study in historic teeth samples. *Molecular Ecology*, **12**, 1087–1093.

Wang, Z., Baker, A. J., Hill, G. E. and Edwards, S. V. (2004). Reconciling actual and inferred population histories in the house finch (*Carpodacus mexicanus*) by AFLP analysis. *Evolution*, **57**, 2852–2864.

Watanabe, S., Minegishi. Y., Yoshinaga, T., Aoyama, J. and Tsukamoto, K. (2004). A quick method for species identification of Japanese eel (*Anguilla japonica*) using real-time PCR: an onboard application for use during sampling surveys. *Marine Biotechnology*, **6**, 566–574.

Whitehead, A. and Crawford, D. L. (2005). Variation in tissue-specific gene expression among natural populations. *Genome Biology*, **6**, R13.

Wilding, C. S., Butlin, R. K. and Gravame, J. (2001). Differential gene exchange between parapatric morphs of *Littorina saxatilis* detected using AFLP markers. *Journal of Evolutionary Biology*, **14**, 611–619.

Wu, K. J., Steding, A. and Becker, C. H. (1993). Matrix-assisted laser desorption time-of-flight mass spectrometry of oligonucleotides using 3-hydroxypicolinic acid as ultraviolet-sensitive matrix. *Rapid Communications in Mass Spectrometry*, **7**, 142–146.

Xiao, W. and Oefner, P. J. (2001). Denaturing high-performance liquid chromatography: a review. *Human Mutation*, **17**, 439–474.

Yeh, S.H., Tsai, C.Y., Kao, J.H. *et al.* (2004). Quantification and genotyping of hepatitis B virus in a single reaction by real-time PCR and melting curve analysis. *Journal of Hepatology*, **41**, 659–666.

Zhang, D. X. (2004). Lepidopteran microsatellite DNA: redundant but promising. *Trends in Ecology and Evolution*, **19**, 507–509.

Zhang, L., Cui, X., Schmitt, K. *et al.* (1992). Whole genome amplification from a single cell: implications for genetic analysis. *Proceedings of the National Academy of Sciences USA*, **89**, 5847–5851.

Theoretical outlook

MARK BEAUMONT

This review addresses a number of current issues in conservation genetics, and highlights possible future trends. The current and future role of statistical model-based inference in population genetics is discussed, particularly in relation to methods that focus on the analysis of haplotype networks. There are a number of current computational issues in model-based methods, for example convergence of Markov chain Monte Carlo (MCMC) with multiple loci, and suggestions for overcoming them are explored. In particular, potential future uses of Approximate Bayesian Computation in a conservation context are discussed. Another issue that is examined is the sensitivity of population genetic modelling to the specific assumptions used. Of particular concern is the potential for many different demographic scenarios to give rise to similar genetic data. A problematic area, but with great relevance to a more detailed dissection of the demographic antecedents of threatened populations, is the development of statistical methods to handle recombination and linkage disequilibrium in linked markers. With the improvement in our ability to inexpensively assay individuals and populations for multiple genetic markers, new directions have become possible in conservation genetics. One area is the use of multilocus genotypes to infer aspects of population structure. Another area is the detection of regions of the genome under adaptive selection. Potential future work with relevance to conservation is discussed, such as pedigree reconstruction purely from genetic data, and the definition of conservation units based on adaptive genetic differences.

INTRODUCTION

If we regard conservation genetics as a subdiscipline of population genetics, then, at least in its theoretical development, the latter is generally regarded as mature (Lewontin 1974). Broadly, the foundations for understanding

Population Genetics for Animal Conservation, eds. G. Bertorelle, M. W. Bruford, H. C. Hauffe, A. Rizzoli and C. Vernesi. Published by Cambridge University Press. © Cambridge University Press 2009.

how gene frequencies respond to drift, selection, immigration and recombination have already been laid down. However, there continue to be novel and, from a conservation perspective, useful, theoretical developments based on these foundations, particularly with regard to debates over the role of genetic factors *per se* in contributing to extinction risk (Whitlock *et al.* 2003). The technical explosion in molecular genetic analysis is more immature – merely 20 years old – and the current ongoing rush in computer intensive statistical methods that aim to build a bridge between data and theory is even younger. This chapter intends to concentrate on current methods of parametric statistical modelling in population genetics and how these impinge on conservation.

GENEALOGICAL MODELLING AND DEMOGRAPHIC HISTORY

One of the major current applications of population genetics in conservation is simply as a tool to recover demographic history (Pearse and Crandall 2004). Often the analysis is performed with a view to better identifying units of conservation (Eizirik *et al.* 2001), or for identifying the timescale and nature of population decline (Goossens *et al.* 2006). Demographic history has been defined by Hey and Machado (2003) as:

> The reproductive history of a population or group of populations. This can include population sizes, sex ratios, migration rates, population-splitting events, variation in reproductive rates and times among organisms, as well as variation over time in all of these quantities.

It has been addressed genetically in three main ways: through the modelling of allele or haplotype frequencies (e.g. Hey and Nielsen 2004), or from analysis of multilocus genotypes (e.g. Pritchard *et al.* 2000), or from the kin structure of populations (e.g. Wang 2004). Gene frequency distributions are modelled either genealogically, using the coalescent, or by a variety of approaches that model diffusions (e.g. Wang and Whitlock 2003; Beaumont and Balding 2004; O'Hara 2005; Williamson *et al.* 2005). As will be noted in this chapter, there is a trend towards the fusion of these three currently separate approaches, and future developments may lead to a single unified approach (albeit joined at the seams via approximations) that can potentially extract all levels of information in the genetic data. Genealogical or frequency-based analyses aim to recover the demographic history – for example, to identify ancient changes in population sizes or to date vicariance events – and will be addressed in this section. Potential future developments in the other techniques will be discussed later in separate sections.

There has been a long tradition of the use of population genetic data to infer demographic history (Cavalli-Sforza and Edwards 1967; Thompson 1973; Slatkin 1981). Methods of inference based on moments and likelihood have both been used. The former approach requires equations that give the expected value of statistics as a function of demographic parameters, whereas the latter involves a formula that gives the probability distribution of the observations of interest. A recent trend in the last 15 years has seen a move towards Bayesian analysis (Wilson and Balding 1998; Beaumont and Rannala 2004), which gives a probability distribution for particular models, their parameter values, and for missing data, conditional on the observed data.

Although these advances have broadened the horizons of population geneticists, and allowed molecular genetic data to be used more efficiently, there are some practical difficulties that seem to have held back their wider usage. A general consideration is that a fair amount of programming and analytical effort is necessary to develop each statistical model, which restricts the number of scenarios available, tempting misapplication by researchers. Other problems include the amount of computational time necessary to carry out the analyses, and the complex nature of the output, which can often be difficult to interpret.

Separately, over this period, with the advent of mitochondrial sequence data, demographic inference based on haplotype networks has become popular (Templeton *et al.* 1995), particularly in human population genetics (Bandelt *et al.* 1999). The idea is that by visualizing the reconstructed trees and applying a number of sophisticated graph-based analyses it is possible to 'read' the demographic history from the trees. In particular, the nested clade phylogeographic analysis (NCPA) method of Templeton and colleagues (Templeton *et al.* 1995; Templeton 2004; see Buhay *et al.* this volume) has been widely used in conservation genetics (e.g. Gottelli *et al.* 2004; Ciofi *et al.* 2006). An attraction of these methods is that they are essentially non-parametric. However a concern is that there is often no strong correlation of individual gene-trees with demography (Machado and Hey 2003). Conceivably, if demographic history always involved a bottleneck at each vicariance event or range expansion, with limited effects of migration, there might be a closer match between gene-trees and demography (Chikhi and Beaumont 2005). The circumstances favouring this possibility could be examined, for example with computer simulations, but it does seem a rather restrictive requirement. Extensions to NCPA that include comparisons of inferences from different genes could allow for some robustness in the face of genealogical diversity (Templeton 2005). Under the cross-validation criterion of Templeton (2002), for example, inferences are regarded as concordant if more than one locus infers the same historical process involving the same locations. There has so far been no examination of the performance of this criterion, either empirically (as performed for single loci in Templeton 2004), or via analysis of simulated data sets. Until recently there has been little simulation-based testing of NCPA, even for single loci (Knowles and Maddison 2002). However, with the advent of an automated version (Panchal 2007), it should become more straightforward to assess the procedure.

It is not the purpose of this article to go into any detail about the validity or otherwise of these network-based approaches (see Panchal and Beaumont 2007, for a more extensive discussion), but to point out that in their apparent deliverables they set a bench-mark against which model-based methods need to be judged. For example a statement that there is evidence of 'restricted gene flow with long distance dispersal' (Templeton *et al.* 1995) could be restated in terms of the posterior probability of such a model given the data. Current model-based methods do not yet approach these aims, but, as techniques improve, it is conceivable that they may do so in the future (see Fagundes *et al.* 2007, for an example). The IM program of Hey and Nielsen (2004), based on the earlier work of Nielsen and Wakeley (2001), is one of the more sophisticated of such models, using MCMC to make inferences about the parameters. This program has appreciable utility in conservation (e.g. Cassens *et al.* 2005), allowing six parameters to be inferred in a two-population setting. To be able address the 'big' questions that NCPA claims to answer, it is necessary to go beyond this and rather than make inferences about particular parameters, we need to make infer-ence about particular models, marginal to (i.e. irrespective of) the parame-ters within the models.

How feasible is it to recover demographic history from genetic data?

Before considering the future development of these approaches, it is worth pausing to consider how much information we can actually glean from genetic data. There is now a tendency with increasing sophistication of computational methods to consider more and more complex models. A question naturally arises whether there is sufficient information in genetic data to enable us to make strong inferences about the past demographic history, and to what extent there is a limit on the amount of information that we can use. Wiuf (2003) examined the situation where one has infinite (non-recombining) sequence data and it is therefore possible to recover the shape of the genealogy exactly. In this case it can be shown that there are no consistent estimators for growth rate, and Wiuf (2003) concluded that this was probably the case for most other parameters of interest as well. This means that as one increases the sample size to infinity the estimates do not converge on the true value. In a recent study Degnan and Rosenberg (2006) show that in models of population divergence with a sufficiently large number of taxa it is possible for the topology of the most probable gene tree to be different from the true population tree. They show that in this case care must be taken when combining the results over multiple loci, other-wise some algorithms are guaranteed in the limit of an infinite number of loci to converge on the wrong topology. More anecdotally, complex coalescent-based models typically have quite wide posterior distributions for many parameters, even when large numbers of loci are considered (Rannala and Yang 2003; Jennings and Edwards 2005).

Results based on the genealogy of structured populations (Nordborg 1997; Wakeley 1998) suggest that under some circumstances the details of the demographic history will become more and more clouded as we con-sider more and more complex ('realistic') scenarios. The idea is that in a structured population with many demes the genealogy can be naturally broken into two phases, termed by Wakeley the 'scattering phase', corre-sponding to the genealogy within a deme, and the 'collecting phase', corresponding to the genealogy of the metapopulation as a whole. It turns out that for quite a large variety of models the genealogy can be well approximated in this way. Examples include: the island model (Wakeley 1999); metapopulation models with migration, colonization and extinction (Wakeley and Aliacar 2001); lattice models with widely spaced samples (Wilkins 2005). The shape of the genealogical structure in all these different cases is very similar: the collecting phase can be described by a standard coalescent genealogy scaled according to an effective population size that depends on the detail of the model, and, for the scattering phase, a distri-bution of lineages that is fairly similar among the different models (slight differences depending on the extent of migration, and timescales of colo-nization). Taken at face value, these ideas would suggest that, at least in populations with a large number of demes and moderate to high gene flow, there is a fundamental limitation on our ability to recover demographic history from genetic data. In future analyses, these results could be used to formulate generic 'null' models with which to compare more detailed scenarios (see also Wakeley 2004). If the posterior probability of more detailed models is not substantially greater than that for such a null model, it is possible to conclude that there is little further information in the data that needs to be explained.

If a method such as BOTTLENECK (Cornuet and Luikart 1996), M_P_VAL (Garza and Williamson 2001) or MSVAR (Beaumont 1999) is applied to a particular population a statistically significant outcome may lead to the conclusion that there has been a bottleneck in the past, and this may result in some recommendation about the population. However, a particular consequence of the considerations of this section for conservation is that it may be difficult to distinguish the presence of bottlenecks from the effects of population structure. In both cases there is a tendency for the genealogy of a sample from a single deme to consist of a more extreme mixture of recent and ancient coalescent events than would be expected for a single stable population at equilibrium (Goossens *et al.* 2006). It is only because the researcher has chosen a method that is specifically designed to detect bottlenecks that she or he interprets the results in terms of a bottleneck. Ideally, and this is discussed in the section 'Improvements to current methods', one needs to compare a number of scenarios to see if they are distinguishable. Similarly, as noted by Wakeley (1999), it is very difficult to determine whether a group of populations has decreased in size over time or whether the migration rate between populations has increased.

Recombination

Broadly, the 'many demes' limit discussed above reflects ignorance of the shape of recent genealogical structure: the mutations necessary to highlight this are assumed only to occur in the collecting phase. If mutations are sufficiently common and informative, then the history of particular lineages could be better resolved. Mitochondrial and Y-chromosome (or equivalent) data are in many ways the closest to this ideal because they provide genealogically useful sequence information without (it is hoped) the complication of recombination. There are a number of caveats, however (Ballard and Whitlock 2004). These markers illuminate only the demographic history of a particular sex, which may differ markedly from that of the other. In addition they represent only a single genealogy, and, as discussed above there is generally not a straightforward mapping from individual gene-genealogies to the underlying demographic history. Also, given that the mtDNA genome contains many completely linked genes, there is much more scope for the action of selection. Indeed, there is evidence that mitochondrial genomes of all organisms have been subject to sufficiently frequent selective sweeps that there is no longer a correlation between the genetic diversity of mitochondrial DNA and that of autosomes across many taxa (Bazin *et al.* 2006).

These observations would point to the need for more autosomal sequence data surveyed from multiple loci. Ideally, it is necessary to additionally model recombination: there is a choice between taking sufficiently short regions that (we hope) may have no recombination, at the price of little genealogical resolution, or the converse. There have been a number of recent studies that have used multiple loci consisting of relatively short regions, typically between 200 and 800 base pairs (Jennings and Edwards 2005; Dolman and Moritz 2006). Even with such short regions there is often strong evidence of recombination, and a significant amount of data may need to be discarded (Dolman and Moritz 2006) because current likelihood-based methods for inferring demographic history assume no recombination (Rannala and Yang 2003; Hey and Nielsen 2004).

A future priority must include improved genealogical modelling of recombination. The need arises not only because the rate of recombination is a nuisance parameter that must be included so that nuclear data can be analysed, but also because the patterns of linkage disequilibrium that arise are potentially informative about demographic history (McVean *et al.* 2007). This is a very hard problem. Currently there are few methods that handle it in a fully likelihood-based setting, and these are restricted to relatively short sequences at one locus (Kuhner *et al.* 2000; Fearnhead and Donnelly 2001). Ingenious approximate methods have been developed, primarily to analyse data generated by the HapMap Consortium (reviewed by McVean *et al.* 2007). One type of approximation is based on composite likelihood (also called 'pseudo-likelihood'), in which the likelihood for a pair of nucleotide sites is computed and then multiplied over all pairs of sites as if they were independent (McVean *et al.* 2004). This is believed to give reasonable point estimates, but with a strong tendency to be over-confident in their precision. An alternative method is based on the so-called product of approximate conditionals (PAC) (Li and Stephens 2003), which will be discussed further below. An outcome of these analyses, supporting more direct evidence from analysis of sperm and deep pedigrees, is that recombination rate varies enormously around the genome, and thus future methods need to assume non-constant recombination rates over the sequences that are analysed. An added complication is that the variation in recombination rates appears, from direct analysis of sperm, to vary between individuals (Tiemann-Boege *et al.* 2006), and hence probably through time and among populations.

FUTURE DEVELOPMENTS

Approximations

Although the current MCMC methods for likelihood-based inference are steadily improving in their sophistication and in the range of scenarios that can be considered (e.g. Hey 2005), it seems doubtful that such techniques can be improved sufficiently rapidly to effectively deal with the volume of data that will soon be obtained. Thus future improvements are likely to involve a certain degree of approximation. The two classes of approximation discussed above are not only used for the analysis of recombinant data. For example, methods for inferring effective population size and admixture (Wang 2003; Wang and Whitlock 2003) use a composite likelihood approximation for multi-allelic data, and are remarkably accurate and much faster in comparison with full-likelihood methods. For microsatellite data a PAC version of the likelihood for scaled mutation rate is very similar to that obtained by alternative techniques (Cornuet and Beaumont 2007). PACs are obtained by deriving an approximate formula for the probability of picking a new allele or haplotype of a given type, conditional on the distribution of types that have already been picked. For the K-allele mutation model, where the type that an allele mutates to is independent of its current state, this conditional probability distribution is known exactly and can be used to give the likelihood. The likelihood is proportional to the product of these conditional probabilities for any sequence in which alleles are picked. Stephens and Donnelly (2000), Fearnhead and Donnelly (2001) and Li and Stephens (2003) suggested ways of approximating this conditional probability distribution for other mutational models and with recombination.

Another class of approximations involves replacing the data with summary statistics calculated from them. These methods can provide approximate likelihood profiles (Weiss and von Haeseler 1998), or can be used for Bayesian inference (Pritchard et al. 1999). The latter has come to be known as approximate Bayesian computation (ABC) (Beaumont et al. 2002; Marjoram et al. 2003; Sisson et al. 2007), and has recently been used to help study some problems in conservation and population management (Estoup et al. 2004; Hamilton et al. 2005; Miller et al. 2005; Chan et al. 2006). Approximations based on summary statistics have also been used to make inferences about recombination rate (Wall 2000; Padhukasahasram et al. 2006), and compare well with composite likelihood methods. Partially linked microsatellites have also been modelled in an analysis of admixture (Excoffier et al. 2005).

The ABC technique is potentially useful for modelling the complex scenarios that frequently arise in applied problems (as in Miller *et al.* 2005, for example). If the summary statistics are jointly sufficient (in the sense that if the true posterior distribution was known, for a particular prior, we could rewrite it equivalently solely in terms of these summary statistics) then the method could potentially be highly accurate. One of the future challenges in the development of the method is to identify useful summary statistics for particular models.

A major advantage of this approximation, particularly for population management, is that it requires little or no analytical treatment to enable it to be implemented. This both widens the scope for applications, and also lowers the potential barrier between primarily empirical researchers and more theoretical folk, by allowing the former to implement their own statistical analyses. All that is needed is some method for simulating 'realistic' data sets that can then be compared, using summary statistics, with empirical data. Scripts written in R are available to then infer parameters in this model (http://www.rubic.rdg.ac.uk/~mab/stuff/ABC_distrib. zip).

Improvements to current methods

The primary computational workhorse for genealogical modelling has been MCMC, and this focus seems likely to continue, particularly with the infusion of ideas and techniques from phylogenetic modelling (Huelsenbeck and Ronquist 2001). The difference between the two fields is merely that in population genetics the genealogy is a nuisance parameter to be integrated out. The key to improved convergence of genealogical MCMC models, particularly with multiple loci, is to be able to update parameters and genealogies jointly (Storz and Beaumont 2002; Rannala and Yang 2003). It is typically quite difficult to devise such updates, and this aspect may currently limit wider application of genealogical MCMC to more complex demographic histories. Ad hoc techniques to improve convergence have been devised, such as the use of 'heated' chains in which the log-likelihood is multiplied by a constant to flatten the likelihood surface (e.g. Hey and Nielsen 2004). There has been much interest in adaptive MCMC methods, but it is generally quite difficult to prove the convergence properties of such schemes, and they do not appear to have been widely used for genealogical problems.

An alternative computational method is importance sampling (Griffiths and Tavaré 1994), in which independent genealogies are simulated from a proposal distribution, conditional on particular parameter values, thereby potentially avoiding the convergence problems that arise with MCMC. These genealogies are given a weight that is proportional to their conditional probability of occurring, given the data. A limitation in importance sampling is that the standard proposal schemes have been developed for stationary models in which the parameters are constant through time; in cases where the populations are growing or diverging from each other through vicariance, the application of standard importance sampling schemes may be inefficient. Thus a potentially fruitful avenue of future research is the development of different schemes that depend on the underlying demographic model (De Iorio and Griffiths 2004).

Importance sampling, as typically used (e.g. Fearnhead and Donnelly 2001), provides an estimate of the likelihood, or likelihood surface. However, a family of methods involving a modification of the approach can allow for Bayesian calculations, as in MCMC, to be straightforwardly carried out. Examples are 'sampling importance resampling' (SIR: Kinas 1996), 'particle filtering', or sequential Monte Carlo methods (Doucet et al. 2001). These involve treating each importance sample (e.g. genealogy) as a particle with an associated weight. Bayesian calculations can then be made by sampling from appropriate distributions. For example, a posterior distribution for the scaled mutation rate θ in a standard coalescent model can be approximated by repeatedly first sampling θ from its prior, then sampling a genealogy using the proposal distribution of Stephens and Donnelly (2000). If n such samples are obtained one can then resample (SIR) m genealogies and values of θ in proportion to the associated importance weights to obtain an approximate posterior distribution for θ. Alternatively, density estimation can be carried out with the n sampled values of θ weighted by their importance weights. Various methods for improving on this approach have been devised. In particular the idea of resampling particles can be used in iterative (and possibly adaptive) schemes leading to progressively more accurate recovery of posterior distributions. Such iterative schemes also lend themselves to improvement of the ABC (Sisson et al. 2007).

For evolution on shorter timescales, inference based on the Wright–Fisher model (Wang 2003; Wang and Whitlock 2003) or diffusion equations (Williamson et al. 2005) may become more widely used. An advantage of these approaches, unlike similar approximations that use coalescent theory (Beaumont 2003; Anderson 2005), is that it is relatively straightforward to incorporate selection. The WinBUGS package (Speigelhalter et al. 1999) provides a particularly useful environment for MCMC-based analysis of the Wright–Fisher model. For example, O'Hara (2005) has used this framework to analyse temporal changes in gene frequency in the scarlet tiger moth (*Callimorpha dominula*), allowing for changes in population size with time and also changes in the selection coefficients with time. Typical WINBUGS programs are very short, compressing into 38 lines the equivalent of several thousand lines of C code.

It should be noted that ideas about how best to formulate and interpret Bayesian models are themselves evolving. Providing that the data are sufficiently informative, Bayesian and frequentist inference using the same likelihood function tend to give similar results (O'Hagen 1994; Gelman *et al.* 1995). However the Bayesian paradigm is sufficiently flexible that it becomes tempting to construct intricate hierarchical models, in which the parameters of the prior distributions, rather than having point values set by the user ('the priors'), themselves have prior distributions (Gelman *et al.* 1995). From this arise a number of choices about how to interpret the results. An important adjunct and alternative to hierarchical models is model choice or model selection in which the marginal likelihood (the probability of data irrespective of parameter values) is compared among models. Again, a number of methods have been proposed for doing this. A problem with Bayesian model choice is that the results are dependent on the priors used for the parameters within each model. From a purist Bayesian perspective, as long as the priors are informative, this does not matter: the Bayesian model choice procedure yields one's 'best guess' of the correct model, given background knowledge and the data. In practice however, given that there is unlikely to be wide agreement on the priors, any application of Bayesian techniques should ideally involve some examination of the sensitivity of the outcome to different priors. Unfortunately, since these methods are fairly computer intensive, this may be difficult to achieve. An as yet relatively undeveloped aspect of Bayesian inference, at least in the context of conservation and management, is decision theory, in which the posterior distribution is weighted by the cost of any particular decision (loss function). It is then possible to estimate parameters and choose among models based on the loss function, and also to identify the decision (Bayes rule) for any particular data set that minimizes the expected loss. A relatively detailed application of this approach to a problem in the management of waterfowl habitat is discussed by Dorazio and Johnson (2003), who use WINBUGS to implement their model. An example of the use of such reasoning in the context of conservation genetics is in the assignment of paternity in the North Atlantic humpback whale (Nielsen *et al.* 2001). It is reasonable to speculate that these approaches will become more common in future.

SPECIES IDENTIFICATION

A good example where technological developments have then stimulated work on statistical modelling is in the area of DNA barcoding (Hebert *et al.* 2003; Hebert *et al.* 2004; Moritz and Cicero 2004; Waugh 2007). The aim, given a database of species designations and corresponding sequence information, is to classify an organism on the basis of a short sequence of its DNA (typically a 648 bp segment of mitochondrial CO1). Polymorphism will often lead to sequences that do not match, and the challenge is to make accurate classifications, taking this into account. If a sequence is substantially diverged from its most closely matching species in the database a decision needs to be made whether it should be assigned to that species or whether a new species has been discovered. For example, it has been suggested that a critical threshold sequence dissimilarity of ten times mean intraspecific dissimilarity should be used as a basis for deciding whether a novel sequence belongs to a new species (Hebert *et al.* 2004). Population genetics theory provides a modelling framework to help make such decisions less arbitrary.

Nielsen and Matz (2006) address the problem that a match may be false. These false positives might be obtained if the true species is not present in that database, or because of shared polymorphisms between species, either because of lineage sorting or from homoplasy. They show how one can compute the false positive rate as a function of the population size scaled by mutation rate, and the time of divergence scaled by population size. They introduce a method for computing whether a given sequence is sufficiently similar that if it were truly from the same population the probability of observing at least as similar a sequence is e.g. 0.95. This is not the same as the probability that the sequence does indeed come from the target population. To perform such a computation they introduce a method for computing the posterior probability that a sequence comes from one of two populations that are compared. This is based on running a modified version of the MCMC-based IM package (Nielsen and Wakeley 2001; Hey and Nielsen 2004) in which the population of origin of the query sequence is 'flipped' between the two populations during the MCMC run. The posterior probability is computed from the proportion of times the sequence is 'resident' in each population. An alternative idea is explored in Matz and Nielsen (2005) in which they perform likelihood ratio tests to test whether the divergence time between the 'population' of the query sequence and that of the reference population is significantly different from 0. Again the likelihoods are computed using the IM program. Tests with both simulated and real data suggest that such methods show promising performance.

The ability of barcoding techniques to discover new species has been recently investigated by Hickerson *et al.* (2006). In this study, by contrast to those discussed above, a false positive constitutes the erroneous identification of a new species. They used two criteria to identify species: the 'ten times' rule, above, and an alternative criterion requiring reciprocal monophyly, which can only be applied when more than one query sequence and more than one reference sequence are available. They suggest that both methods are only likely to have an acceptable error rate if population divergence times are of the order of 1 million generations.

An alternative approach to species discovery has been suggested by Pons *et al.* (2006). They use a model of lineage branching that is a mixture of a coalescent process and a Yule process. They make the assumption that up to a time *T* before the present there is a Yule process in which lineages branch at a constant rate, and that after time *T*, up until the present, there is a coalescent process. The critical time *T* is estimated by maximum likelihood, and species are then defined as those that derive from a single lineage crossing this critical time. Although there is scope for improvement in the model, for example, relaxing the need for only one *T*, this approach points to one interesting way forwards in species discovery.

The DNA barcoding idea touches on the often heated debate concerning species concepts, which is significant for conservation (Mace 2004). Barcoding, depending on the precise criterion used, presumably conforms to some variant of the phylogenetic species concept (Cracraft 1989). The utility of such a concept is open to debate (Hey *et al.* 2003). As these authors have noted, species are rather like droughts: we all know what they are, but it is difficult to encapsulate them in an all-encompassing definition. Biologists tacitly tend to apply what might be called the 'Humpty Dumpty species concept': a species is whatever you want it to be in order to address the problem in hand. There appears to be little prospect of regaining the certainties expressed in Mayr (1963).

MULTILOCUS GENOTYPIC METHODS

One of the most widely used groups of techniques in conservation genetics are those based on the information contained in multilocus genotypes. At the heart of these techniques lies a calculation of the probability of obtaining the genotype of an individual, given what can be called 'background frequencies' – the allele frequencies in some notional random mating population from which the alleles in the individual were drawn. The range of application in conservation biology has been significant; for example, in the analysis of individual movement of bears (Paetkau *et al.* 1995); quantifying hybridization between domestic animals and their wild counterparts (Beaumont *et al.* 2001; Randi *et al.* 2001), including estimation of when the process began (Verardi *et al.* 2006); identifying the origin of illegal ivory (Wasser *et al.* 2004); analysing the effects of barriers such as roads or rivers on dispersal (Goossens *et al.* 2005; Coulon *et al.* 2006).

Broadly, we can see that in the development of these methods there has been an increasing sophistication in the modelling of the background frequencies. Initially point estimates were used (Paetkau *et al.* 1995), using data from known prior groupings. A subsequent development was to incorporate uncertainty into the estimates of the background frequencies (Rannala and Mountain 1997), but still based on data for groups defined a priori. This was followed by then allowing uncertainty in the grouping of individuals (Pritchard *et al.* 2000; Dawson and Belkhir 2001; Corander *et al.* 2003), followed by increasing complexity in the specification of the priors for these background frequencies, also taking into account the non-independence of partially linked loci (Falush *et al.* 2003). A major preoccupation has been the development of the methodology as a clustering algorithm for grouping individuals, and some effort has been expended in deciding how many clusters, or groups, are present in the data. Obviously this is a rather model-dependent concept. Presumably if there are strong prior beliefs that the individuals under analysis do indeed come from some number of discrete populations, then the answers obtained on the posterior distribution of the number of contributing populations will be useful. Frequently, however, it may be the case that the methods are trying to convert a continuum into discrete categories. Recent studies have attempted to model spatial distribution of gene frequency (Wasser *et al.* 2004; Guillot *et al.* 2005; François *et al.* 2006).

A potential limitation in the future development of these techniques is that correlations between relatives due to common ancestry are typically ignored. The likelihoods are computed on the assumption that the alleles within individuals and the individuals themselves are (conditionally) independent. Thus, for example, two individuals that are the descendants of a single immigrant will be assumed to have arisen from two immigration events, with a concomitant inflation of the real immigration rate. Further progress in this area may well depend on work in pedigree reconstruction discussed below. Whether the computational cost will make this worthwhile remains to be seen.

RELATEDNESS AND PEDIGREES

Analysis of kin structure in endangered and managed populations has always been an important component of the conservation geneticist's toolbox, for example in identifying potential mates (Russello and Amato 2004) or in situations where endangered taxa are threatened by hybridization (Daniels *et al.* 2001). Relatedness has been used to estimate genetic components of variance of phenotypic traits in outbred populations (e.g. Mousseau *et al.* 1998), and thus can potentially be used in conservation to quantify the amount of phenotypically important genetic variation in endangered populations (Storfer 1996; Carvajal-Rodríguez *et al.* 2005; Thomas 2005). Primary concerns have been estimation of coefficients of pairwise relatedness (Queller and Goodnight 1989; Ritland 1996a), paternity assignment (Marshall *et al.* 1998), and pairwise likelihood-based tests for discriminating between different degrees of relationship (Goodnight and Queller 1999). However, recently we have seen the development of methods for identifying sib-relationships (Painter 1997; Thomas and Hill 2002; Wang 2004), and, almost equivalently, identifying paternities and maternity of a sample of individuals (Emery *et al.* 2001; Hadfield *et al.* 2006). These typically involve the partitioning a sample of individuals into groups according to their relatedness. In this regard the methods are similar to the multilocus techniques (e.g. Structure and Partition) described above. The difference is that the latter do not include any genealogical structure, but assume that alleles are independently drawn from particular gene frequency distributions.

The techniques used in the analysis of kin structure are, of course, similar to those in forensic analysis and can be used to quantify the probability that two samples are identical matches to the same individual. This then leads to the ability to determine census size from faecal and hair samples, using the same statistical techniques involved in classical mark–release–recapture studies (Schwartz *et al.* 1998). In analysing these data there are a number of uncertainties: in the identification of individuals, in the genotyping, and in the underlying mark–release–recapture model used to estimate population census sizes (Bellemain *et al.* 2005; Petit and Valiere 2006). Given that genotyping techniques are becoming more and more sensitive there is scope for very detailed surveillance of natural populations of rare and elusive species. It would seem that the time is ripe for considerably more statistical development, both to further address the areas of uncertainty given above, but also to allow for far more detailed inference about the interactions between individuals of a known degree of relationship, and their spatial movements in complex landscapes.

Genotyping errors seriously bedevil all methods that attempt to infer degrees of relationship (Marshall *et al.* 1998). This appears to be in contrast to assignment methods, where moderate levels of genotyping error may not so strongly affect the outcome (Hauser *et al.* 2006). One approach is the development of heuristic methods to allow researchers to pick up genotyping errors as the data are generated (McKelvey and Schwartz 2005). The assumption is that problematic loci and/or individuals will then be re-genotyped until all detectable discrepancies disappear. However, it is likely that there will always be a need to incorporate the possibility of error when modelling genotypic data. Marshall *et al.* (1998) used a model of genotyping error in paternity analysis where alleles were assumed to have been replaced (at rates of e.g. 0.01) by others chosen in proportion to the marginal allele frequencies. A more detailed model, including the possibility of allelic dropout, has been considered by Wang (2004). Further developments along these lines can be envisioned, and applied to other markers.

It is tempting to see that here has been, at least superficially, a parallel progression of ideas in assignment methods that identify the population to which a given individual might belong (Paetkau *et al.* 1995), and in relatedness and paternity testing, which identify the kin-group to which an individual might belong (Painter 1997). With the application of methods of computational statistics such as MCMC it has been possible to consider more and more complex models. The logical endpoint is to attempt to recover the pedigree of a population from the genotypic data of individuals. Described as a 'Holy Grail' (Blouin 2003), if such a goal were achievable, it would then unify the many currently apparently disparate methods of population genetic analysis. A recent major advance has been made by Gasbarra *et al.* (2005, 2007), who have devised an MCMC scheme for generating samples from the posterior distribution of pedigrees back to some arbitrary time point in the past, given the multilocus genotypes of a number of individuals. A related approach for partially selfing species has been devised by Wilson and Dawson (2007). It is surely an attractive idea that as such methods become routinely available and are extended to more complex demographic scenarios, many of the problems that are currently tackled approximately and disparately through coalescent modelling and population assignment will be more comprehensively addressed within this single coherent framework. However, it is likely that the computational complexity inherent in sampling across the space of pedigrees is sufficiently challenging that it might impose limitations on what can be done, and the difficulty in sampling from the ancestral recombination graph, given even quite modest sequence data, is a salutary example.

NATURAL SELECTION AND ADAPTATION

The ease with which it is possible to develop large numbers of markers, particularly with amplified fragment length polymorphisms (AFLPs), has led to a renewed interest in the possibility of identifying genomic regions under locally adaptive selection (Wilding *et al.* 2001; Beaumont 2005; Storz 2005; Vasemagi *et al.* 2005; Bonin *et al.* 2006; Savolainen *et al.* 2006). From the point of view of conservation, the most interesting questions are not necessarily concerned with the identification of a potential region of the genome that may be influenced by selection, but with some general assay of adaptive divergence (Luikart *et al.* 2003; Beaumont and Balding 2004). In principle, if one could be confident in the assay for 'adaptive' genetic variation, then it might be possible to identify, for example, suitable populations for translocations of individuals with a view to avoiding outbreeding depression, or to identify populations that are unique in their possession of local adaptations (Bonin *et al.* 2007).

Genome scan methods have recently been widely used in the analysis of the human genome (Nielsen 2005), where densely spaced markers and linkage information are available. This has become the subject of intense research into methodological development, particularly where linkage information is used. One approach to performing such analyses would be through the use of structured coalescent modelling of selection, following the method of Kaplan *et al.* (1989). This has been successfully applied for likelihood-based inference of non-recombining markers (Coop and Griffiths 2004), and also, through the use of an ABC approach (Przeworski 2003) on recombining markers. A difficulty with the explicit analysis of selection is that it requires a priori identification of the nucleotide site under selection, whereas in most situations we would need to make inferences marginal to (i.e. summed over) potential sites under selection. Furthermore, a posteriori analyses of selection at a particular sequence previously identified as 'interesting' from some more general scan need to take this ascertainment into account. The analysis of individual regions also needs to take into account unknown demography, and thus needs to include data from multiple other loci, either as priors or as part of an all-encompassing analysis. Exact likelihood-based inference (at least, within reasonable Monte Carlo error) is very difficult to obtain for recombining sequences, and there is a need to recourse to approximations. It is possible that PAC-likelihood approximations (Li and Stephens 2003) may be useful for such analyses in the future.

For general scans, a good example is the study of Nielsen *et al.* (2005), who have developed a method in which the likelihood for a model of a selective sweep can be computed from the allele frequencies at a given single nucleotide polymorphism (SNP) locus. This likelihood can be multiplied across all SNPs in a local region of the genome (this is an example of composite likelihood, because an assumption is made that the SNPs are independent when they are in fact partially linked). The likelihood is maximized for the key parameters describing the distance of the SNP from the centre of the selective sweep, and the strength of the sweep. The ratio of this maximized composite likelihood to that under a purely neutral model can be used as a score statistic to identify regions of the genome that are under selection. By contrast many non-human applications have worked with unlinked and unmapped markers, although Campbell and Bernatchez (2004) have strongly recommended the mapping of markers as an adjunct to attempting to identify regions under potential selection.

Future theoretical developments need to consider explicit models of natural selection. The selective sweep model considered by Nielsen *et al.* (2005) is a useful approach, particularly for humans, but many loci that are under selection in populations of interest from the point of view of conservation may not conform to this idealization very closely. Many polymorphisms may be better understood at the metapopulation level as being under a form of balancing selection caused by local selection of alternative alleles among demes (Petry 1983; Charlesworth *et al.* 1997). In this case evidence of selection may best come from comparisons among populations in gene frequencies. Wright (1935, 1949) derived a number of stationary distributions for loci under selection in the presence of migration and selection. In the purely neutral case the stationary distribution is a Dirichlet distribution, and this has formed the basis for a Bayesian approach (Beaumont and Balding 2004) in which selection is modelled as a reduction in effective migration rates at affected loci. This can be justified from earlier studies, where the effective migration rate is given as:

$$m' = \frac{rm}{r + hs}$$

(Petry 1983; Barton and Bengtsson 1986; Charlesworth *et al.* 1997), where r is the recombination rate between the neutral marker and the selected marker, m is the neutral migration rate, h is the degree of dominance and s is the selection coefficient. These approaches are suitable for unlinked, unmapped markers, of the type frequently encountered in conservation genetics. It is possible to envisage that in the not-too-distant future it will be relatively inexpensive to obtain large numbers of at least approximately mapped markers for any organism of interest. In this case, more information about selection will come from an analysis of the patterns of linkage disequilibrium, as in current analyses of the HapMap data discussed above. However, rather than base the composite likelihood functions on selective sweep models as in Nielsen *et al.* (2005), patterns of gene frequencies among populations may be better modelled using the approximation of Petry (1983) for local selection with Sewall Wright's equations. A problem with the use of these analytical approximations is that they are an idealisation, ignoring the effects of demographic history and mutation process (Beaumont 2005), which may give spurious signals of selection. The ABC framework may allow for complex demographic histories and selection to be jointly inferred in future studies.

A related set of studies for the identification of the effect of selection is that on quantitative traits. The quantity Q_{ST} (Spitze 1993) has been defined as $Q_{ST} = V_B/(V_B + 2 V_A)$, where V_B is the between-population component of variance in the trait and V_A is the additive genetic variance within populations (for reviews see Merila and Crnokrak 2001; McKay and Latta 2002; see also Bonin and Bematchez this volume). The expected Q_{ST} for neutral loci under an additive model is the same as that of F_{ST}. This has stimulated a number of studies to relate the two by comparing the Q_{ST}s measured for various traits with F_{ST}s from presumably neutral markers. High levels of Q_{ST} relative to F_{ST} would suggest local adaptation with respect to the morphological traits (Whitlock 1999). A fairly extensive literature (Goudet and Buchet 2006 and citations therein) suggests that this prediction is relatively robust to departures from additivity and inbreeding. Studies on morphological characters could complement analyses based on genetic markers to enable a more integrated study of local adaptation in populations, and thereby help inform decision-making for conservation and management. Mapping of quantitative trait loci (QTLs) may be useful here, and the method has been applied to identify regions of the genome that are responsible for local adaptation in managed populations, particularly forest trees (González-Martínez *et al.* 2006). As these authors point out, QTL mapping is quite challenging for a number of reasons, such as the variation of the effects due to genetic background and environment. A difficulty with morphological traits in Q_{ST} analyses is that it is often difficult to estimate heritabilities and thereby obtain V_A. Heritabilities have been estimated using measures of relatedness (Ritland 1996b; Mousseau *et al.* 1998), and in the long run it may be possible to integrate methods for inferring pedigrees (Gasbarra *et al.* 2007) from molecular markers with the Q_{ST} studies. Indeed, given that some of the markers may themselves be highly differentiated because of their linkage to the QTLs affecting the Q_{ST}s, one can envision a fully integrated morphological and molecular genetic analysis, similar to that of linkage disequilibrium mapping of QTLs in outbred populations (Zhao *et al.* 2007).

CONCLUSIONS

This review has attempted to map out, at least from the perspective of this author, how current areas of research in statistical genetics could potentially impinge on conservation genetics in the future. It is reasonable to assume that it will soon be possible to routinely genotype individuals of any organism of interest for very large numbers of markers. The ability to do so will potentially give conservation biologists information on how any given population of interest came to be there, its uniqueness in terms of adaptation to the local environment, and its current life history, all of which has relevance to any management programme. The individuals concerned may never even be observed, but sampled through their spoor. From a theoretical perspective it seems that two main features stand out. One is that most methods of analysis will necessarily involve a fair amount of approximation, and many types of approximation are currently being developed. Even though computer speed is increasing exponentially, approximations are needed because, in population genetic analysis, we can typically only write likelihoods for particular genealogical histories or gene-frequency trajectories, and the number of such histories increases much faster than exponentially with linear increases in the amount of data. The second feature, in antagonism to the first, and in a similar state of development to that of genealogical inference in the mid-1990s, is that we are now in a position to infer the pedigrees of individuals from their genotypic information alone. Once we can do this deep enough into the past then many problems of interest – detection of selective sweeps, linkage analysis, QTL analysis, detection of immigrants and current dispersal rates, analysis of local adaptation, inferring demographic history – can be encompassed in one single framework. However, at the moment, inferring pedigrees involves Monte Carlo sampling of a very large space, and is probably not tractable for most problems of interest. Thus a future challenge will be to develop good approximations on pedigrees, and if this is successful genetic analysis will certainly become a very powerful tool for conservation.

REFERENCES

Anderson, E. C. (2005). An efficient Monte Carlo method for estimating N_e from temporally spaced samples using a coalescent-based likelihood. *Genetics*, **170**, 955–967.

Ballard, J. W. O. and Whitlock, M. C. (2004). The incomplete natural history of mitochondria. *Molecular Ecology*, **13**, 729–744.

Bandelt, H.-J., Forster, P. and Röhl, A. (1999). Median-joining networks for inferring intraspecific phylogenies. *Molecular Biology and Evolution*, **16**, 37–48.

Barton, N. H., and Bengtsson, B. O. (1986). The barrier to genetic exchange between hybridizing populations. *Heredity*, **56**, 357–376.

Bazin E., Glémin, S. and Galtier, N. (2006). Population size does not influence mitochondrial diversity in animals. *Science*, **312**, 570–572.

Beaumont, M. A. (1999). Detecting population expansion and decline using microsatellites. *Genetics*, **153**, 2013–2029.

Beaumont, M. A. (2003). Estimation of population growth or decline in genetically monitored populations. *Genetics*, **164**, 1139–1160.

Beaumont, M. A. (2005). Adaptation and speciation: what can F_{ST} tell us? *Trends in Ecology and Evolution*, **20**, 435–440.

Beaumont, M. A. and Balding, D. J. (2004). Identifying adaptive genetic divergence among populations from genome scans. *Molecular Ecology*, **13**, 969–980.

Beaumont, M. A. and Rannala, B. (2004). The Bayesian revolution in genetics. *Nature Reviews Genetics*, **5**, 251–261.

Beaumont, M. A., Barratt, E. M., Gottelli, D. *et al.* (2001). Genetic diversity and introgression in the Scottish wildcat. *Molecular Ecology*, **10**, 319–336.

Beaumont, M. A., Zhang, W. and Balding, D. J. (2002). Approximate Bayesian computation in population genetics. *Genetics*, **162**, 2025–2035.

Bellemain, E., Swenson, J. E., Tallmon, D., Brunberg, S. and Taberlet, P. (2005). Estimating population size of elusive animals with DNA from hunter-collected feces: four methods for brown bears. *Conservation Biology*, **19**, 150–161.

Blouin, M. S. (2003). DNA-based methods for pedigree reconstruction and kinship analysis in natural populations. *Trends in Ecology and Evolution*, **18**, 503–511.

Bonin, A., Miaud, C., Taberlet, P. and Pompanon, F. (2006). Explorative genome scan to detect candidate loci for adaptation along a gradient of altitude in the common frog (*Rana temporaria*). *Molecular Biology and Evolution*, **23**, 773–783.

Bonin, A., Nicole, F., Pompanon, F., Miaud, C. and Taberlet, P. (2007). Population adaptive index: a new method to help measure intraspecific genetic diversity and prioritize populations for conservation. *Conservation Biology*, **21**, 697–708.

Campbell, D. and Bernatchez, L. (2004). Generic scan using AFLP markers as a means to assess the role of directional selection in the divergence of sympatric whitefish ecotypes. *Molecular Biology and Evolution*, **21**, 945–956.

Carvajal-Rodriguez A., Rolan-Alvarez, E. and Caballero A. (2005). Quantitative variation as a tool for detecting human-induced impacts on genetic diversity. *Biological Conservation*, **124**, 1–13.

Cassens, I., Van Waerebeek, K., Best, P. B. *et al.* (2005). Evidence for male dispersal along the coasts but no migration in pelagic waters in dusky dolphins (*Lagenorhynchus obscurus*). *Molecular Ecology*, **14**, 107–121.

Cavalli-Sforza, L. L. and Edwards, A. W. F. (1967). Phylogenetic analysis, models and estimation procedures. *Evolution*, **32**, 550–570.

Chan, Y. L., Anderson, C. N. K. and Hadly, E. A. (2006). Bayesian estimation of the timing and severity of a population bottleneck from ancient DNA. *PLoS Genetics*, **2**, 451–460.

Charlesworth, B., Nordborg, M. and Charlesworth, D. (1997). The effects of local selection, balanced polymorphism and background selection on equilibrium patterns of genetic diversity in subdivided populations. *Genetical Research Cambridge*, **70**, 155–174.

Chikhi, L. and Beaumont, M. A. (2005). Modelling human genetic history. In *The Encyclopedia of Genetics, Genomics, Proteomics and Bioinformatics*, ed. M. J. Dunn, L. B. Jorde, P. F. R. Little and S. Subramaniam. New York: John Wiley, Vol. 1, pp. 11–31.

Ciofi, C., Wilson, G. A., Beheregaray, L. B. *et al.* (2006). Phylogeographic history and gene flow among giant Galapagos tortoises on southern Isabela Island. *Genetics*, **172**, 1727–1744.

Coop, G., and Griffiths, R. C. (2004). Ancestral inference on gene trees under selection. *Theoretical Population Biology*, **66**, 219–232.

Corander, J., Waldmann, P. and Sillanpaa, M. J. (2003). Bayesian analysis of genetic differentiation between populations. *Genetics*, **163**, 367–374.

Cornuet, J. M. and Beaumont, M. A. (2007). A note on the accuracy of PAC-likelihood inference with microsatellite data. *Theoretical Population Biology*, **71**, 12–19.

Cornuet, J. M. and Luikart, G. (1996). Description of power analysis of two tests for detecting recent population bottlenecks from allele frequency data. *Genetics*, **144**, 2001–2004.

Coulon, A., Guillot, G., Cosson, J. F. *et al.* (2006). Genetic structure is influenced by landscape features: empirical evidence from a roe deer population. *Molecular Ecology*, **15**, 1669–1679.

Cracraft, J. (1989). Speciation and its ontology: the empirical consequences of alternative species concepts for understanding patterns and processes of differentiation. In *Speciation and its Consequences*, ed. D. Otte and J. A. Endler. Sunderland: Sinauer Associates, pp. 28–59.

Daniels, M. J., Beaumont, M. A., Johnson, P. J., Balharry, D. and Macdonald D. W. (2001). Ecology and genetics of wild living cats in the north east of Scotland and the implications for the conservation of the wildcat. *Journal of Applied Ecology*, **38**, 146–161.

Dawson, K. J. and Belkhir, K. (2001). A Bayesian approach to the identification of panmictic populations and the assignment of individuals. *Genetical Research Cambridge*, **78**, 59–77.

Degnan, J. H. and Rosenberg, N. A. (2006). Discordance of species trees with their most likely gene trees. *PLoS Genetics*, **2**, 762–768.

De Iorio, M. and Griffiths, R. C. (2004). Importance sampling on coalescent histories. I. *Advances in Applied Probability*, **36**, 417–433.

Dolman, G. and Moritz, C. (2006). A multilocus perspective on refugial isolation and divergence in rainforest skinks (*Carlia*). *Evolution*, **60**, 573–582.

Dorazio, R. M. and Johnson, F. A. (2003). Bayesian inference and decision theory: a framework for decision making in natural resource management. *Ecological Applications*, **13**, 556–563.

Doucet, A., Godsill, S. and Andrieu, C. (2001). *Sequential Monte Carlo Methods in Practice*. New York: Springer-Verlag.

Eizirik, E., Kim, J. H., Menotti-Raymond, M. *et al.* (2001). Phylogeography, population history and conservation genetics of jaguars (*Panthera onca*, Mammalia, Felidae). *Molecular Ecology*, **10**, 65–79.

Emery, A. M., Wilson, I. J., Craig, S., Boyle, P. R. and Noble, L. R. (2001). Assignment of paternity groups without access to parental genotypes: multiple mating and developmental plasticity in squid. *Molecular Ecology*, **10**, 1265–1278.

Estoup, A., Beaumont, M. A., Sennedot, F., Moritz, C. and Cornuet, J. M. (2004). Genetic analysis of complex demographic scenarios: the case of spatially expanding populations in the cane toad, *Bufo marinus*. *Evolution*, **58**, 2021–2036.

Excoffier, L., Estoup, A. and Cornuet, J. M. (2005). Bayesian analysis of an admixture model with mutations. *Genetics*, **169**, 1727–1738.

Fagundes, N. J. R., Ray, N., Beaumont, M. *et al.* (2007). Statistical evaluation of alternative models of human evolution. *Proceedings of the National Academy of Sciences USA*, **104**, 17 614–17 619.

Falush, D., Stephens, M. and Pritchard, J. K. (2003). Inference of population structure using multilocus genotype data: linked loci and correlated allele frequencies. *Genetics*, **164**, 1567–1587.

Fearnhead, P. and Donnelly, P. (2001). Estimating recombination rates from population genetic data. *Genetics*, **159**, 1299–1318.

Francois, O., Ancelet, S. and Guillot, G. (2006). Bayesian clustering using hidden Markov random fields in spatial population genetics. *Genetics*, **174**, 805–816.

Garza, J. C., and Williamson, E. G. (2001). Detection of reduction in population size using data from microsatellite loci. *Molecular Ecology*, **10**, 305–318.

Gasbarra, D., Sillanpaa, M. J. and Arjas, E. (2005). Backward simulation of ancestors of sampled individuals. *Theoretical Population Biology*, **67**, 75–83.

Gasbarra, D., Pirinen, M., Sillanpaa, M. J., Salmela, E. and Arjas, E. (2007). Estimating genealogies from unlinked marker data: a Bayesian approach. *Theoretical Population Biology*, **72**, 305–322.

Gelman, A., Carlin, J. B., Stern, H. S. and Rubin, D. B. (1995). *Bayesian Data Analysis*. London: Chapman and Hall.

González-Martínez, S. C, Krutovsky, K. V. and Neale, D. B. (2006). Forest-tree population genomics and adaptive evolution. *New Phytologist*, **170**, 227–238

Goodnight, K. F. and Queller, D. C. (1999). Computer software for performing likelihood tests of pedigree relationship using genetic markers. *Molecular Ecology*, **8**, 1231–1234.

Goossens, B., Chikhi, L., Jalil, M. F. *et al.* (2005). Patterns of genetic diversity and migration in increasingly fragmented and declining orang-utan (*Pongo pygmaeus*) populations from Sabah, Malaysia. *Molecular Ecology*, **14**, 441–456.

Goossens, B., Chikhi, L., Ancrenaz, M. *et al.* (2006). Genetic signature of anthropogenic population collapse in orang-utans. *PLoS Biology*, **4**, 285–291.

Gottelli, D., Marino, J., Sillero-Zubiri, C. and Funk, S. M. (2004). The effect of the last glacial age on speciation and population genetic structure of the endangered Ethiopian wolf (*Canis simensis*). *Molecular Ecology*, **13**, 2275–2286.

Goudet, J. and Buchi, L. (2006). The effects of dominance, regular inbreeding and sampling design on Q_{ST}, an estimator of population differentiation for quantitative traits. *Genetics*, **172**, 1337–1347.

Griffiths, R. C. and Tavaré, S. (1994). Simulating probability distributions in the coalescent. *Theoretical Population Biology*, **46**, 131–159.

Guillot, G., Estoup, A., Mortier, F. and Cosson, J. (2005). A spatial statistical model for landscape genetics. *Genetics*, **170**, 1261–1280.

Hadfield, J. D., Richardson D. S. and Burke, T. (2006). Towards unbiased parentage assignment, combining genetic, behavioural and spatial data in a Bayesian framework. *Molecular Ecology*, **15**, 3715–3730.

Hamilton, G., Currat, M., Ray, N. *et al.* (2005). Bayesian estimation of recent migration rates after a spatial expansion. *Genetics*, **170**, 409–417.

Hauser, L., Seamons, T. R., Dauer, M., Naish, K. A. and Quinn, T. P. (2006). An empirical verification of population assignment methods by marking and parentage data, hatchery and wild steelhead (*Oncorhynchus mykiss*) in Forks Creek, Washington, USA. *Molecular Ecology*, **15**, 3157–3173.

Hebert, P. D. N., Cywinska, A., Ball, S. L. and deWaard, J. R. (2003). Biological identification through DNA barcodes. *Proceedings of the Royal Society Series B*, **270**, 313–321.

Hebert, P. D. N., Penton, E. H., Burns, J. M., Janzen, D. H. and Hallwachs, W. (2004). Ten species in one: DNA barcoding reveals cryptic species in the neotropical skipper butterfly *Astraptes fulgerator*. *Proceedings of the National Academy of Sciences USA*, **101**, 14 812–14 817.

Hey, J. (2005). On the number of new world founders, a population genetic portrait of the peopling of the Americas. *PLoS Biology*, **3**, 965–975.

Hey, J. and Machado, C. A. (2003). The study of structured populations: new hope for a difficult and divided science. *Nature Reviews Genetics*, **4**, 535–543.

Hey, J. and Nielsen, R. (2004). Multilocus methods for estimating population sizes, migration rates and divergence time, with applications to the divergence of *Drosophila pseudoobscura* and *D. persimilis*. *Genetics*, **167**, 747–760.

Hey, J., Waples, R. S., Arnold, M. L., Butlin, R. K. and Harrison, R. G. (2003). Understanding and confronting species uncertainty in biology and conservation. *Trends in Ecology and Evolution*, **18**, 597–603.

Hickerson, M. J., Meyer, C. P. and Moritz, C. (2006). DNA barcoding will often fail to discover new animal species over broad parameter space. *Systematic Biology*, **55**, 729–739.

Huelsenbeck, J. P. and Ronquist, F. (2001). MRBAYES: Bayesian inference of phylogenetic trees. *Bioinformatics*, **17**, 754–755.

Jennings, W. B. and Edwards, S. V. (2005). Speciational history of Australian grass finches (*Poephila*) inferred from thirty gene trees. *Evolution*, **59**, 2033–2047.

Kaplan, N. L., Hudson, R. R. and Langley, C. H. (1989). The 'hitchhiking effect' revisited. *Genetics*, **123**, 887–899.

Kinas, P. G. (1996). Bayesian fishery stock assessment and decision making using adaptive importance sampling. *Canadian Journal of Fisheries and Aquatic Sciences*, **53**, 414–423.

Knowles, L. L. and Maddison, W. P. (2002). Statistical phylogeography. *Molecular Ecology*, **11**, 2623–2635.

Kuhner, M. K., Yamato, J. and Felsenstein, J. (2000). Maximum likelihood estimation of recombination rates from population data. *Genetics*, **156**, 1393–1401.

Lewontin, R.C. (1974). *The Genetic Basis of Evolutionary Change*. New York: Columbia University Press.

Li, N. and Stephens, M. (2003). Modeling linkage disequilibrium and identifying recombination hotspots using single-nucleotide polymorphism data. *Genetics*, **165**, 2213–2233.

Luikart, G., England, P. R., Tallmon, D., Jordan, S. and Taberlet, P. (2003). The power and promise of population genomics, from genotyping to genome typing. *Nature Reviews Genetics*, **4**, 981–994.

Mace, G. M. (2004). The role of taxonomy in species conservation. *Philosophical Transactions of the Royal Society Series B*, **359**, 711–719.

Machado, C. A. and Hey, J. (2003). The causes of phylogenetic conflict in a classic *Drosophila* species group. *Proceedings of the Royal Society Series B*, **270**, 1193–1202.

Marjoram, P., Molitor, J., Plagnol, V. and Tavaré, S. (2003). Markov chain Monte Carlo without likelihoods. *Proceedings of the National Academy of Sciences USA*, **100**, 15324–15328.

Marshall, T. C., Slate, J., Kruuk, L. E. B. and Pemberton, J. M. (1998). Statistical confidence for likelihood-based paternity inference in natural populations. *Molecular Ecology*, **7**, 639–655.

Matz, M. V. and Nielsen, R. (2005). A likelihood ratio test for species membership based on DNA sequence data. *Philosophical Transactions of the Royal Society Series B*, **360**, 1969–1974.

McKay, J. and Latta, R. (2002). Adaptive population divergence, markers, QTL and traits. *Trends in Ecology and Evolution*, **17**, 285–291.

McKelvey, K. S. and Schwartz, M. K. (2005). DROPOUT, a program to identify problem loci and samples for noninvasive genetic samples in a capture–mark–recapture framework. *Molecular Ecology Notes*, **5**, 716–718.

McVean, G. A. T. (2007). Linkage disequilibrium, recombination and selection. In *The Handbook of Statistical Genetics*, 3 edn, ed. D. J. Balding, M. Bishop and C. Cannings. New York: John Wiley, pp. 909–944.

McVean, G. A. T., Myers, S. R., Hunt, S. *et al.* (2004). The fine-scale structure of recombination rate variation in the human genome. *Science*, **304**, 581–584.

Mayr, E. (1963). *Animal Species and Evolution*. Cambridge, Mass.: Belknap Press of Harvard University Press.

Merila, J. and Crnokrak, P. (2001). Comparison of genetic differentiation at marker loci and quantitative traits. *Journal of Evolutionary Biology*, **14**, 892–903.

Miller, N., Estoup, A., Toepfer, S. *et al.* (2005). Multiple transatlantic introductions of the Western Corn Rootworm. *Science*, **310**, 992.

Moritz, C. and Cicero, C. (2004). DNA barcoding: promise and pitfalls. *PLoS Biology*, **2**, 1529–1531.

Mousseau, T. A., Ritland, K. and Heath, D. D. (1998). A novel method for estimating heritability using molecular markers. *Heredity*, **80**, 218–224.

Nielsen, R. (2005). Molecular signatures of natural selection. *Annual Review of Genetics*, **39**, 197–218.

Nielsen, R. and Matz, M. (2006). Statistical approaches for DNA barcoding. *Systematic Biology*, **55**, 162–169.

Nielsen, R. and Wakeley, J. (2001). Distinguishing migration from isolation: a Markov chain Monte Carlo approach. *Genetics*, **158**, 885–896.

Nielsen, R., Mattila, D. K., Clapham, P. J. and Palsboll, P. (2001). Statistical approaches to paternity analysis in natural populations and applications to the North Atlantic humpback whale. *Genetics*, **157**, 1673–1682.

Nielsen, R., Williamson, S., Kim, Y. *et al.* (2005). Genomic scans for selective sweeps using SNP data. *Genome Research*, **15**, 1566–1575.

Nordborg, M. (1997). Structured coalescent processes on different time scales. *Genetics*, **146**, 1501–1514.

O'Hagan, A. (1994). *Kendall's Advanced Theory of Statistics*, vol. 2B, *Bayesian Inference*. London: Arnold.

O'Hara, R. B. (2005). Comparing the effects of genetic drift and fluctuating selection on genotype frequency changes in the scarlet tiger moth. *Proceedings of the Royal Society Series B*, **272**, 211–217.

Padhukasahasram, B., Wall, J. D., Marjoram P. and Nordborg, M. (2006). Estimating recombination rates from single-nucleotide polymorphisms using summary statistics. *Genetics*, **174**, 1517–1528.

Paetkau, D., Calvert, W., Stirling, I. and Strobeck, C. (1995). Microsatellite analysis of population structure in Canadian polar bears. *Molecular Ecology*, **4**, 347–354.

Painter, I. (1997). Sibship reconstruction without parental information. *Journal of Agricultural, Biological, and Environmental Statistics*, **2**, 212–229.

Panchal, M. (2007). The automation of Nested Clade Phylogeographic Analysis. *Bioinformatics*, **23**, 509–510.

Panchal, M. and Beaumont, M. A. (2007). The automation and evaluation of nested clade phylogeographic analysis. *Evolution*, **61**, 1466–1480.

Pearse, D. E. and Crandall, K. A. (2004). Beyond F_{ST}: analysis of population genetic data for conservation. *Conservation Genetics*, **5**, 585–602.

Petit, E. and Valiere, N. (2006). Estimating population size with non-invasive capture–mark–recapture data. *Conservation Biology*, **20**, 1062–1073.

Petry, D. (1983). The effect on neutral gene flow of selection at a linked locus. *Theoretical Population Biology*, **23**, 300–313.

Pons, J., Barraclough, T. G., Gomez-Zurita, J. *et al.* (2006). Sequence-based species delimitation for the DNA taxonomy of undescribed insects. *Systematic Biology*, **55**, 595–609.

Pritchard, J. K., Seielstad, M. T., Perez-Lezaun, A. and Feldman, M. W. (1999). Population growth of human Y chromosomes: a study of Y chromosome microsatellites. *Molecular Biology and Evolution*, **16**, 1791–1798.

Pritchard, J. K., Stephens, M. and Donnelly, P. (2000). Inference of population structure using multilocus genotype data. *Genetics*, **155**, 945–959.

Przeworski, M. (2003). Estimating the time since the fixation of a beneficial allele. *Genetics*, **164**, 1667–1676.

Queller, D. C. and Goodnight, K. F. (1989). Estimating relatedness using genetic markers. *Evolution*, **43**, 258–275.

Randi, E., Pierpaoli, M., Beaumont, M., Ragni, B. and Sforzi, A. (2001). Genetic identification of wild and domestic cats (*Felis silvestris*) and their hybrids using Bayesian clustering methods. *Molecular Bology and Evolution*, **18**, 1679–1693.

Rannala, B. and Mountain, J. L. (1997). Detecting immigration by using multilocus genotypes. *Proceedings of the National Academy of Sciences USA*, **94**, 9197–9201.

Rannala, B. and Yang, Z. (2003). Bayes estimation of species divergence times and ancestral population sizes using DNA sequences from multiple loci. *Genetics*, **164**, 1645–1656.

Ritland, K. (1996a). Estimators for pairwise relatedness and individual inbreeding coefficients. *Genetical Research*, **67**, 175–185.

Ritland, K. (1996b). A marker-based method for inferences about quantitative inheritance in natural populations. *Evolution*, 50, 1062–1073.

Russello, M. and Amato, G. (2004). Ex situ management in the absence of pedigree information. *Molecular Ecology*, 13, 2829–2840.

Savolainen, V., Anstett, M.-C., Lexer, C. *et al.* (2006). Sympatric speciation in palms on an oceanic island. *Nature*, 441, 210–213.

Schwartz, M. K., Tallmon, D. A. and Luikart, G. (1998). Review of DNA-based census and effective population size estimators. *Animal Conservation*, 1, 293–299.

Sisson, S. A., Fan, Y. and Tanaka, M. M. (2007). Sequential Monte Carlo without likelihoods. *Proceedings of the National Academy of Sciences USA*, 104, 1760–1765.

Slatkin, M. (1981). Estimating levels of gene flow in natural populations. *Genetics*, 99, 323–335.

Speigelhalter, D. J., Thomas, A., and Best, N. G. (1999). *WinBUGS v. 1.2 User Manual*. Cambridge: MRC Biostatistics Unit.

Spitze, K. (1993). Population structure in *Daphnia obtusa*: quantitative genetics and allozyme variation. *Genetics*, 135, 367–374.

Stephens, M. and Donnelly, P. (2000). Inference in molecular population genetics (with discussion). *Journal of the Royal Statistical Society B*, 62, 605–655.

Storfer, A. (1996). Quantitative genetics, a promising approach for the assessment of genetic variation in endangered species. *Trends in Ecology and Evolution*, 11, 343–348.

Storz, J. F. (2005). Using genome scans of DNA polymorphism to infer adaptive population divergence. *Molecular Ecology*, 14, 671–688.

Storz, J. F. and Beaumont, M. A. (2002). Testing for genetic evidence of population expansion and contraction: an empirical analysis of microsatellite DNA variation using a hierarchical Bayesian model. *Evolution*, 56, 154–166.

Templeton, A. R. (2002). Out of Africa again and again. *Nature*, 416, 45–51.

Templeton, A. R. (2004). Statistical phylogeography: methods of evaluating and minimizing inference errors. *Molecular Ecology*, 13, 789–809.

Templeton, A. R. (2005). Haplotype trees and modern human origins. *Yearbook of Physical Anthropology*, 48, 33–59.

Templeton, A. R., Routman, E. and Phillips, C. A. (1995). Separating population structure from population history: a cladistic analysis of the geographical distribution of mitochondrial DNA haplotypes in the tiger salamander, *Ambystoma tigrinum*. *Genetics*, 140, 767–782.

Thomas, S. C. (2005). The estimation of genetic relationships using molecular markers and their efficiency in estimating heritability in natural populations. *Philosophical Transactions of the Royal Society Series B*, 360, 1457–1467.

Thomas, S. C. and Hill, W. G. (2002). Sibship reconstruction in hierarchical population structures using Markov chain Monte Carlo techniques. *Genetical Research*, 79, 227–234.

Thompson, E. A. (1973). The Icelandic admixture problem. *Annals of Human Genetics*, 37, 69–80.

Tiemann-Boege, I., Calabrese, P., Cochran, D. M., Sokol, R. and Arnheim, N. (2006). High-resolution recombination patterns in a region of human chromosome 21 measured by sperm typing. *PLoS Genetics*, 2, 682–692.

Vasemagi, A., Nilsson, J. and Primmer, C. R. (2005). Expressed sequence tag-linked microsatellites as a source of gene-associated polymorphisms for detecting signatures of divergent selection in Atlantic salmon (*Salmo salar* L.). *Molecular Biology and Evolution*, **22**, 1067–1076.

Verardi, A., Lucchini, V. and Randi, E. (2006). Detecting introgressive hybridization between free-ranging domestic dogs and wild wolves (*Canis lupus*) by admixture linkage disequilibrium analysis. *Molecular Ecology*, **15**, 2845–2855.

Wakeley, J. (1998). Segregating sites in Wright's island model. *Theoretical Population Biology*, **53**, 166–175.

Wakeley, J. (1999). Nonequilibrium migration in human history. *Genetics*, **153**, 1863–1871.

Wakeley, J. (2004). Recent trends in population genetics: More data! More math! Simple models? *Journal of Heredity*, **95**, 397–405.

Wakeley, J. and Aliacar, N. (2001). Gene genealogies in a metapopulation. *Genetics*, **159**, 893–905.

Wall, J. D. (2000). A comparison of estimators of the population recombination rate. *Molecular Biology and Evolution*, **17**, 156–163.

Wang, J. L. (2003). Maximum-likelihood estimation of admixture proportions from genetic data. *Genetics*, **164**, 747–765.

Wang, J. L. (2004). Sibship reconstruction from genetic data with typing errors. *Genetics*, **166**, 1963–1979.

Wang, J. L. and Whitlock, M. C. (2003). Estimating effective population size and migration rates from genetic samples over space and time. *Genetics*, **163**, 429–446.

Wasser, S. K., Shedlock, A. M., Comstock K. *et al.* (2004). Assigning African elephant DNA to geographic region of origin: applications to the ivory trade. *Proceedings of the National Academy of Sciences USA*, **101**, 14847–14852.

Waugh, J. (2007). DNA barcoding in animal species, progress, potential and pitfalls. *BioEssays*, **29**, 188–197.

Weiss, G. and von Haeseler, A. (1998). Inference of population history using a likelihood approach. *Genetics*, **149**, 1539–1546.

Whitlock, M. C. (1999). Neutral additive variance in a metapopulation. *Genetical Research Cambridge*, **74**, 215–221.

Whitlock, M. C., Griswold, C. and Peters, A. D. (2003). Compensating for the meltdown: the critical effective size of a population with deleterious and compensatory mutations. *Annales Zoologici Fennici*, **40**, 169–183.

Wilding, C. S., Butlin, R. K. and Grahame, J. (2001). Differential gene exchange between parapatric morphs of *Littorina saxatilis* detected using AFLP markers. *Journal of Evolutionary Biology*, **14**, 611–619.

Wilkins, J. F. (2005). A separation-of-timescales approach to the coalescent in a continuous population. *Genetics*, **168**, 2227–2244.

Williamson, S. H., Hernadez, R., Fledel-Alon, A. *et al.* (2005). Simultaneous inference of selection and population growth from patterns of variation in the human genome. *Proceedings of the National Academy of Sciences USA*, **102**, 7882–7887.

Wilson, I. J. and Balding, D. J. (1998). Genealogical inference from microsatellite data. *Genetics*, **150**, 499–510.

Wilson, I. J. and Dawson, K. J. (2007). A Markov chain Monte Carlo strategy for sampling from the joint posterior distribution of pedigrees and population parameters under a Fisher–Wright model with partial selfing. *Theoretical Population Biology*, **72**, 436–458.

Wiuf, C. (2003). Inferring population history from genealogical trees. *Journal of Mathematical Biology*, **46**, 241–264.

Wright, S. (1935). Evolution in populations in approximate equilibrium. *Journal of Genetics*, **30**, 257–266.

Wright, S. (1949). Adaptation and selection. In *Genetics, Paleontology, and Evolution*, ed. G. L. Jepson, G. G. Simpson and E. Mayr. Princeton: Princeton University Press, pp. 365–389.

Zhao, H. H., Fernando, R. L. and Dekkers, J. C. M. (2007). Power and precision of alternate methods for linkage disequilibrium mapping of quantitative trait loci. *Genetics*, **175**, 1975–1986.

Software index

†indicates a list of software within the text

API-CALC 182
ARLEQUIN 94, 106, 273, 277

BAYES 31
BAYESASS 279
BAYESASS+ 25, 28, 30, 37–39, 40
BEAST 50, 51, 67, 68, 205
BOTTLENECK 350

CAPWIRE 185
CERVUS 183

DELRIOUS 183

GEMINI, Genotyping Errors and Multitube
    Approach for Individual
    Identification 181
GENECAP 182
GENECLASS 149, 183, 276, 283
GENEPOP 273
GENETIX 273
GEODIS 88–92, 279
    input 89, 90, 94
GIMLET 180, 182

HYBRIDLAB 39

IDENTIX 183
IM 51, 67, 94, 348, 356

KINSHIP 183

LAMARC 47, 50, 59, 61, 67
LAM-MPI 62, 63

MESQUITE 94
MICRO-CHECKER 182
MIGRATE ch.3, 43
    analysis 55–67
    assumptions 50–51
    central probability 48
    common mistakes 55
    comparison of BI and ML approaches
        58–59
    comparison of two migration models 62,
        63–67
    computer systems 62
    default values 55, 56
    effect of gene flow 57–58
    example data set 53
    heating 60–61
    input 47
    replication 60–61
    runtime 60, 61–62
    summary 69
    with many loci 62
    with many populations 62
MISMATCH DISTRIBUTIONS 94
MODELTEST 47
MPICH2 62, 63–67
M_P_Val 350
MRBAYES 274
MSVAR 350

NETWORK 107, 111, 115, 116
NEWHYBRIDS 25, 28, 30, 36–37, 38,
    39, 40

OPENMPI 62, 63–64

PAPA 281
parentage software[†] 184
PAUP* 47, 111, 115, 274
PCA-Gen 279
PROC MIXED 254

R2D2 248
Relatedness 183
Repmaker 336

SIMCOAL 205
simdata_nh 39
SPECTRONET 107
spip 39

Structure 25, 28, 30, 33, 34, 35, 36, 38, 39,
    40, 149, 183, 274, 303
  with admixture 31–32
  with admixture and prior population
    information 33–36, 37, 38
    limitations 35
  without admixture 31

TCS 84, 86, 107, 111, 113, 115, 116, 118, 274,
    277, 279

VORTEX 238

WinBUGS 354, 355

Species index

(in alphabetical order of common name)
*indicates a list of species within the text

amphibians 2, 172, 260
apes, great 178
arthropods 2
axolotl, Mexican (*Ambystoma mexicanum*) 333

bacteria 212, 324
bear 178, 185
   cave 335
   North American brown (*Ursus arctos*)
     206, 217
   polar (*Ursus maritimus*) 294
   short-faced (*Arctodus simus*) 217
beetle, northeastern beach tiger (*Cicindela
   dorsalis dorsalis*) 207
birds 2, 168, 178
   species studied non-invasively using
     faecal samples 171–172*
bison, Beringian (*Bison* cf. *priscus*)
   217, 248
bivalve, freshwater (*Potamilus inflatus*) 81
bonobo (*Pan paniscus*) 183
bustard, great (*Otis tarda*) 183
butterflies 2, 323

cat, sabretooth 209
cetaceans 294, 295
chamois
   Alpine (*Rupicapra rupicapra*) 161
   Pyrenean (*Rupicapra pyrenaica*) 161
chicken, greater prairie (*Tympanuchus
   cupido*) 213
chimpanzee (*Pan troglodytes*) 168, 185
cod, Atlantic (*Gadus morhua*) 469
cow, snake-eating 211
coyote (*Canis latrans*) 81, 185, 207

crane, whooping (*Grus americana*) 208, 213
crayfish
   freshwater Tasmanian (*Astacopsis
   gouldi*) 82
   obligate cave (*Orconectes* spp.) 84–86, 94

deer 172
   red (*Cervus elaphus*) 160
   roe (*Capreolus capreolus*) 157
   white-tailed (*Odocoileus virginianus*) 156
dodo (*Raphus cucullatus*) 209
dolphin 172, 294, 299, 303–305
   bottlenose (*Tursiops* spp.) 183, 299, 300,
     302–303, 310, 311
   common (*Delphinus* spp.) 302, 305
   dusky (*Lagenorhynchus obscurus*) 105, 111,
     305
   Franciscana (*Pontoporia blainvillei*) 300
   spotted (*Stenella attenuata*) 301–302
   spotted (*Stenella frontalis*) 301–302
   striped (*Stenella coeruleoalba*) 299
duck
   koloa (*Anas wyvilliana*) 208
   Laysan (*Anas laysanensis*) 208

eagle
   short-toed (*Circaetus gallicus*) 178
   Hawaiian (*Haliaeetus* spp.) 210
elk, Irish (*Megaloceros giganteus*) 209

felids 178
fish 2, 47, 172, 178, 260, 332
   lake whitefish (*Coregonus clupeaformis*)
     127–128
   marine 68

species studied non-invasively using
faecal samples 172*
razorback sucker (*Xyrauchen texanus*)
151–156
fly, fruit (*Drosophila* spp.) 5, 251
frog
common (*Rana temporaria*) 53,
135, 236
African clawed (*Xenopus laevis*) 333
fox, red (*Vulpes vulpes*) 180

gnatcatcher, California (*Polioptila
californica*) 229
goose
Canada (*Branta canadensis*) 213
giant Hawaiian 210
gopher, pocket (*Thomomys talpoides*) 216

human (*Homo sapiens*) 47, 217

ibex, Alpine (*Capra ibex ibex*) 157
iguana, Galápagos island (marine)
(*Conolophus subcristatus*) 209, 281
insects 2
invertebrates 2, 10

kite, Cape Verde (*Milvus milvus
fasciicauda*) 230

langur, Hanuman (*Semnopithecus
entellus*) 183
lizard complex (*Liolaemus alongatus-kriegi*)
83, 98–99

mammals
Beringian 216
marine ch.13, 2, 168, 178, 310
terrestrial 168
species studied non-invasively using
hairs 171*
species studied non-invasively using
faecal samples 171*
mammoth 335
microorganisms 335
moa 210
moa-nalos 210
mollusc 2
moth, scarlet tiger (*Callimorpha dominula*) 355
mouse
laboratory 251, 259
meadow jumping (*Zapus hudsonius*)
83–84

Preble's meadow jumping (*Zapus
hudsonius preblei*) 83–84, 230–233
rock pocket (*Chatodipus intermedius*)
129–132

nene (*Branta sandvicensis*) 213–215

otter
European (*Lutra lutra*) 159
sea (*Enhydra lutris*) 294
oyster, pearl (*Pinctada margaritifera
cumingii*) 156

panda, giant (*Ailuropoda melanoleuca*) 185
parrot, St Vincent (*Amazona
guildingii*) 178
pinnipeds 294, 295, 299, 308
po'ouli (*Melamprosops phaeosoma*) 211
porpoise
Burmeister's (*Phocena spinipinnis*) 305
harbour (*Phocoena phocoena*) 310

quagga (*Equus quagga*) 203

reptiles 2, 168
rodents 216

saddleback, New Zealand (*Philesturnus
carunculatus rufusater*) 157
salamander 89
Eastern tiger (*Ambystoma tigrinum*) 333
salmon ch.11, 28, 244, 245
Atlantic (*Salmo salar*) 138–139, 246
Lake Saimaa (*Salmo salar* m. *sebago*) 245
seal
Antarctic fur (*Arctocephalus gazella*) 296
harbour (*Phoca vitulina*) 300, 310
northern elephant (*Mirounga
angustirostris*) 213
southern elephant (*Mirounga
leonina*) 299, 300–301, 308, 309,
310, 311
sea lion (*Otaria flavescens*) 295
shark, great white (*Carcharodon
carcharias*) 211
sirenians 294
skink, grand (*Oligosoma grande*) 183
snail, land (*Candidula unifasciata*)
86, 95
snakes 172
spiders 82
starling, Mascarene 211

tit, great (*Parus major*) 234
toads 82
tortoise ch.12
  giant (genus *Dipsochelys*) 209
  giant Galápagos (*Geochelone nigra*)
    ch.12, 269
trout, brown (*Salmo trutta*) 259
tuco-tuco, South American (*Ctenomys*
    sociabilis) 216
turtle
  painted (*Chrysemys picta*) 185
  sea 172
  Yunnan box (*Cuora yunnanensis*) 211

ungulates 229

vertebrates 2, 178
viruses 212
vole, montane (*Microtus montanus*) 216
vulture, bearded (*Gypaetus barbatus*) 213

wallaby
  brush-tailed (*Petrogale penicillata*) 180
  rock (*Petrogale lateralis*), 183
warbler
  large-billed reed (*Acrocephalus*
    orinus) 210
  reed (*Acrocephalus*
    arundinaceus) 183
waterfowl 355

whale 172, 185
  blue (*Balaenoptera musculus*) 294
  fin (*Balaenoptera physalus*) 308
  grey (*Eschrichtius robustus*) 306
  humpback (*Megaptera novaeangliae*) 68,
    306–307, 355
  killer (*Orcinus orca*) 295, 296, 305–306,
    309, 311
  long-finned pilot (*Globicephala*
    melas) 298
  minke (*Balaenoptera acutorostrata*)
    307–308
  North Atlantic (northern) right
    (*Eubalaena glacialis*) 213, 306
  North Pacific right (*Eubalaena japonica*)
    306
  northern bottlenose (*Hyperoodon*
    ampullatus) 299
  southern right (*Eubalaena australis*) 306
  sperm (*Physeter catadon*) 298, 309
wild boar (*Sus scrofa*) 157
wolf
  Ethiopian (*Canis simensis*) 211
  gray/grey (*Canis lupus*) 150, 162, 178, 207,
    208, 306
  Indian (*Canis* spp.) 211
  marsupial (*Thylacinus cynocephalus*) 209
  red (*Canis rufus*) 207

yeast 332

see also
Table 5.1 (**species** with datasets used for comparison of network methods)
Table 7.1 (**species** for which translocation plans have been evaluated)
Table 8.1 (**species** studied using non-invasive genetic techniques
Table 13.1 (**marine mammals** and F_{ST} values)

Species index

(in alphabetical order of *Latin* name)

Acrocephalus arundinaceus	reed warbler	183
Acrocephalus orinus	large-billed reed warbler	210
Ailuropoda melanoleuca	giant panda	185
Amazona guildingii	St Vincent parrot	178
Ambystoma mexicanum	Mexican axolotl	333
Ambystoma tigrinum	Eastern tiger	333
Anas laysanensis	Laysan duck	208
Anas wyvilliana	koloa	208
Arctocephalus gazelle	Antarctic fur seal	296
Arctodus simus	short-faced bear	217
Astacopsis gouldi	freshwater Tasmanian crayfish	82
Balaenoptera acutorostrata	minke whale	307–308
Balaenoptera musculus	blue whale	294
Balaenoptera physalus	fin whale	308
Bison cf. priscus	Beringian steppe bison	217, 248
Branta canadensis	Canada goose	213
Branta sandvicensis	nene	213–215
Candidula unifasciata	land snail	86, 95
Canis latrans	coyote	81, 185, 207
Canis lupus	gray wolf	150, 162, 178, 207, 208, 306
Canis rufus	red wolf	207
Canis simensis	Ethiopian wolf	211
Canis spp.	Indian wolf	211
Capra ibex ibex	Alpine ibex	157
Capreolus capreolus	roe deer	157
Carcharodon carcharias	great white shark	211
Cervus elaphus	red deer	160
Chatodipus intermedius	rock pocket mouse	129–132
Chrysemys picta	painted turtle	185
Cicindela dorsalis dorsalis	northeastern beach tiger beetle	207
Circaetus gallicus	short-toed eagle	178
Conolophus subcristatus	Galpagos island (marine) iguana	209, 281
Coregonus clupeaformis	lake whitefish	127–128

Ctenomys sociabilis	South American tuco-tuco	216
Cuora yunnanensis	Yunnan box turtle	211
Delphinus spp.	common dolphin	302, 305
Dipsochelys spp.	giant tortoise	209
Drosophila spp.	fruit fly	5, 251
Enhydra lutris	sea otter	213, 294
Equus quagga	quagga	203
Eschrichtius robustus	grey whale	306
Eubalaena australis	southern right whale	306
Eubalaena glacialis	North Atlantic (northern) right whale	213, 306
Eubalaena japonica	North Pacific right whale	306
Gadus morhua	Atlantic cod	39
Geochelone nigra	giant Galpagos tortoise ch.12,	269
Globicephala melas	long-finned pilot whale	298
Grus americana	whooping crane	208, 213
Gypaetus barbatus	bearded vulture	213
Haliaeetus spp.	Hawaiian eagle	210
Homo sapiens	human	47, 217
Hyperoodon ampullatus	northern bottlenose whale	299
Lagenorhynchus obscurus	dusky dolphin	105, 111, 305
Liolaemus alongatus-kriegi	lizard complex	83, 98–99
Lutra lutra	European otter	159
Megaloceros giganteus	Irish elk	209
Megaptera novaeangliae	humpback whale	68, 306–307, 355
Melamprosops phaeosoma	po'ouli	211
Microtus montanus	montane vole	216
Milvus milvus fasciicauda	Cape Verde kite	230
Mirounga angustirostris	northern elephant seal	48
Mirounga leonine	southern elephant seal	299, 300–301, 308, 309, 310, 311
Odocoileus virginianus	white-tailed deer	156
Oligosoma grande	grand skink	183
Orcinus orca	killer whale	295, 296, 305–306, 309, 311
Orconectes spp.	obligate cave	84–86, 94
Otaria flavescens	sea lion	295
Otis tarda	great bustard	183
Pan paniscus	bonobo	183
Pan troglodytes	chimpanzee	168, 185
Parus major	great tit	234
Petrogale lateralis	rock wallaby	183
Petrogale penicillata	brush-tailed wallaby	180

Philesturnus carunculatus rufusater	New Zealand saddleback	157
Phoca vitulina	harbour seal	300, 310
Phocena spinipinnis	Burmeister's porpoise	305
Phocoena phocoena	harbour porpoise	310
Physeter catadon	sperm whale	298, 309
Pinctada margaritifera cumingii	pearl oyster	156
Polioptila californica	California gnatcatcher	229
Pontoporia blainvillei	Franciscana dolphin	300
Potamilus inflatus	freshwater bivalve	81
Rana temporaria	common frog	53, 135, 236
Raphus cucullatus	dodo	209
Rupicapra pyrenaica	Pyrenean chamois	161
Rupicapra rupicapra	Alpine chamois	161
Salmo salar	Atlantic salmon	138–139, 246
Salmo salar m. *sebago*	Lake Saimaa salmon	245
Salmo trutta	brown trout tortoise	259
Semnopithecus entellus	Hanuman langur	183
Stenella attenuate	spotted dolphin	301–302
Stenella coeruleoalba	striped dolphin	299
Stenella frontalis	Atlantic spotted dolphin	301–302
Sus scrofa	wild boar	157
Thomomys talpoides	pocket gopher	216
Thylacinus cynocephalus	marsupial wolf	209
Tursiops spp.	bottlenose dolphin	183, 299, 300, 302–303, 310, 311
Tympanuchus cupido	greater prairie chicken	213
Ursus arctos	North American brown bear	206, 217
Ursus maritimus	polar bear	294
Vulpes vulpes	red fox	180
Xenopus laevis	African clawed frog	333
Xyrauchen texanus	razorback sucker	151–156
Zapus hudsonius	meadow jumping mouse	83–84
Zapus hudsonius preblei	Preble's meadow jumping mouse	83–84, 230–233

see also

Table 5.1 (**species** with datasets used for comparison of network methods)
Table 7.1 (**species** for which translocation plans have been evaluated)
Table 8.1 (**species** studied using non-invasive genetic techniques
Table 13.1 (**marine mammals** and F_{ST} values)

Subject index

α, level of admixture 31
γ, prior distribution of θ 30
ζ, prior distribution for π 30
Θ, mutation scaled population size 52
θ, allele frequency 26–27, 30
λ, scaled mutation rate 354
ν, prior probability of immigrant ancestry, migration rate 33, 35
   matrix of individual migration rate 38
   see also migration, rate
π, proportion or frequency 28, 30, 37
16S gene 82, 84, 86, 94
454 sequence technology 331, 333, 335
   see also pyrosequencing
ABC see Approximate Bayesian Computation
abundance 9
acceptance ratios 61–62
action plan 238, 321
activities, anthropogenic 288
adaptability, reduced 212
adaptation
   cryptic 134
   genetic 215
   local 364
admixture 28, 31, 32, 33, 35, 36, 37, 38, 39, 159, 259, 352
adNA see ancient DNA
AFLP-PCR see amplified fragment length polymorphism PCR
AFLPs see amplified fragment length polymorphisms
age 244
aggressiveness 246, 258
   and genetic diversity 249–262
AIC see Akaike's information criterion
akaike's information criterion (AIC) 65
allele frequency, estimation of 26–27, 37

allelic dropout 180, 360
allocation, categorical 182
alloparental care 296
allozyme 241, 245, 307
amelogenin 178
AMOVA see analysis of molecular variance
amplified fragment length polymorphism PCR (AFLP-PCR) 329–331
amplified fragment length polymorphisms (AFLPs) 132, 133, 168, 178, 236, 324, 361
analysis
   Bayesian 347
   data 323
   data, with MIGRATE 45, 51, 55–63
   diet/dietary 170, 180, 183, 215, 296, 303
   habitat pattern 229
   landscape 229
   linkage 364
   morphological, integrated with molecular genetic analysis 364
   spatial 229
   statistical parsimony 107–108, 274
   transcriptome 331
   see also approach, multidisciplinary, restriction fragment length polymorphism analysis
analysis of molecular variance (AMOVA) 149
analysis of variance (ANOVA) 251
ancestry, recent immigrant 33, 34
ancient DNA (aDNA) ch.9, 9, 202, 203, 325, 327, 334, 335
   analysis 205–206, 233
   applications 203, 206–217
   laboratory methods 203–205
   sources 203
   systematics and forensics 209–212
ANOVA see analysis of variance

approach
  coalescent-based 8, 43
  complementarity 237
  multidisciplinary 8, 287, 289
  multi-tubes 181
approximate Bayesian approach 336, 345
approximate Bayesian computation (ABC)
  352, 354, 361, 363
approximate likelihood profiles 352
approximations 352–353, 361
  composite likelihood 352
assignment
  genetic, of individuals 270, 271, 272, 276,
    281–287, 288, 330
  population 25, 28–30, 162, 170, 178,
    287, 360
association study 129
assumptions
  of F-statistics 42
  of MIGRATE 50–51
  of probability model 26
  of structure with admixture and prior
    population information 35
  violated in MIGRATE 51
augmentation 82
  demographic 288
avian pox 212

backcrossing 37
balls-in-barrels see model, conceptual
barrier 89, 358
Bayesian
  information criterion (BIC) 66
  see analysis, estimation, method,
    paradigm, reconstruction, skyline
    plots, specification
behaviour 7, 167, 245, 249
  breeding 309
  feeding 303
  of methods 39, 40
BI 49 see inference, Bayesian
BIC see Bayesian information criterion
biodiversity 9, 210, 234, 238, 321
  assessment 321
  loss 1
  preservation 1
bioinformatics 336
biological significance 150, 233
biologically misleading 228
biology
  conservation 3, 4, 99, 133, 203, 323
  wildlife 3

blood 203, 271
bones 203, 208, 209, 210, 214, 217, 272
bottleneck 9, 50, 68, 205, 212–215, 259, 278,
  289, 296, 312, 347, 350
  during a translocation 158
  recent 53
  short 53
  very sudden, very recent 53
boundaries, population
  population 32
  species 84, 96–99
branch length 48
breeding grounds, fidelity to 306–307
breeding programme 270
breeding success see reproductive success
breeding system 9
buccal swabs 172

c, sample correction factor 66
candidate gene/loci 228, 229, 236
  analysis of 128–129
captive breeding/breeders 7, 281, 288
captive individuals 271, 288
capture–recapture 182, 185
carcasses 172
carnivore 180
caviar 172–175, 211
census
  size 359
  population 170
Centro di Ecologia Alpina 10
change
  climate 205, 234
  environmental 95, 212, 213, 215–217, 238
  morphological 215
characters, multistate 107
CHD see chromo-helicase-DNA-binding
chromo-helicase-DNA-binding 178
cladogram 88
class, genealogical 36, 37, 39
  see also Z, genealogical class
cloning 5
coalescence 8, 42, 45, 46–47, 98
coalescent 46, 47, 67–69, 346
  multiple-merger 47
  n- 46
  serial 205
  see also approach, coalescence, model
Cohesion Species Concept 97, 98
COI see cytochrome oxidase I
colonization 92, 277, 279, 287
competition 311

competitive ability 246, 258
  and genetic variation 246
computation time 336, 347
computer cluster
  to run MIGRATE 62–63
confidence intervals
  maximum likelihood estimates 49
  MCMC-assisted ML analysis 62, 63–64
connectivity 8, 131
conservation 80, 130, 135, 138, 139, 141, 178,
    215, 216, 244, 260, 269, 271, 295, 312
  see biology, decision, plan, priorities, tools,
    units
constant 43, 44
constraints, environmental 295
contact zone 39, 300
contamination 179, 204, 334
control region 83, 161, 208, 213, 230, 272,
    283, 300, 302, 310
  see also mitochondrial DNA
convergence diagnostic
  Gelman–Rubin 60
coprolites 203, 215
corridors, wildlife 9
cost see expense
crisis discipline 3
cross-contamination 179
cultural transmission 304, 312
cytochome oxidase I (COI) 81, 356
cytochome oxidase III 207
cytochrome b gene 83, 156, 161, 209, 232, 272

d^2 243, 247–248, 253, 255–261
D_a, number of connections that need to
    be added to yield a single network
    115, 116
DAGs
  see directed acyclic graphs (DAGs)
data 8, 25, 31, 42, 162, 229, 323
  availability, transferability, sharing 323
  co-dominant 42
  DNA sequence 48, 68, 80, 81, 84, 106, 107,
    111, 133, 141, 178, 271, 323, 335, 350
  to estimate [v] 36
  false 179
  features affecting behaviour of 39, 40
  genetic 8, 9, 25, 35, 42, 68,
    78–79, 348
  genomic 8, 141, 323
  -handling 336
  missing 53
  molecular 6, 238

multilocus genetic 182
  no 35
  observed 27, 30, 35
  phenotypic 133
  quantitative 141
  real 39
  required 31, 32, 34, 37, 38
  RNA sequence 48
  run conditions for MIGRATE examples
    in text 78–79
  simulated 26, 37, 39, 40, 46, 111
  to weight genealogies 43
  your own 26, 27, 35, 39, 40
  see also expressed sequence tag,
    microsatellites, mitochondrial DNA,
    single nucleotide polymorphism
data sets 73, 83, 89, 91, 107, 115, 238, 336
D_b, sum of the connections that are different
    between the two networks 115, 116
D_c, within clade distance 91
decisions, conservation 149, 205, 207
decline
  population 213, 216, 346
  species 2
degenerate-oligonucleotide-primed PCR
    (DOP) 333
degradation, environmental 244, 262
deme 38, 349
demographic see augmentation, information,
    trend, fluctuation, history, processes
demographic events, past 8
denaturing high-performance liquid
    chromatography (DHPLC) 335
designation, species 170
destruction, habitat 7, 202
developmental instability 213
DHPLC see denaturing high-performance
    liquid chromatography
diagnosis, character-based 229
differentiation, genetic 4, 31, 39, 160, 279,
    300, 310
  between coastal and pelagic populations
    301
diffusion approximation/equations 46,
    354–355
direct effect hypothesis 242
directed acyclic graphs (DAGs) 25
disease 7, 170, 212, 327, 335
dispersal 5, 9, 95, 167, 168, 210, 215, 287,
    301, 303, 308, 309, 310, 351, 358, 364
  effective rate 9
  sex-specific rate 9, 303, 307, 308–311

distribution, spatial
of gene frequency 358
divergence 272–273
adaptive 139
phenotypic 126, 138
time 51
diversification
evolutionary 279
morphological 210
diversity ch.6, 129, 233, 287
allelic 213
and behaviour 244
and fitness 243–244
biological 3, 4
genetic 6–7, 8, 9, 131, 212–215, 216, 238,
241–242, 244, 247–248, 270, 273–,
335, 350
loss of 212
low 245, 307
neutral genetic 123, 129, 228, 233
reduction in 295
species 210
see also biodiversity, variation
D_n, nested clade distance 91
DNA
amplification 5, 7, 8
barcoding 9, 321, 356, 357
-chip 325
copy number 179
damage 204
degraded 181, 334
entire genomic, gDNA 333
exogenous 181
extraction 179, 180, 272
-fingerprinting 5
isolation 335
polymerase slippage 179
polymorphism 128
quantitation 181
sequences 6
sequencing 5
template 325
see also data, mitochondrial DNA
dominance 258
DOP see degenerate-oligonucleotide-
primed PCR
drift see genetic drift

ecology 3, 7, 178
wildlife 9
economy-of-scale 322
ecotype 281, 296

egg membranes 172
eggs 172
eggshells 172
electrophoresis
allozyme/enzyme 4–5, 42
gel/capillary 325, 327
Endangered Species Act (ESA 1973) 207,
229, 231, 232
endemic 229
environmental decay experiment 180–181
eruption, volcanic 216, 278
EST see expressed sequence tag
estimation
Bayesian 274
problem of 26, 27
ESU see evolutionarily significant unit
ethical principles 3
evolution 7
convergent 302, 303
evolutionary conservation genetics 8
evolutionarily significant unit (ESU) 123, 142,
149, 170, 206–207, 228, 229, 232, 233
adaptive definition 123
neutral definition 123
exchangeability 227
ecological and genetic 227, 230, 233
genetic 234
exclusion 182
principle 161
exonuclease 5′ assay 326
see TaqMan assay
expansion/contraction, population 312
expense 322, 325
experimental design 234
exploitation 270, 307
expressed sequence tag (EST) 323, 332
expression array 324
extinction 1–2, 6, 7, 215
crisis 322
rates 1, 2
risk 346
species 2, 3

F, inbreeding parameter 38
F84 8 see model, Felsenstein
faeces/faecal samples 168, 171, 180, 334, 359
DNA extraction of 176
storage of 173–175
false allele lengths 179
feathers 171–172, 178
collection and storage 175
DNA extraction 176

feeding grounds 307–308
fidelity to 306–307
filters, data 182
fin 254
shark 211
fitness 6, 125, 148, 151, 159, 212, 213,
    241–242, 288
of endangered species 269
hybrid 161
of offspring 260
traits 246
fixation, chance 298
fluctuating asymmetry 213
fluctuation, demographic 215, 305
foraging 302, 303, 305, 306, 312
index 256
range 301
forces, evolutionary 132
forensics 170, 211–212, 359
fossil record 1
fossils 210, 216
founder 245, 261, 270, 282
effect 157, 206
event 302
fractional allocation 183
fragment length polymorphism 324
see also amplified fragment length
    polymorphism, microsatellites
fragmentation
geographic 210
habitat 9, 38, 92, 170, 207, 215, 217
F_{ST} 46, 48, 51, 67–69, 126, 160, 310, 330, 363
assumptions 50
methods 42
F-statistics 4, 42, 80
fully connected graph 112

gametes, many per individual 47
gDNA see DNA, entire genomic
gene flow 9, 32, 42, 50, 51, 82, 83, 92, 130,
    132, 150, 168, 170, 178, 183, 202, 205,
    207, 230, 279, 288, 296, 298, 301,
    302, 303, 305, 309, 350, 360
gene inheritance 36, 38
gene pool 288
gene tree 47
see also genealogy/genealogies
Genealogical Concordance Species Concept
    97, 98, 114
genealogical exclusivity 81
genealogy/genealogies 43, 47, 48, 51, 105, 349
probability of 46

general effect hypothesis 242, 258
generalized linear mixed models
    (GLMMs) 254
genetic distance 296
genetic diversity see diversity, genetic
genetic drift 5, 9, 123, 126, 132, 139, 157, 159,
    205, 206, 228, 241, 272, 288, 305,
    310, 346
genetic stock identification 28
see also assignment, population
genetic variability 6, 132
see also heterozygosity
genetic variance 126
genome era 336
genome project 331
genome scan 132–134, 228, 236, 238, 361
pitfalls 134
genomes, viral 170
genomic survey ch.6, 8
genomics
functional 135
landscape 228, 229
population 132, 134, 323
genotypes, multilocus 31, 32, 34, 37, 38, 39,
    346, 360
genotyping errors 360
geography 80
geology 210
glacial cycles 96
glaciation events 96
GLMMs see generalized linear mixed
    models

h, degree of dominance 362
habitat 281, 300, 302, 306
missing 85
hair roots 168
hairs 171, 359
collection and storage 172–173
DNA extraction 175
haplotypes
age 86
divergent 160
extinct 84
missing 87
unsampled interior 84
Hardy–Weinberg equilibrium
assumption of 27
departures from 38, 160, 182
harvesting ch.7, 68
illegal 9
heterosis 151, 241, 251

heterozygosity/heterozygous 6, 8, 213, 246,
    257, 261
    and fitness 241, 242–243
    see also genetic variability
heterozygosity–heterozygosity correlation
    values 258
hiatus, between genetic data and
    management 322
high throughput 321, 322
history
    demographic 205, 237, 346–348, 351, 358,
        364
    evolutionary 80, 84, 105, 131, 211, 271
history, population 51–55
HKY see model, Hasegawa–Kishino–Yano
H_{EST}, estimated heterozygosity 248
H_{OBS}, number of heterozygous loci/total
    number of loci 247
homoplasy 107
hot spots 229
housekeeping gene 328
Humpty Dumpty species concept 357
hunting 211, 213, 295, 296
hybrid ch.2, 25, 31, 283
    see also Q, expected proportion of ancestry
hybrid swarm 32
hybrid zone 32, 37, 82
hybridization 8, 25, 37, 161, 170, 207, 330,
    358, 359
hydrolytic deamination 204

i, regular number of repeat units 247
ice age, last 245
identification
    individual 168, 182
    species 9, 81, 168, 333, 356–357
identity disequilibrium hypothesis see
    general effect hypothesis
immigration/immigrants 346, 364
    rate 51
implanted transponder tags 156
importance, conservation ch.10, 8, 128, 287
improved-primer extension
    preamplificaion (I-PEP) 333, 334
inbreeding 4, 7, 8, 9, 38, 182, 202, 206,
    212, 241, 244, 245, 251, 259,
    260, 289
    depression 148, 150, 241, 303
    prevention 149
index
    genetic diversity 248
    hybrid 32

inference 30, 31, 48, 83
    Bayesian 30, 48, 49–50, 352
    exact likelihood-based 361
    key 92–95
    likelihood-based 352
    maximum likelihood 48, 49, 50
    model-based 25, 35, 38, 40
    population genetic 47
    of biogeographic patterns 95–96
    of coalescence theory 42
    of parameters 46, 48
    phylogenetic 330
    phylogeographic 83
    problem 29, 30, 37
    statistical model-based 26, 345
influenza, 212, 246
information, demographic 182–185
integral 43, 44
integration 44, 45, 324
integration-function 43, 44, 45
    height 43
    offspring 46
    steepness 43
internal relatedness (IR) 243
interpretation 323
introduction 202
    see also reintroduction
introgression 37
introgressive exchange 9
invasive species 93
I-PEP see improved-primer extension
    preamplificaion
IR see internal relatedness
island biogeography 5
isolation
    genetic 311
    geographic 298
isolation by distance 92, 95, 183, 300
I-T distance 91
IUCN, International Union for
    Conservation of Nature 2
    Red List 2
ivory 172, 211, 358

k
    populations 29
    number of parameters 66
K, number of (sub)populations 28, 31–33
    see also subpopulation
kin associations 295–299
kin recognition 252
kin selection 295, 359

kin-group 360
kinship 168
kin-structure 346
kit, commercial 325, 329, 334

[l], locus 29
L, number of loci 28
landscape genetics 8, 228
landscapes, adaptive 126
last glacial maximum 206, 217, 300, 305
learning 303, 304, 311, 312
learning sample 30, 31, 32, 37
    see also training sample
legal action 170
life stage 244, 259
likelihood ratio test (LRT) 356
    to compare two migration models using
        MIGRATE 62, 63–64
likelihood
    composite 351, 362
    maximum 4, 274
    see inference, likelihood
lineage
    diversification 210
    merge 46
lineages-through-time plots 8
local effect hypothesis 242, 255
loci
    adaptive ch.6, 124, 255
    agent 242
    electrophoretic 53
    how many 39
    multiple unlinked 68
    nuclear 204
    outlier 132
    see also markers, microsatellites,
        mitochondrial DNA, QTLs, SNPs
loss, habitat 1
LRT *see* likelihood ratio test

M, individuals sampled 27
maintenance 202
major histocompatibility complex (MHC)
    168, 178, 228, 236
majority consensus vectors 106
malaria 212
MALDI-TOF *see* matrix-assisted laser
    desorption/ionization time-of-flight
    mass spectrometry
management 9, 10, 28, 68, 81, 82, 151, 227,
    228, 234, 238, 269, 287, 288, 295,
    321, 322, 352, 353, 355

inappropriate 162
plan 150, 159, 271, 289, 307
managers 238, 322
Mantel tests 183
marker screening strategy 124
markers 229, 321, 323, 361
    adaptive genetic 227
    DNA 241
    generic 323
    genetic 148
    large numbers 364
    linked to genomic regions under
        selection 330
    kilo- 323
    molecular 132, 168, 177–178, 300, 323
    neutral 131, 150, 159, 236, 327
    non-neutral 336
    nuclear ch.2-6 ch.15, 7, 8, 296, 298, 307
    outlier 236
    sex-specific 330
    random 132
    see also loci
Markov chain Monte Carlo (MCMC) ch.3,
    30, 43–45, 205, 274, 345, 348, 352,
    353, 360
mate choice 9
maternal effects 126
maternity 359
mates, potential 359
matrix-assisted laser desorption/ionization
    time-of-flight mass spectrometry
    (MALDI-TOF) 328
maturation 246
maximum parsimony (MP) 104, 274, 283
MCMC *see* Markov chain Monte Carlo
MCMC-chain 49
MCMCMC *see* metropolis-coupled MCMC
MDA *see* multiple displacement
    amplification
measurements, morphological 83
meat 172
median vectors *see* majority consensus vectors
median-joining network (MJN) 107, 111,
    113, 115
megafauna 2, 217
melanism 129
melanocortin-1-receptor gene, Mc1r 129
meta-analyses 260
metagenomics 331, 334–335
metallothionein 139
metapopulation 4, 80, 349
    dynamics 81

method
approximate 351
assignment 183, 360
autocorrelated 43
Bayesian 30, 205, 279
genetic distance-based 229
laboratory ch.14, 325
likelihood 205, 351
median-joining 84
model-based ch.2, 25
molecular ch.14, 6, 309
multilocus genotypes 357–358
parallel sequencing 322
for searching for adaptive loci 125
sequential Monte Carlo 354
statistical 6
method-of-moments estimator (MME) 243
Metropolis algorithm 43
Metropolis-coupled MCMC (MCMCMC) 61
Metropolis–Hastings–Green algorithm 43
MHC see major histocompatibility complex
microarray 135, 137, 139, 178, 324, 325,
332–333, 336
microbes 47, 335
microbial communities 10
microsatellites 8, 83, 159, 160, 161, 162,
168, 178, 179, 180, 207, 213, 228,
230, 242, 243, 245, 247, 248, 252,
254, 255, 271, 272, 273, 281, 300,
302, 309, 310, 323, 324, 328, 330,
334, 352
see also short tandem repeats
migrant 33, 35, 38, 51, 277
migration 4, 5, 28, 33, 35, 38, 82, 123, 126,
160, 178, 246, 262, 279, 295, 300,
306, 310, 347, 361
matrix 38
model 38
rate 25, 35, 38, 43, 45, 51, 68, 281, 346
migration matrix 38
see also v, prior probability of immigrant
ancestry
Mindel 53
minimum spanning network (MSN) 84,
106, 108, 111
minimum spanning tree (MST) 106
minisatellites 168, 277
minisequencing 324–326
mismatch
analysis 8
curve 277
distribution 277

mitochondrial DNA (mtDNA) 5, 7, 68, 157,
159, 161, 162, 168, 177, 204, 207,
229, 230, 232, 236, 271, 272, 273,
283, 296, 298, 302, 304, 307, 309,
310, 325, 330, 347, 350
correlation of diversity with autosomes 351
mixing, mechanical 39
MJN see median-joining network
ML 49
see also inference, likelihood
MME see method-of-moments estimator
model
for allozyme data 48
Bayesian 355
Brownian motion 48
Cannings 46
coalescent 42, 43, 45, 86–88, 202, 349
conceptual 26–28
for electrophoretic markers 48
Felsenstein 47
gene flow 46
of genotyping error 360
graphical 26–30
Hasegawa–Kishino–Yano (HKY) 47
infinite sites 47, 94
island 349
of lineage branching 357
metapopulation 349
for microsatellite markers 48
mixture 28–30, 36, 37
mutation 45, 46, 47–49
no mutation 47
population 46
probability 25, 26
recombination 46
sampling 29
selection 46, 94
selective sweep 362
sequence 48
single-step mutation 48
speciation 46
statistical 25
substitution 47
Wright–Fisher population 46,
354–355
modelling
coalescent 360, 361
genealogical 346–348
of allele or haplotype frequencies 346
parametric statistical 346
simulation 8
molecular tracking 170

monitoring 148, 149, 150, 159, 202
  ecological 333
  non-invasive 167, 172
morphotypes 302
movement
  individual 168
  source-sink 279
  spatial 360
MP *see* maximum parsimony
mRNA 138
MSN *see* minimum spanning network
MST *see* minimum spanning tree
mtDNA *see* mitochondrial DNA
MU *see* units, management
multiple displacement amplification
  (MDA) 177, 333, 334
multiplex(ing) 325
  PCR 176–177
  SNPs 325
mutation 8, 46, 126, 135, 350
  adaptive 129
  artificial point 179
  candidate 128–129
  neutral 46
  rates 4, 51, 205, 310
  see also model

n
  number of generations 33
  number of loci 247
  number of samples 66
NADH dehydrogenase 1 (ND1) 82
NADH dehydrogenase 5 (ND5) 272
NADH dehydrogenase 6 (ND6) 272
National Environmental Policy Act (1969) 208
N_c, census size 68
NCA *see* nested clade analysis
NCPA *see* nested clade phylogeographic
  analysis
ND1 *see* NADH dehydrogenase 1
ND5 *see* NADH dehydrogenase 5
ND6 *see* NADH dehydrogenase 6
N_e, effective population size 68
  see also size
neighbour-joining 274, 283
nested clade analysis (NCA) 80
  see also nested clade phylogeographic
  analysis (NCPA)
nested clade distance (D_n) 91
nested clade phylogeographic analysis
  (NCPA) ch.4, 80, 81, 279, 347, 348
  applications 99

nesting design 87–88
network
  95% confidence interval 85
  approach 84, 104–105, 347
  construction ch.5, 84
    comparison of methods, empirical data
    sets 117
    comparison of methods, simulated
    sequenced data 111–113
  diagrams 84–86
  unresolved 86–88
  which method 117–118
neutral sweep 47
node 26
non-invasive sampling ch.8, 4, 9, 325, 326,
  327, 334
  analysis 181–182
  applications 168–171
  limitations 179–180
  requirements 181
  sources 171–172
  see also sample (non-invasive), hair,
  faeces, urine, etc.
non-parametric permutation approaches 156
number of subpopulations 32
numerical integration methods 43

O, subpopulation 33
on-line appendices 336
organismal transcriptomes 137
  see also microarray
outbreeding
  avoid 149
  depression 148, 149, 151, 361
outgroup 84
  probability 86
over-diagnosis 228
overexploitation 1
overfishing 244, 262
overhunting 82, 202, 213
owl pellets 203

PAC *see* product of approximate
  conditionals
PAI *see* population adaptive index
pairwise distances 89
paleogenomics 331
paleomicrobiology 212
paradigm, Bayesian 30
parameter 8, 27, 42, 44
  biased estimates with *F*-statistics 42
  coalescence-based 68

derived for MIGRATE 47
driving 49
estimate 42, 43
multiple 44, 45
mutation model 47
overestimated 47, 48
site rate variation 47, 48
parasites 212, 302
parentage 9, 168, 178, 182–171, 271
analysis 281–283
see also relatedness
parental reconstruction 183
parsimony 84
limit 108
statistical 84
particle filtering 354
paternity 168, 355, 359, 360
pathogens 183, 327, 330
pattern 132
biogeographical 81
demographic 271, 277
of differentiation 132
dispersal 9, 168, 183
effect of glaciations and uplift of
mountain ranges on 93
evolutionary 82
of gene flow 269
of gene inheritance 35, 36, 38
of genetic diversity 132, 159
of genetic variation 149, 151, 178
of genotypes 39
geographic 80
haplotype similarity 277
kinship 168
of linkage disequilibrium 132, 351
macrogeographic 7
of mate choice 9
movement 168
'natural' 150
phylogeographic 269
of seasonal movement 312
PCA see principal component analysis
PCR see polymerase chain reaction
(PCR)
Pearson product moment correlation
analysis 254
pedigree 9, 133, 345, 350, 358, 359–361, 364
penises 211
PEP see primer extension preamplification
phenotype matching hypothesis 252
phenotypic plasticity 126, 215,
234, 281

philopatry 295, 298, 299, 309, 310,
311, 312
phylogenetic species concept 207
phylogenetic(s) 46, 47, 51
analysis 234, 277
approach 229
techniques 80
trees 47
phylogeography 80, 129, 170
what it does not provide 80
pilot studies 180–181
plan
conservation 132–134, 307
translocation 152
plants, vascular 2
plate 27
Pleistocene 216, 217
glaciations 93
poaching, detection 172
policy 1, 289, 322, 335
pollen 233
pollution 1, 139, 333
polymerase chain reaction (PCR) 5–6, 168,
178, 179, 204, 242
inhibition 179
product purification 327
replicate 182
see also real-time polymerase chain
reaction
population
captive 10
closed 26, 27
dynamics 272
endangered 359
expanding 277
fragmented 38
growing 53
shrinking 53
small 5, 7
non-anadromous 245
source 9, 183
threatened 345
wild 10
population adaptive index (PAI) 134–135,
227, 236–238
calculation of 134–136
Population Genetics for Animal
Conservation Workshop
(PGAC) 9–10
population of origin ch.2, 25, 28, 270,
271, 284
see also W, population of origin

population size 5, 43, 45, 46, 47, 50, 68, 170,
    185, 206, 216, 259, 288, 307, 309,
    310, 346
  constant 67
  effective 4, 9, 47, 67, 68, 170, 202, 215,
    271, 309, 352
  historical/past 68, 205
  randomly fluctuating 53, 67
  very small 46, 149
  viable 9
posterior distribution 30, 35, 349
posterior probability 31, 35
potential
  adaptive 7
  evolutionary 148
predictions 260
  behaviour of methods 40
prey
  choice 303
  resources 303
primer extension preamplification
    (PEP) 333
principal component analysis (PCA) 279
prior distribution 30, 31, 35, 49–50
prior population information 33–35, 37, 38
priorities, conservation 211, 289
probability distribution, expectation 44
probability-landscape 44
processes
  demographic 160, 269
  evolutionary 82
product of approximate conditionals (PAC)
    351, 352, 361
profile
  expression 327
  gene expression 326, 332, 336
profile likelihoods 49, 62, 63
project design 84
protection 68, 81, 142, 287, 288
pseudo-likelihood 351
  see also likelihood
pyrosequencing 331–332, 333, 334, 335

Q, expected proportion of ancestry 31, 32, 36
  -barrel 32
  prior probability distribution for 33, 35
Q_{ST} 363
Q_{ST}–F_{ST} comparison 125–127
QTLs see quantitative trait loci
quantitative trait loci (QTLs) ch.6, ch.11,
    ch.14, 7, 330, 363
  analysis 124, 364

r, recombination rate 362
radiocarbon dating 209
random mating 51
range
  alteration of 215
  expansion 92, 95, 279
  prior 207–209
real-time polymerase chain reaction
    (RT-PCR) 326–328
recalcitrant genome 323
recent common ancestor/ancestry 38,
    46, 68
reciprocal transplant experiments 126
recolonization 162
recombination 346, 350–351
  rate 351, 352
reconstruction, Bayesian 160
reduced median network (RMN) 106–107
regional genealogical exclusivity 83
reintroduction ch. 7, ch. 12, 9, 82, 148, 159,
    207, 208, 209, 288, 289
  future guidelines 162
  of offspring 281
relatedness 9, 168, 183, 359–361
  see also parentage
relationship, degrees of 359
relevance of conservation genetics 322
reproduction strategy 244
reproductive capacity 288
reproductive skew 295
reproductive success 282, 288, 311, 312
  uneven 47
  variance in 69
resequencing 331
reserve design 10
resistance 139
resource allocation/specialization 1, 298
resource competition hypothesis 308
restocking ch.7, 9, 148, 156
restoration 262, 270, 288
  ecosystem 148, 202
  genetics 10, 270
restriction fragment length polymorphism
    (RFLP) analysis 5, 327
results, non-congruent 322, 323
RMN see reduced median network
RNA 181, 272
RT-PCR see real-time polymerase chain
    reaction

S, sampled subpopulation 33
s, selection coefficient 362

sample(s) 46, 160, 271–272, 323
  age 180
  effect of number of 46
  fresh 180
  genetic, issues 82, 83–84
  geographic, issues 83–84
  how large 39
  large 47
  museum 150, 151, 203, 207, 209, 213, 214,
    232, 233, 271, 272, 277
  for network analysis 82
  non-invasive ch.8, 167, 168
  random 46
  small 81, 205
  storage of 172–175
  thorough 82
sampling importance resampling (SIR) 354
sampling, importance 353–354
  see also non-invasive sampling
satellite tag 308
scales 172
scrimshaw 172
selection 5, 51, 132, 138, 139, 140, 141, 202,
    228, 327, 345, 346, 361–364
  against hybrids 37
  artificial 138
  balancing 141
  co-efficients 51, 126
  directional 141
  divergent natural 125
  factors that can mimic patterns produced
    by 134
  inadvertent 139
  negative 148
  on loci 51
  relaxation of 151
selective sweep 47, 68, 214, 350, 364
sequencing
  automated 7, 325, 329
  direct 334
  Sanger 324
  rapid 321
sex 244
  determination/identification 168, 170
sex chromosomes 178
sex ratio 9, 346
sex-linked nuclear DNA 210
short tandem repeats (STRs) 7
  see also microsatellites
simulate
  admixture 39
  mechanical mixing 39

multilocus genotype data 39, 40
  scenario 39
simulation program, genetic 39
single nucleotide polymorphisms
  (SNPs) 8, 47, 132, 178, 321, 323,
    324, 325, 326, 328, 330, 331, 334,
    335, 362
SIR see sampling importance resampling
site fidelity 299, 308
size
  family 309
  sample 46, 47
  see also population size
skin 172, 208
skin mucus 172
skyline plots, Bayesian 205
SNPs see Single Nucleotide
  Polymorphisms
sociality 295, 304, 305, 312
software 326, 336
  see also Software Index
soil cores 203, 215
Southern blotting 5
spatial distribution, population 4
spatial genetics 8
Spearman rank correlation analysis 254
specialization, ecological 210
speciation 210, 333
  processes of 97
  sympatric 296
species
  acquatic 90
  candidate 99
  captive 8, 260
  critically endangered 2
  endangered 2, 5, 81, 167, 172, 211, 212,
    233, 244
  extinct 210
  genealogical 98–99
  gregarious 311
  hunting/fishing 160
  introduced 1
  protected and elusive 334
  social ch.13, 311
  terrestrial vs riparian 94–95
  threatened 2, 9, 133, 289, 321, 323
  vulnerable 2
  with low genetic variation 149
  see also animal species index,
    reintroduction, translocation
specification, Bayesian 30
sperm 283

statistical significance 233
step clade 88
stochastic 202
  events 288
  factors 1
  simulation tools 238
STR *see* short tandem repeats
strain 327
stress, environmental 333
structure
  adaptive 130
  population 5, 9, 32, 39, 156, 167,
      170, 288, 294, 295, 300, 302,
      303, 306, 310, 312, 330, 345,
      350, 360
  social 167, 303
studies, simulation 39
subpopulation 31–33, 34, 35
  cryptic 31
  see also K, number of (sub)populations, *S*,
      sampled subpopulation
subspecies 83, 229, 230, 232
substitution 46
summary statistics 353
  methods 42
support interval
  for MCMC-assisted ML analysis
      62–63
  see also confidence interval
survey, multilocus 132
syndrome, behavioural 259
synergy 1, 3–12
system, population 262

T, generation 33
tandem repeats DNA 5
  see also DNA-fingerprinting
TaqMan assay 326, 334
taxonomic status 8, 271
taxonomy 269–270, 274–281
TBR branch swapping heuristic search
      option 115
teeth 172, 181, 203, 216
temperature increase, global 215
template switching 204
theoretical foundations 4
time-forward process 46
toe pads 203
tools, conservation 4
trafficking in endangered species 211
training sample 30
  see also learning sample

trait
  adaptive 128, 228
  life-history 244
  phenotypic 128
  polygenic 140
  quantitative 363
transcription profiling 135–138
transition–transversion ratio 47
translocations ch.7, 148, 159, 208, 234,
      236, 361
  detecting 160–162
  examples 162
  future guidelines 162
  genetic analyses 151–159
  genetic effects 162
  illegal 160
  IUCN definition 150
  monitoring 151–159
  plan 148, 149, 150
  success 150
trees
  bifurcating 80
  consensus 104
  phylogenetic 81, 82
trend, demographic 8
tuberculosis 212

U S Endangered Species Act (ESA) *see*
      Endangered Species Act
UMP *see* union of maximum parsimonious
      trees
union of maximum parsimonious
      trees (UMP) 108–111, 113,
      115, 118
units
  conservation 81, 287, 345, 346
  management, MU 105, 149, 207, 232,
      233, 271, 288
unscaled function-evaluations 44
uplift of mountain ranges
  effect 93
urine 172, 334
  collection and storage 175
  DNA extraction 176

value, adaptive 133
variability
  intra-population 330
  maintenance of long term 67
variable 27, 35
  demographic 7
  ecological 7

environmental 229
genetic 25
posterior distribution 35
standard 4
unobserved 35
variant
adaptive 228
unique 149
variation
adaptive genetic ch.6, 140–142,
    150, 159, 227, 228, 232,
    238, 323
decline in 159
genetic 135, 149, 150, 156, 159, 202, 205,
    269, 288, 359
increased levels 10
intraspecific 97
maintenance 159
    highly reduced 160
neutral genetic 227
phenotypic 126
site rate 51
supplement 149
viability 270
vicariance 287, 288

VIE *see* Visible Implant Fluorescent
    Elastomer tags
Visible Implant Fluorescent Elastomer tags
    (VIE) 253

W, population of origin 28
    see also population of origin
wadges 172
    collection and storage 175
    DNA extraction 176
watersheds 82
WGA *see* whole genome amplification
whole genome amplification (WGA) 177,
    333–334
Wiens–Penkrot protocol 98
wildlife forensics *see* forensics
Würm 53

X-linked 178

Y, allelic type 26, 29
Y-chromosome 178, 350

Z, genealogical class 37
zoo 241